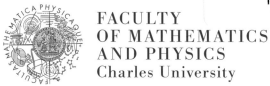

FACULTY
OF MATHEMATICS
AND PHYSICS
Charles University

Jiří Hořejší

# Lectures on
# QUANTUM FIELD THEORY

MatfyzPress
Publishing House

2024

Published by MatfyzPress Publishing House, Faculty of Mathematics and Physics of Charles University, Sokolovská 83, 186 75 Prague 8

Distributed in arrangement with Karolinum Press.

First edition

A catalogue record for this book is available from the National Library of the Czech Republic.

ISBN 978-80-7378-500-0 (MatfyzPress Publishing House)
ISBN 978-80-7378-508-6 (MatfyzPress Publishing House, pdf)
ISBN 978-80-246-5791-2 (Karolinum Press)
ISBN 978-80-246-5792-9 (Karolinum Press, pdf)

# Contents

# Preface

This work covers the material of a two-semester course of quantum field theory (QFT) that I taught for more than 20 years at the Charles University and Czech Technical University in Prague. For years, I was reluctant to write up such a set of lecture notes, since the current literature in this area is quite rich and there are dozens of books on the subject. However, eventually I was forced to do it, because of the pandemy of the infamous coronavirus that has broken out in spring 2020. I comment on this in more detail below. Conceptually, my approach is traditional, starting with several introductory chapters on the relativistic quantum mechanics. Then, after a brief interlude on the classical field theory, one proceeds to the quantization of free fields and to some elementary examples of field interactions, the basic tool being the Dyson perturbation expansion of the $S$-matrix in the interaction representation. The pragmatic aim of the first half of the text (chapters 1–25) is to arrive at the basic techniques for calculations of Feynman diagrams in the lowest perturbative order, as well as for the computation of the particle decay rates and scattering cross sections. This is just the matter that should be ideally explained during the first (winter) semester, since a part of the curriculum in the second (summer) semester, at least for some students, is a course on the standard model of particle physics, where a Feynman diagram calculation is an everyday occurrence. The second half (chapters 26–50) represents topics to be explained during the second semester and the main theme here is quantum electrodynamics at the level of one-loop diagrams, including techniques of regularization of ultraviolet divergences and renormalization. In this way, the whole material of the present lecture notes is divided into 50 chapters and each of them corresponds, roughly, to a 90 min. lecture (the total number of QFT lectures in a given academic year is about fifty). I would like to stress that the text is really intended to have the character of lecture notes, which means that, among other things, some explicit calculations are shown here in greater detail than in most of the representative monographs and textbooks, so as to make the life of a QFT beginner easier. Throughout the text one also encounters numerous hints to possible independent calculations, addressed to interested diligent readers; some of the problems in question may also serve as appropriate topics for tutorials. Admittedly, readers that are not quite fond of performing independent calculations may find the repeated offers of problems left to them as "instructive exercises" somewhat disturbing (or even annoying); anyway, there are just about three dozen of such hints in the whole text, i.e. less than one per chapter on average.

As I have indicated above, these lecture notes have been written under rather special circumstances, during the protracted coronavirus (COVID-19) crisis in 2020 and 2021. It was a situation that people of my generation have experienced never before, so let me add some personal recollection (which is, admittedly, somewhat emotional). The outbreak of the pandemy was officially announced in March 2020. Thus, on Wednesday, March 11, the personal attendance of students in the lecture rooms was banned "until further notice" and I decided to write immediately the text of a lecture scheduled for Thursday, to be able to send it to students via e-mail. Such a procedure seemed to me more efficient than a system of videoconferences or so, and I hoped also that the students' opinion would coincide with that of the aspiring student in Goethe's Faust, expressed in a dialogue with Mephistopheles, namely, "You won't need to tell me twice! I think, myself, it's very helpful, too, that one can take back home, and use, what someone's penned in black and white".[1] In any case, it is obvious that a carefully written text is more durable than lectures presented on a blackboard and erased immediately after the classes. Thus I went on in this manner, sticking to the maxim "nulla dies sine linea", till the end of May when the semester terminates. When the summer semester and the students' exams were over, I returned to the material of the envisaged next winter semester and continued writing down the relevant lectures so as to have a complete set (in musical terms, "da capo al fine"). In the meantime, I had to put together a collection of lectures for another course, aimed at a more advanced audience (25 chapters as well). In this way, the whole work has been basically completed in May 2021, with the nasty virus still around. Then there followed a period of transforming the manuscript full of handwritten formulae into a user-friendly electronic file, as well as gradual detailed proofreading of the text, mostly during the academic year 2021/2022. This was largely finished in autumn 2022, when the pandemy was fading away, but was overshadowed by even more tragic events — of course, I have in mind the absurd criminal war that Russia started against Ukraine.

When I started writing the lecture notes, in the gloomy atmosphere of the covid calamity on the rise, it came to my mind that there is a famous work of the world literature that was created under similar circumstances and survived over centuries. Yes, you guessed right; it is the Decameron by Giovanni Boccaccio. Its origin is widely known. It represents a collection of one hundred tales told by a group of ten young people who escaped from Florence, where the epidemic of plague broke out in 1348, and stayed in a hideout in the countryside to avoid the dangerous infection. Concerning my text, I have also written the lecture notes partly in a hideout (the "home office"). These consist of only fifty tales told by myself (not young anymore), concerning topics not so easily accessible to a general public and I certainly do not expect that my opus will become so famous as the Boccaccio's Decameron, or that it could survive through centuries. Nevertheless, I believe that it may have an appropriate (though inevitably limited) lifetime and may be useful for at least some students and other potentially interested scientifically minded readers. My primary aim has been to make it a comprehensible and digestible introduction to the rather difficult subject of quantum field theory, which, among others, forms a basis of the contemporary particle physics.

One last remark is perhaps in order here. In view of the above-mentioned origin of these lecture notes, it is to be expected that most of the potential readers will be university students fluent in Czech. Thus, I could not resist the temptation to include, occasionally, some notes concerning the Czech equivalents of the international English terminology, or even some elements of a common literary folklore. Hopefully, this might add some cheering moments to the serious scholarly style of the whole opus.

## Acknowledgements

From what I have written above it might seem that I should thank the malicious coronavirus in the first place, for stimulating me to write up these lecture notes. But I will not, taking into account that, apart from the positive impact mentioned above, this dangerous invisible bug did also so much harm to so many people all over the world. Needless to say, my acknowledgements are aimed in a completely different, genuinely positive, direction. In particular, I recognize the work of my younger colleagues who conducted and supervised, during the previous years, the

---

[1] A translation into English by A. S. Kline, 2003. In Czech (in the classic translation by O. Fischer) it reads: "Toť praktické, i heleď me se! To tělem duší při tom jsem. Neb co je černé na bílém, to vesele se domů nese."

tutorials related to my lectures. They are, in alphabetical order: Karol Kampf, Karel Kolář, Jiří Novotný and Martin Zdráhal. Further, I appreciate questions and comments that the students made throughout the years; this certainly led to many improvements of the style and contents of the lectures. Actually, I have also received some useful remarks from other colleagues; for instance, Walter Grimus from Vienna University has drawn my attention to the fact that the frequently cited "Lorentz condition" in electromagnetism is in fact "Lorenz condition". Finally, my great thanks are due to Tomáš Husek and Tomáš Kadavý, who recast my manuscript in LaTeX and thus made it ready for publication; the whole work matured to its present form in spring 2023.

Prague, May 2023                                                                                J. Hořejší

# Conventions, notations and units

Unless stated otherwise, we use the natural system of units, in which $\hbar = c = 1$ (note that Peskin and Schroeder call it "God-given" units in their book [14]). Obviously, within such a system, the time and length have the same dimension, the energy, momentum and mass have the same dimension, inverse length has the dimension of a mass, etc. The passage from the economical natural system to ordinary units is quite straightforward. To this end, one may use the commonly known approximate values of the Planck constant $\hbar$ and the "conversion constant" $\hbar c$, namely

$$\hbar = 6.58 \times 10^{-22}\,\text{MeV s}\,, \qquad \hbar c = 197\,\text{MeV fm}\,,$$

where $1\,\text{fm} = 10^{-13}\,\text{cm}$ (fm stands for "fermi" or "femtometer"). Numerical values of observable quantities (such as decay rates or scattering cross sections) are then converted into ordinary units by setting

$$1\,\text{MeV}^{-1} = 6.58 \times 10^{-22}\,\text{s}\,,$$

or

$$1\,\text{MeV}^{-1} = 197\,\text{fm}\,.$$

While the natural system of units is universally accepted in the literature concerning quantum field theory and particle physics, there are three other conventions that may differ in various books, so one must emphasize what is our particular choice (to avoid any misunderstanding when comparing our results with other books or papers). First, the metric of the flat spacetime used throughout the present text is defined by

$$g_{\mu\nu} = g^{\mu\nu} = \text{diag}(+1, -1, -1, -1)\,.$$

In other words, the metric we are employing here has the signature $(+ - - -)$. Let us remark that such a choice seems to be prevalent in current literature; for instance, among the books that we cite in the list of relevant literature, only [13] and [18] use the metric with the inverse signature $(- + + +)$. Anyway, one should keep in mind that there is no question of which metric is "right" or "wrong"; its choice is just a matter of convention. Note also that readers specialized mostly in relativity and gravitation should not worry about our notation $g_{\mu\nu}$ for the metric tensor, which they got used to employ for the general case of curved Riemann space (and distinguish the case of the flat spacetime by using the symbol $\eta_{\mu\nu}$ or so). The notation used here is a common practice in the literature concerning relativistic quantum theory and particle physics, since in this area one is dealing just with flat spacetime (nevertheless, $\eta_{\mu\nu}$ is employed conventionally e.g. in the books [11, 13] or [29]).

Second, another important convention is that for the fifth Dirac gamma matrix $\gamma_5$. Here we use the definition

$$\gamma_5 = i\gamma^0\gamma^1\gamma^2\gamma^3\,.$$

Again, this choice seems to be prevalent in the literature (note that within our list, the books [7] and [13] define $\gamma_5$ with opposite sign).

4

Finally, our convention for the fully antisymmetric Levi-Civita tensor is such that

$$\varepsilon_{0123} = +1 \, .$$

In this case, one must admit that this is a minority choice, since the option prevalent in current literature is $\varepsilon^{0123} = +1$ (which corresponds to the sign change in contrast to our convention). So, the reader must be careful when comparing our formulae in Appendix C and elsewhere (see in particular (C.11)) with those presented in other textbooks. Note that the convention employed here agrees with the classic books by Bjorken and Drell [1, 2].

# Chapter 1

# Klein–Gordon and Dirac equations: brief history

The best known equation of quantum mechanics is undoubtedly the Schrödinger equation, which for a particle moving in an external field reads

$$i\hbar \frac{\partial \psi}{\partial t} = \left( -\frac{\hbar^2}{2m}\Delta + V(\vec{x}) \right)\psi , \qquad (1.1)$$

where $\Delta$ is the Laplace operator, $\Delta = \vec{\nabla}^2$, and $V(\vec{x})$ is the potential energy corresponding to an external force. The wave function $\psi = \psi(\vec{x}, t)$ has the familiar interpretation: $|\psi(\vec{x}, t)|^2$ represents the probability density for the particle localization at the point $\vec{x}$ and time $t$. Erwin Schrödinger published it in 1926 (and subsequently won the Nobel Prize in 1933). Let us consider first Eq. (1.1) for a free particle, i.e. for $V = 0$. There is a simple "correspondence principle" that may serve as a recipe for recovering the Schrödinger equation. Denoting the energy as $E$ and momentum as $\vec{p}$, one may observe the correspondence

$$\begin{aligned} E &\longleftrightarrow i\hbar \frac{\partial}{\partial t} , \\ \vec{p} &\longleftrightarrow -i\hbar \vec{\nabla} , \end{aligned} \qquad (1.2)$$

which leads from the usual non-relativistic relation between kinetic energy and momentum

$$E = \frac{\vec{p}^2}{2m} \qquad (1.3)$$

to the Schrödinger equation

$$i\hbar \frac{\partial \psi}{\partial t} = -\frac{\hbar^2}{2m}\vec{\nabla}^2\psi . \qquad (1.4)$$

Let us stress as emphatically as possible that the correspondence (1.2) does not represent a derivation of the Schrödinger equation. This cannot be derived, it can only be postulated; this is what the founding fathers of quantum theory did. The meaning of the correspondence (1.2) is that it guarantees recovering the right relation between the energy and momentum (1.3) when the operators in (1.2) act on an appropriate wave function $\psi$, in particular the plane wave

$$\psi(\vec{x}, t) \propto e^{-\frac{i}{\hbar}(Et - \vec{p}\cdot\vec{x})} . \qquad (1.5)$$

In the same year when the non-relativistic equation (1.4) or (1.1) was postulated, a pertinent relativistic version was considered (preferably as a quantum mechanical equation for

6

the electron). In that case, one has to use as a motivating hint the relation between the energy and momentum valid in special relativity, i.e.

$$E^2 = c^2\vec{p}^2 + m^2c^4 \,. \tag{1.6}$$

Then, using the correspondence (1.2), one gets immediately

$$-\hbar^2\frac{\partial^2\psi}{\partial t^2} = \left(-\hbar^2c^2\vec{\nabla}^2 + m^2c^4\right)\psi \,, \tag{1.7}$$

and this can be recast in a more elegant form

$$\left(\frac{1}{c^2}\frac{\partial^2}{\partial t^2} - \Delta + \frac{m^2c^2}{\hbar^2}\right)\psi = 0 \,. \tag{1.8}$$

Now, the differential operator in Eq. (1.8) is the familiar d'Alembert operator

$$\Box = \frac{1}{c^2}\frac{\partial^2}{\partial t^2} - \Delta \,, \tag{1.9}$$

and thus we end up with

$$\left(\Box + \frac{m^2c^2}{\hbar^2}\right)\psi(x) = 0 \,. \tag{1.10}$$

The equation (1.10) had been formulated in 1926 independently by several theorists: Erwin Schrödinger (who subsequently rejected it), Oskar Klein, Walter Gordon, and Vladimir Fock (or, better Fok: the name reads Фок in Russian). So, although it is apparently an equation with many parents, it is universally called the **Klein–Gordon equation**.

A remark is perhaps in order here. The constant appearing in Eq. (1.10) is the square of inverse of the Compton wavelength and one might wonder why it happens to be there, when (1.10) clearly has nothing to do with the famous Compton process (the photon scattering on a charged particle). The answer is guessed easily on dimensional grounds: the d'Alembert operator $\Box$ has, obviously, the dimension of inverse length squared, and any possible additive constant (with the same dimension) must be made of the fundamental constants of a relativistic quantum theory, i.e. $c$ and $\hbar$ and, eventually, the relevant mass $m$. The combination $\hbar/mc$ is then the only possibility how to form a constant with the dimension of length (it is a refreshing simple exercise to show that such a combination of $c$, $\hbar$ and $m$ is indeed unique).[2]

For convenience, let us now pass to the natural system of units with $\hbar = 1$, $c = 1$. Using the standard relativistic covariant notation, one then has

$$\left(\Box + m^2\right)\psi(x) = 0 \,, \tag{1.11}$$

with $\Box = \partial_\mu\partial^\mu$. The simplest solutions of Eq. (1.11) have the form of plane waves; for their description one may use two linearly independent exponentials

$$\begin{aligned}\psi_{(+)}(x) &= \text{const. } e^{-ip\cdot x} \,, \\ \psi_{(-)}(x) &= \text{const. } e^{ip\cdot x} \,,\end{aligned} \tag{1.12}$$

where $p \cdot x = p_0x_0 - \vec{p}\cdot\vec{x}$ (the logic of the chosen notation will become clear shortly). Inserting (1.12) into (1.11), one gets the condition

$$p^2 = m^2 \,, \tag{1.13}$$

---

[2] In fact, sticking to the traditional terminology, $\hbar/mc$ is the Compton wavelength divided by $2\pi$.

i.e. $p_0^2 = \vec{p}^2 + m^2$. Without loss of generality, one may choose $p_0 > 0$,

$$p_0 = \sqrt{\vec{p}^2 + m^2}\,. \tag{1.14}$$

So, as expected, one recovers the correct relation between the energy and momentum of a particle with the mass $m$. Using the correspondence (1.2), one sees that the solution $\psi_{(+)}(x)$ describes a state with positive energy $E = p_0$ and momentum $\vec{p}$, while $\psi_{(-)}(x)$ carries negative energy $E = -p_0$ and momentum $-\vec{p}$. In any case, the four-component quantity $p$ satisfying (1.13) is rightly called the four-momentum of the particle with the mass $m$. Thus we have encountered, for the first time, the problem of a wave function for the free particle with negative energy; we will see that this is a generic feature of the equations of relativistic quantum mechanics.

In fact, there is another difficulty inherent in the Klein–Gordon equation. If one wants to implement the probabilistic interpretation of the wave function $\psi$, one should derive first a pertinent continuity equation connecting the probability density (for particle localization) and the density of probability current. Let us first remind the reader how one proceeds in the case of non-relativistic Schrödinger equation (1.4) (we are going to use the natural system of units, i.e. set $\hbar = 1$). We have the equations for $\psi$ and $\psi^*$,

$$i\frac{\partial \psi}{\partial t} = -\frac{1}{2m}\vec{\nabla}^2\psi\,, \tag{1.15}$$

$$-i\frac{\partial \psi^*}{\partial t} = -\frac{1}{2m}\vec{\nabla}^2\psi^*\,. \tag{1.16}$$

Multiplying Eq. (1.15) by $\psi^*$ and (1.16) by $\psi$, and subtracting the two equations, one gets immediately

$$i\frac{\partial}{\partial t}\left(\psi\psi^*\right) = -\frac{1}{2m}\left(\psi^*\,\vec{\nabla}^2\psi - \psi\,\vec{\nabla}^2\psi^*\right) = -\frac{1}{2m}\vec{\nabla}\left(\psi^*\,\vec{\nabla}\psi - \psi\,\vec{\nabla}\psi^*\right).$$

Thus one obtains the familiar result

$$\frac{\partial}{\partial t}\rho_{\text{Sch.}} + \vec{\nabla}\cdot\vec{j}_{\text{Sch.}} = 0\,, \tag{1.17}$$

with

$$\rho_{\text{Sch.}} = \psi\psi^* = |\psi|^2\,, \qquad \vec{j}_{\text{Sch.}} = \frac{1}{2mi}\left(\psi^*\,\vec{\nabla}\psi - \psi\,\vec{\nabla}\psi^*\right). \tag{1.18}$$

For the Klein–Gordon equation (1.11) one may try to proceed in a similar manner. To begin with, (1.11) is recast as

$$\frac{\partial^2\psi}{\partial t^2} = \vec{\nabla}^2\psi - m^2\psi\,, \tag{1.19}$$

and the same equation holds for $\psi^*$. Next, using the multiplication and subtraction trick as above, one gets first

$$\frac{\partial}{\partial t}\left(\psi^*\frac{\partial \psi}{\partial t} - \psi\frac{\partial \psi^*}{\partial t}\right) = \vec{\nabla}\left(\psi^*\,\vec{\nabla}\psi - \psi\,\vec{\nabla}\psi^*\right). \tag{1.20}$$

In order to make the left-hand side of Eq. (1.20) real, one has to include a factor of $i$; for getting quantities with the same dimension as in the case of the Schrödinger equation, one may write finally

$$\frac{\partial}{\partial t}\rho_{\text{KG}} + \vec{\nabla}\cdot\vec{j}_{\text{KG}} = 0\,,$$

8

where

$$\rho_{KG} = \frac{i}{2m} \left( \psi^* \frac{\partial \psi}{\partial t} - \psi \frac{\partial \psi^*}{\partial t} \right),$$

$$\vec{j}_{KG} = \frac{1}{2mi} \left( \psi^* \vec{\nabla} \psi - \psi \vec{\nabla} \psi^* \right). \tag{1.21}$$

Obviously, in contrast to (1.18), the would-be probability density $\rho_{KG}$ in (1.21) is not *a priori* positive. In particular, it is easy to see that for $\psi_{(+)}$ shown in (1.12) the expression for $\rho_{KG}$ is positive, while for $\psi_{(-)}$ one gets a negative value of $\rho_{KG}$. This, of course, is a serious flaw. On the top of that, it has soon become clear that the Klein–Gordon equation is not viable as an equation for the electron, because it cannot incorporate a description of intrinsic angular momentum, the spin (note that the concept of electron spin appeared on the physical stage in 1925, when it was introduced by George Uhlenbeck and Samuel Goudsmit — surprisingly, they have never received the Nobel Prize for it).

Despite the above-mentioned difficulty with the interpretation of the probability density, Klein–Gordon equation, as an equation of relativistic quantum mechanics, does have some limited applicability for the description of spinless particles (for more details, see e.g. the book [1]). However, we will exploit this equation fully later on, within the framework of field theory.

Anyway, it is clear that a topical question that certainly resonated in minds of quantum theorists in the second half of 1920s was: So, what is the right relativistic quantum equation for electron? The problem was resolved in 1928 by Paul Dirac. His solution was, at that time, quite astonishing and this historical breakthrough is thus worth recapitulating here (for the original paper, see [32]).

As we have already stressed, a major flaw of the Klein–Gordon equation is the non-positivity of the would-be probability density in (1.21). It is clear what is the source of this inherent feature of the $\rho_{KG}$: the equation (1.19) is of the second order in time and thus a time derivative emerges necessarily in (1.21). Thus, it is desirable to have an equation that would be of the first order with respect to time. To ensure the relativistic covariance, it should also be of the first order in space variables (time and space coordinates are treated on an equal footing in Lorentz transformations). In any case, one has to maintain the relativistic relation between energy and momentum (1.6) (for a moment, we come back to ordinary units). For his purpose, Dirac took the square root of (1.6) by linearizing it as follows,

$$E = c\alpha^j p^j + \beta mc^2, \tag{1.22}$$

where $\alpha^j$, $j = 1, 2, 3$, and $\beta$ are some constant coefficients (summation over the index $j$ is understood here, so $\alpha^j p^j$ can also be written as $\vec{\alpha} \cdot \vec{p}$). Now, employing the correspondence (1.2) one arrives at the equation

$$i\hbar \frac{\partial \psi}{\partial t} = \left( -i\hbar c\alpha^j \nabla^j + \beta mc^2 \right) \psi. \tag{1.23}$$

The consistency condition for such an equation is that upon squaring it, one should recover the Klein–Gordon equation (which corresponds trivially to the energy–momentum relation (1.6)). Before squaring Eq. (1.23) one must clarify a simple point: If one has an equation $A\psi = B\psi$ with $A$, $B$ being some operators, it does not imply automatically that $A^2\psi = B^2\psi$. Indeed, if $A$ and $B$ do not commute, the latter identity is not guaranteed. However, if $AB = BA$, then obviously $A\psi = B\psi \Rightarrow A^2\psi = AB\psi = BA\psi = B^2\psi$. Eq. (1.23) clearly corresponds to the case $[A, B] = 0$, since the time derivative commutes with $\nabla^j$ on the right-hand side. One may now

9

square Eq. (1.23) with confidence; the only caveat is that one must not assume *a priori* that the coefficients $\alpha^j$, $\beta$ commute (they cannot be ordinary numbers). Thus, one gets

$$-\hbar^2\frac{\partial^2\psi}{\partial t^2} = \left[-\hbar^2c^2(\vec{\alpha}\cdot\vec{\nabla})(\vec{\alpha}\cdot\vec{\nabla}) - i\hbar c\cdot mc^2(\vec{\alpha}\beta + \beta\vec{\alpha})\cdot\vec{\nabla} + \beta^2m^2c^4\right]\psi. \qquad (1.24)$$

In Eq. (1.24) one has

$$(\vec{\alpha}\cdot\vec{\nabla})(\vec{\alpha}\cdot\vec{\nabla}) = \alpha^j\alpha^k\nabla^j\nabla^k = \frac{1}{2}\{\alpha^j,\alpha^k\}\nabla^j\nabla^k + \frac{1}{2}\left[\alpha^j,\alpha^k\right]\nabla^j\nabla^k, \qquad (1.25)$$

but the last term in (1.25) vanishes, since $\nabla^j\nabla^k = \nabla^k\nabla^j$. In order to turn Eq. (1.24) into the form of the Klein–Gordon equation, the coefficients $\alpha^j$ must obviously satisfy the identities

$$\{\alpha^j,\alpha^k\} = 2\delta^{jk},$$
$$\{\beta,\alpha^j\} = 0, \qquad (1.26)$$
$$\beta^2 = 1.$$

So, it is clear that $\alpha^j$ and $\beta$ must be matrices rather than ordinary numbers, as we have rightly anticipated before. Moreover, the equation (1.23) has a "Schrödinger-like" form; the operator on its right-hand side could be interpreted as a Hamiltonian that should be Hermitian (self-adjoint). It means that one should impose an additional constraint on $\alpha^j$ and $\beta$, namely

$$\left(\alpha^j\right)^\dagger = \alpha^j, \qquad \beta^\dagger = \beta. \qquad (1.27)$$

Now the question is, what can be matrices satisfying (1.26) and (1.27). First of all, it is not difficult to show that the dimension of such matrices must be even. Indeed, (1.26) means, in particular, that

$$\alpha^j\alpha^k = -\alpha^k\alpha^j \qquad \text{for} \quad j \neq k, \qquad (1.28)$$

and

$$\left(\alpha^j\right)^2 = 1 \qquad \text{for} \quad j = 1,2,3. \qquad (1.29)$$

Let us now consider the determinants of the matrix products in (1.28). One has

$$\det\alpha^j \det\alpha^k = \det(-1)\det\alpha^k \det\alpha^j = (-1)^d \det\alpha^j \det\alpha^k, \qquad (1.30)$$

where $d$ is the dimension of matrices in question. Obviously, $\det\alpha^j \neq 0$ because of (1.29), so (1.30) implies $(-1)^d = 1$, i.e. $d$ is even.

The simplest choice would be $d = 2$, but it does not work; the point is that there are not four mutually anticommuting $2\times 2$ matrices. Indeed, for $\alpha^j$, $j = 1,2,3$, one could take the Pauli matrices $\sigma^j$, but then there is no non-trivial $\beta$ that would anticommute with them. Proving this statement independently is left to the reader as an instructive algebraic exercise.

The next try is $d = 4$ and we will see shortly that it does work. Needless to say, it then also means that the wave function $\psi$ in Eq. (1.23) has four components. Before showing an explicit example of $4 \times 4$ matrices satisfying (1.26) and (1.27), let us mention another general property of the matrices in question. It is easy to show that matrices $\alpha^j$ and $\beta$ are traceless,

$$\text{Tr}\,\alpha^j = 0, \qquad j = 1,2,3,$$
$$\text{Tr}\,\beta = 0. \qquad (1.31)$$

Let us prove e.g. the first identity (1.31). Since $\beta^2 = 1$, one may write

$$\text{Tr}\,\alpha^j = \text{Tr}(\beta^2\alpha^j) = \text{Tr}(\beta\alpha^j\beta) = -\text{Tr}(\alpha^j\beta^2) = -\text{Tr}\,\alpha^j,$$

10

so that indeed $\mathrm{Tr}\,\alpha^j = 0$. Note that we have utilized just the trace cyclicity and the anticommutation property $\beta\alpha^j = -\alpha^j\beta$. The second identity (1.31) can be proved in the similar way, employing the same trick with e.g. $\left(\alpha^1\right)^2 = 1$.

Finally, let us display an explicit example of the $4 \times 4$ matrices satisfying (1.26) and (1.27). They are

$$\alpha^j = \begin{pmatrix} 0 & \sigma^j \\ \sigma^j & 0 \end{pmatrix}, \qquad j = 1, 2, 3, \qquad \beta = \begin{pmatrix} 1 & 0 \\ 0 & -1 \end{pmatrix}, \tag{1.32}$$

where $\sigma^j$ are the familiar Pauli matrices

$$\sigma^1 = \begin{pmatrix} 0 & 1 \\ 1 & 0 \end{pmatrix}, \qquad \sigma^2 = \begin{pmatrix} 0 & -i \\ i & 0 \end{pmatrix}, \qquad \sigma^3 = \begin{pmatrix} 1 & 0 \\ 0 & -1 \end{pmatrix}, \tag{1.33}$$

and $1$ stands for the $2 \times 2$ unit matrix. It is straightforward to verify that the matrices (1.32) have indeed the required properties. The representation (1.32) is used frequently in practical calculations and is called the **standard representation**.

In the next chapter we will see that the magic Dirac's trick of taking the square root of the energy–momentum relation (1.6) in terms of $4 \times 4$ matrix coefficients leads indeed to a successful description of the electron. The great leap from the simple kinematical relation (1.6) to the deep quantum equation with rich physical contents makes the Dirac equation one of the most remarkable achievements of the 20th century physics. Note that Dirac received the Nobel Prize in 1933 together with E. Schrödinger. Many historical details concerning the Dirac's discovery can be found in the book [9].

# Chapter 2

# Physical contents of Dirac equation: preliminary discussion

As we have noted in the preceding chapter, the prime motivation for finding an alternative to the Klein–Gordon equation was the requirement that the probability defined in terms of a quantum mechanical wave function should be positive. So, let us now examine this problem for the Dirac equation; for convenience, we return to the natural units. Eq. (1.23) then reads

$$i \frac{\partial \psi}{\partial t} = -i \vec{\alpha} \cdot \vec{\nabla} \psi + \beta m \psi \tag{2.1}$$

(we will use the standard representation (1.32) in what follows). Let us recall that $\psi$ is a four-component wave function that is conventionally written as a column

$$\psi(x) = \begin{pmatrix} \psi_1(x) \\ \psi_2(x) \\ \psi_3(x) \\ \psi_4(x) \end{pmatrix}. \tag{2.2}$$

Upon Hermitian conjugation of Eq. (2.1) one has

$$-i \frac{\partial \psi^\dagger}{\partial t} = i \vec{\nabla} \psi^\dagger \vec{\alpha} + m \psi^\dagger \beta, \tag{2.3}$$

where $\psi^\dagger = (\psi_1^*, \psi_2^*, \psi_3^*, \psi_4^*)$, and we have utilized the hermiticity property (1.27) of $\vec{\alpha}$ and $\beta$. Multiplying Eq. (2.1) by $\psi^\dagger$ from the left and (2.3) by $\psi$ from the right, and taking then the difference of the two equations, one gets immediately

$$\frac{\partial}{\partial t} (\psi^\dagger \psi) + \vec{\nabla} (\psi^\dagger \vec{\alpha} \psi) = 0, \tag{2.4}$$

which is the anticipated continuity equation. Thus we may identify the probability density and the probability current as

$$\rho_{\text{Dirac}} = \psi^\dagger \psi, \qquad \vec{j}_{\text{Dirac}} = \psi^\dagger \vec{\alpha} \psi. \tag{2.5}$$

The positivity of the $\rho_{\text{Dirac}}$ is obvious, since

$$\psi^\dagger \psi = |\psi_1|^2 + |\psi_2|^2 + |\psi_3|^2 + |\psi_4|^2. \tag{2.6}$$

This is an expected result, due to the fact that the Dirac equation (2.1) is, in a sense, "square root of Klein–Gordon equation"; more precisely, it is an evolution equation of the 1st order in time, having the form

$$i \frac{\partial \psi}{\partial t} = H \psi, \tag{2.7}$$

where $H$ is the Dirac Hamiltonian

$$H = -i\vec{\alpha} \cdot \vec{\nabla} + \beta m \,. \tag{2.8}$$

Thus, the time evolution is generated by an operator of energy, as it should be, in accordance with the general principles of quantum theory.

A next issue is the angular momentum. Let us start with orbital angular momentum, defined in the standard way as $\vec{L} = \vec{x} \times \vec{p}$, where $\vec{p}$ is the (linear) momentum $\vec{p} = -i\vec{\nabla}$. As we know, $\vec{L}$ commutes with the non-relativistic Hamiltonian in the Schrödinger equation (1.4). For the Dirac Hamiltonian (2.8) one gets, employing the canonical commutation relation $[x^j, p^k] = i\delta^{jk}$,

$$[H, \vec{L}] = -i(\vec{\alpha} \times \vec{p}) \,. \tag{2.9}$$

Let us remark that the vector product in (2.9) is defined formally as usual, i.e.

$$(\vec{\alpha} \times \vec{p})^j = \varepsilon^{jkl} \alpha^k p^l \,.$$

So, apparently, there is something missing, since any decent angular momentum should be an integral of motion for the free particle, i.e. the corresponding operator should commute with the Hamiltonian. In other words, the fact that $[H, \vec{L}] \neq 0$ is a hint that we are on the right track towards the electron spin. A good candidate for such an additional ingredient of the full angular momentum is guessed quite easily. Let us consider the $4 \times 4$ matrices

$$\vec{S} = \frac{1}{2}\vec{\Sigma} \,, \qquad \vec{\Sigma} = \begin{pmatrix} \vec{\sigma} & 0 \\ 0 & \vec{\sigma} \end{pmatrix} \,, \tag{2.10}$$

and recall that the Pauli matrices have the commutation relations

$$[\sigma_j, \sigma_k] = 2i\varepsilon_{jkl}\sigma_l \,. \tag{2.11}$$

This means that the matrices $\vec{S}$ defined by (2.10) satisfy

$$[S_j, S_k] = i\varepsilon_{jkl}S_l \,, \tag{2.12}$$

which, of course, is a set of commutation relations for components of an angular momentum. Needless to say, the matrices $\vec{S}$ possess eigenvalues $\pm 1/2$ (because $(\sigma_j)^2 = 1$ for $j = 1, 2, 3$). Now we may evaluate the commutator $[H, \vec{S}]$. Clearly, $\vec{S}$ commutes with the diagonal matrix $\beta$ (see (1.32)). Concerning the commutator involving $\vec{\alpha}$, one gets first

$$[\alpha^j, \Sigma^k] = \begin{pmatrix} 0 & 2i\varepsilon^{jkl}\sigma^l \\ 2i\varepsilon^{jkl}\sigma^l & 0 \end{pmatrix} \,,$$

so that

$$[H, \Sigma^k] = 2i(\vec{\alpha} \times \vec{p})^k \,. \tag{2.13}$$

Summarizing the results of our simple algebraic exercise, we have

$$\begin{aligned} [H, \vec{L}] &= -i(\vec{\alpha} \times \vec{p}) \,, \\ [H, \vec{S}] &= i(\vec{\alpha} \times \vec{p}) \,, \end{aligned} \tag{2.14}$$

and thus

$$[H, \vec{J}] = 0 \,, \tag{2.15}$$

13

with

$$\vec{J} = \vec{L} + \vec{S}. \tag{2.16}$$

Thus, in such a straightforward manner we have recovered the electron spin as a part of the conserved total angular momentum (2.16).

Let us now recall the problem of negative energy solutions of the Klein–Gordon equation, mentioned in the preceding chapter (cf. (1.12)). One may wonder whether the Dirac equation suffers an analogous difficulty. For clarifying this point, we are going to consider the solution of Eq. (2.1) in the form of a plane wave involving the usual factor $\exp\left[-i(Et - \vec{p} \cdot \vec{x})\right]$. To make our discussion as simple as possible, we will restrict ourselves to the case of a particle at rest, i.e. set $\vec{p} = 0$. Eq. (2.1) is then reduced to

$$i\frac{\partial \psi}{\partial t} = \beta m \psi . \tag{2.17}$$

Taking into account the block diagonal structure of the matrix $\beta$ ((1.32), it is useful to split the $\psi$ as

$$\psi = \begin{pmatrix} \varphi \\ \chi \end{pmatrix}, \tag{2.18}$$

where $\varphi$ and $\chi$ are two-component column vectors. Eq. (2.17) is then recast as

$$i\frac{\partial \varphi}{\partial t} = m\varphi , \tag{2.19}$$

$$i\frac{\partial \chi}{\partial t} = -m\chi . \tag{2.20}$$

Thus, two linearly independent solutions of Eq. (2.19) may be written e.g. as

$$\varphi_{(1)} = e^{-imt} \begin{pmatrix} 1 \\ 0 \end{pmatrix}, \qquad \varphi_{(2)} = e^{-imt} \begin{pmatrix} 0 \\ 1 \end{pmatrix}, \tag{2.21}$$

and similarly for (2.20),

$$\chi_{(1)} = e^{imt} \begin{pmatrix} 1 \\ 0 \end{pmatrix}, \qquad \chi_{(2)} = e^{imt} \begin{pmatrix} 0 \\ 1 \end{pmatrix}. \tag{2.22}$$

In this way, we obtain a set of four independent solutions of Eq. (2.1)

$$\psi_{(1)} = e^{-imt} \begin{pmatrix} 1 \\ 0 \\ 0 \\ 0 \end{pmatrix}, \quad \psi_{(2)} = e^{-imt} \begin{pmatrix} 0 \\ 1 \\ 0 \\ 0 \end{pmatrix}, \quad \psi_{(3)} = e^{imt} \begin{pmatrix} 0 \\ 0 \\ 1 \\ 0 \end{pmatrix}, \quad \psi_{(4)} = e^{imt} \begin{pmatrix} 0 \\ 0 \\ 0 \\ 1 \end{pmatrix}. \tag{2.23}$$

Obviously, $\psi_{(1)}$ and $\psi_{(2)}$ correspond to the positive rest energy $E = m$, while $\psi_{(3)}$ and $\psi_{(4)}$ carry negative energy $E = -m$ (they are also characterized by the two possible spin projections to the third axis, up and down ($\pm 1/2$)). It is interesting to notice that in the considered case, the existence of the negative energy solutions is a consequence of the specific structure of the matrix $\beta$. If $\beta$ were $4 \times 4$ unit matrix, we would have only a solution with positive energy. But, alas, $\beta$ can never be the unit matrix because of the required anticommutation relations (1.26). As we have already noted in the preceding chapter, the appearance of negative energy solutions is a generic feature of the equations of relativistic quantum mechanics. We will discuss the plane-wave solutions of Dirac equation in detail later on.

The last topic that we are going to discuss here is a derivation of the spin magnetic moment of the electron. Soon after the birth of relativistic quantum mechanics this was indeed one of the most remarkable achievements of the Dirac theory, so it certainly deserves a detailed exposition.

To this end, one has to start with the Dirac equation for the electron in an external electromagnetic field. Using the scalar potential $\phi$ and vector potential $\vec{A}$, one may write the relevant equation as

$$i\frac{\partial \psi}{\partial t} = \left[\vec{\alpha} \cdot (-i\vec{\nabla} - e\vec{A}) + e\phi + \beta m\right] \psi . \tag{2.24}$$

Note that the form (2.24) represents the so-called **minimal electromagnetic interaction** and satisfies certainly the requirement of gauge invariance (invariance under gauge transformations of the potentials $\phi$ and $\vec{A}$). In fact, it is not the most general choice, but coincides with the recipe to be employed later on, in quantum electrodynamics. More comments on a possible extension of the gauge invariant electromagnetic interaction within the framework of Dirac equation are deferred to the Chapter 13.

Our ultimate goal is to get the non-relativistic two-component **Pauli equation**, from which one can extract easily the value of the magnetic moment in question. For this purpose, we will separate upper and lower components of the wave function $\psi$ as

$$\psi = \begin{pmatrix} \widetilde{\varphi} \\ \widetilde{\chi} \end{pmatrix} . \tag{2.25}$$

Then, denoting

$$-i\vec{\nabla} - e\vec{A} = \vec{\pi} , \tag{2.26}$$

Eq. (2.24) is recast as a pair of coupled two-component equations

$$\begin{aligned} i\frac{\partial \widetilde{\varphi}}{\partial t} &= (\vec{\sigma} \cdot \vec{\pi})\widetilde{\chi} + (e\phi + m)\widetilde{\varphi} , \\ i\frac{\partial \widetilde{\chi}}{\partial t} &= (\vec{\sigma} \cdot \vec{\pi})\widetilde{\varphi} + (e\phi - m)\widetilde{\chi} . \end{aligned} \tag{2.27}$$

Throughout our calculation we will have in mind a situation close to the non-relativistic limit; thus, it is convenient to factorize in the wave function a part corresponding to the rest energy. (cf. (2.21)), i.e. introduce the Ansatz

$$\begin{pmatrix} \widetilde{\varphi} \\ \widetilde{\chi} \end{pmatrix} = e^{-imt} \begin{pmatrix} \varphi \\ \chi \end{pmatrix} . \tag{2.28}$$

Inserting (2.28) into Eq. (2.27) one gets, after a simple manipulation,

$$i\frac{\partial \varphi}{\partial t} = (\vec{\sigma} \cdot \vec{\pi})\chi + e\phi\varphi , \tag{2.29a}$$

$$i\frac{\partial \chi}{\partial t} = (\vec{\sigma} \cdot \vec{\pi})\varphi + e\phi\chi - 2m\chi . \tag{2.29b}$$

We consider weak fields, in particular $e\phi \ll m$, as well as a small kinetic energy; the latter assumption may be expressed, technically, as

$$\frac{\partial \chi}{\partial t} \ll m\chi .$$

Thus, in Eq. (2.29b) we will neglect $\partial \chi / \partial t$ and $e\phi\chi$ in comparison with $2m\chi$. Consequently, the function $\chi$ can be approximately written as

$$\chi \doteq \frac{1}{2m}(\vec{\sigma} \cdot \vec{\pi})\varphi . \tag{2.30}$$

Using the last expression in Eq. (2.29a), we have

$$i \frac{\partial \varphi}{\partial t} = \frac{1}{2m} (\vec{\sigma} \cdot \vec{\pi})(\vec{\sigma} \cdot \vec{\pi}) \varphi + e\phi\varphi \,. \tag{2.31}$$

To work out the right-hand side of Eq. (2.31), one may utilize the familiar identity for Pauli matrices

$$\sigma_j \sigma_k = \delta_{jk} \cdot \mathbb{1} + i\varepsilon_{jkl}\sigma_l \,. \tag{2.32}$$

From (2.32) one then gets

$$(\vec{\sigma} \cdot \vec{\pi})(\vec{\sigma} \cdot \vec{\pi}) = \vec{\pi}^2 + i\vec{\sigma} \cdot (\vec{\pi} \times \vec{\pi}) \,. \tag{2.33}$$

One must treat the vector product carefully, since $\vec{\pi}$ is a differential operator. So, one has to evaluate it by letting it act on an arbitrary test function $f$; one obtains, after some manipulations,

$$(\vec{\pi} \times \vec{\pi})^j f = ie(\vec{\nabla} \times \vec{A})^j f \,,$$

so that

$$\vec{\pi} \times \vec{\pi} = ie(\vec{\nabla} \times \vec{A}) = ie\vec{B} \,, \tag{2.34}$$

where $\vec{B}$ is the magnetic field (the reader is encouraged to reproduce independently the result (2.34)). In total, we thus have

$$(\vec{\sigma} \cdot \vec{\pi})(\vec{\sigma} \cdot \vec{\pi}) = (-i\vec{\nabla} - e\vec{A})^2 - e\vec{\sigma} \cdot \vec{B} \,.$$

The two-component equation (2.31) thus becomes

$$i \frac{\partial \varphi}{\partial t} = \left[ \frac{1}{2m}(\vec{p} - e\vec{A})^2 + e\phi - \frac{e}{2m}\vec{\sigma} \cdot \vec{B} \right] \varphi \,, \tag{2.35}$$

and this is the anticipated Pauli equation. Obviously, the last term in the square brackets represents an interaction of magnetic moment with magnetic field $\vec{B}$. Since the Pauli matrices have eigenvalues $\pm 1$, one may conclude that the value of the magnetic moment in question is $e/(2m)$ (i.e. one **Bohr magneton**). Note that Wolfgang Pauli formulated Eq. (2.35) in 1927 as a phenomenological description of the electron moving in an external field; he then used the empirically known value of the spin magnetic moment. The derivation described above is actually a **prediction** of the relevant value, made on the basis of a more fundamental equation (though restricted to the minimal electromagnetic interaction). This is why the result (2.35) obtained as a non-relativistic approximation of Dirac equation is extolled as a true achievement.

One more remark is in order here. Magnetic moment of a particle is usually characterized also by its **gyromagnetic ratio**, which is the ratio of the magnetic moment to the angular momentum. It is a well-known fact that for the orbital motion, the gyromagnetic ratio is equal to $e/(2m)$ (this holds both in classical and in quantum theory). For the spin magnetic moment we obviously have the gyromagnetic ratio $e/m$, since the magnitude of the spin projection is $1/2$. Thus, the spin magnetic moment of electron does not obey the "normal" rule and differs from it by a dimensionless factor called simply **g-factor**, here equal to 2. The g-factor has become a usual way of description of intrinsic magnetic moment of subatomic particles.

The above-described elegant derivation of the electron spin magnetic moment, in particular the natural explanation of the "anomalous" value $g = 2$ for the g-factor was certainly a great success of the Dirac theory in 1928. In fact, even more remarkable was the continuation of this success story some 20 years later. It turned out that quantum electrodynamics (QED) leads to a tiny correction to the Dirac's prediction. The correction is of relative order of one per-mille; it was found experimentally in 1947 and subsequently calculated theoretically by Julian Schwinger, one of the founding fathers of modern QED. This achievement corroborated strongly the QED as the relevant model of quantum field theory capable to describe the most subtle electromagnetic phenomena. We will discuss this topic in detail in Chapter 50.

# Chapter 3

# Covariant form of Dirac equation. Fun with $\gamma$-matrices

The main topic of this and the following chapter is the relativistic invariance of the Dirac equation. Before proceeding to this extensive theme, let us return briefly to the Klein–Gordon equation. In that case, the relativistic invariance is almost obvious, since the d'Alembert operator $\Box = \partial^\mu \partial_\mu$ has the structure of a scalar product in the four-dimensional spacetime. So, when passing from one Lorentz reference frame to another, with coordinates transformed as $x' = \Lambda x$ (with $\Lambda$ being the matrix of a Lorentz transformation), one can get along with a trivial transformation of the wave function

$$\psi'(x') = \psi(x). \tag{3.1}$$

As we will see in the next chapter, in the case of Dirac equation the situation is much more interesting, i.e. far from trivial. For a proper assessment of this problem, it is convenient to recast first the Dirac equation in a form that is more symmetric with respect to spacetime coordinates; this is what is meant by the term "covariant form" in the title of this chapter.

For a moment, let us use the ordinary units with $\hbar \neq 1$, $c \neq 1$. As we know, Dirac equation reads

$$i\hbar \frac{\partial \psi}{\partial t} = -i\hbar c \, \vec{\alpha} \cdot \vec{\nabla} \psi + \beta mc^2 \psi. \tag{3.2}$$

Introducing the usual notation $x_0 = ct$ and taking into account that $\nabla^j$ is defined, conventionally, as $\partial / \partial x^j = \partial_j$, one may rewrite Eq. (3.2) as

$$i\hbar c \frac{\partial \psi}{\partial x_0} = -i\hbar c \, \vec{\alpha} \cdot \vec{\nabla} \psi + mc^2 \beta \psi, \tag{3.3}$$

and this subsequently becomes

$$i\hbar \beta \partial_0 \psi + i\hbar \beta \alpha^j \partial_j \psi - mc\psi = 0. \tag{3.4}$$

If we now denote

$$\gamma^0 = \beta, \qquad \gamma^j = \beta \alpha^j, \tag{3.5}$$

then Eq. (3.4) can be rewritten as

$$i\gamma^\mu \partial_\mu \psi - \frac{mc}{\hbar} \psi = 0. \tag{3.6}$$

One may notice the appearance of the inverse Compton length in the second term; this, of course, was to be expected on dimensional grounds. Thus, in natural units, Eq. (3.6) reads

$$(i\gamma^\mu \partial_\mu - m)\psi(x) = 0, \tag{3.7}$$

and this is the promised "covariant form" of Dirac equation, which will be our staple food from now on. The Dirac matrices $\gamma^\mu$ will be called simply gamma matrices, or $\gamma$-matrices in what follows. Notice that Eq. (3.7) seems to look covariant, since the term $\gamma^\mu \partial_\mu$ has, at first sight, the form of a scalar product in Minkowski spacetime; however, $\gamma^\mu$, $\mu = 0, 1, 2, 3$, are fixed $4 \times 4$ matrices to be used in any reference frame, so one should not jump to conclusions at this point. Anyway, a most economical form of Eq. (3.7), utilizing the scalar product symbol, is perhaps

$$i\gamma \cdot \partial\psi = m\psi \,, \tag{3.8}$$

and this is precisely what is engraved in the commemorative marker in Dirac's honour in Westminster Abbey (it is there since 1995).

So, from now on, we will work with the set of matrices $\gamma^\mu$ introduced in (3.5); it is also convenient to employ formally the rule for raising and lowering the Lorentz indices and define $\gamma_\mu = g_{\mu\nu}\gamma^\nu$, i.e.

$$\gamma_0 = \gamma^0 \,, \qquad \gamma_j = -\gamma^j \,. \tag{3.9}$$

To begin with, we should rewrite the anticommutation relations (1.26) in terms of $\gamma^\mu$. This is an easy exercise; we have

$$\{\gamma^0, \gamma^j\} = \{\beta, \beta\alpha^j\} = \beta^2\alpha^j + \beta\alpha^j\beta = 0 \,,$$
$$\{\gamma^j, \gamma^k\} = \{\beta\alpha^j, \beta\alpha^k\} = \beta\alpha^j\beta\alpha^k + \beta\alpha^k\beta\alpha^j = -\{\alpha^j, \alpha^k\} = -2\delta^{jk} \cdot \mathbb{1} \,,$$

and, of course,

$$\{\gamma^0, \gamma^0\} = 2\beta^2 = 2 \cdot \mathbb{1} \,,$$

where $\mathbb{1}$ denotes the $4 \times 4$ unit matrix. Thus, we can summarize the above results as

$$\{\gamma^\mu, \gamma^\nu\} = 2g^{\mu\nu} \cdot \mathbb{1} \tag{3.10}$$

(for brevity, we will usually omit $\mathbb{1}$ when using (3.10)). In mathematics, the anticommutation relations (3.10) are known to correspond to generators of the so-called **Clifford algebra**. Note that (3.10) means, in particular,

$$(\gamma^0)^2 = \mathbb{1} \,, \qquad (\gamma^j)^2 = -\mathbb{1} \,. \tag{3.11}$$

Further, let us see what becomes of the hermiticity relations (1.27). Obviously, one has

$$(\gamma^0)^\dagger = \gamma^0 \,, \qquad (\gamma^j)^\dagger = (\beta\alpha^j)^\dagger = \alpha^j\beta = -\gamma^j \,. \tag{3.12}$$

So, taking into account (3.10), (3.12) may be summarized as

$$(\gamma^\mu)^\dagger = \gamma^0\gamma^\mu\gamma^0 \,. \tag{3.13}$$

This last relation is one of the identities that will be used very frequently in our forthcoming calculations.

It is highly useful to introduce a fifth $\gamma$-matrix, denoted traditionally as $\gamma_5$, which is proportional to the product $\gamma_0\gamma_1\gamma_2\gamma_3$. The salient feature of such a matrix product is that it anticommutes with any $\gamma^\mu$, $\mu = 0, 1, 2, 3$ (the reader is encouraged to check this statement independently). Note that we fix the definition of $\gamma_5$ conventionally as

$$\gamma_5 = i\gamma^0\gamma^1\gamma^2\gamma^3 \,. \tag{3.14}$$

18

So, we have

$$\{\gamma_5, \gamma^\mu\} = 0, \qquad \mu = 0, 1, 2, 3. \tag{3.15}$$

Other basic properties of the $\gamma_5$ shown in (3.14) are

$$(\gamma_5)^2 = \mathbb{1}, \qquad (\gamma_5)^\dagger = \gamma_5 \tag{3.16}$$

(again, proving (3.16) is left to the reader as an easy exercise).

The rest of this chapter is devoted to a rather detailed discussion of various properties of the gamma matrices; it is just an algebra, no physics. So, this is the would-be "fun" mentioned in the heading (though, admittedly, the "fun" in the present context is a matter of personal taste — obviously, the trick here was to lure the reader into studying an otherwise somewhat boring subject).

First of all, it is useful to master some simple formulae for traces of products of $\gamma$-matrices. We already know (see (1.31)) that traces of the original Dirac matrices $\alpha^j$ and $\beta$ are zero; this finding is easily reproduced for the $\gamma$-matrices as well. With the matrix $\gamma_5$ at hand, it is straightforward to prove that the trace of a product of odd number of $\gamma$-matrices is zero; symbolically,

$$\mathrm{Tr}(\text{odd} \#) = 0. \tag{3.17}$$

The proof can be left to the reader as an instructive exercise. Hint: Use $\gamma_5^2 = \mathbb{1}$, the anticommutation property (3.15) and trace cyclicity. By the way, using a similar trick, one can prove as well that

$$\mathrm{Tr}\, \gamma_5 = 0 \tag{3.18}$$

(to this end, one may use e.g. $\gamma_0^2 = \mathbb{1}$).

For products of an even number $n = 2k$ of $\gamma$-matrices one gets a series of formulae that have quite uniform structure. Let's start with $n = 2$. One has certainly $\mathrm{Tr}(\gamma_\mu \gamma_\nu) = \mathrm{Tr}(\gamma_\nu \gamma_\mu)$, so that, employing (3.10),

$$\mathrm{Tr}(\gamma_\mu \gamma_\nu) = \frac{1}{2}\, \mathrm{Tr}(\gamma_\mu \gamma_\nu + \gamma_\nu \gamma_\mu) = \frac{1}{2} \cdot 2 g_{\mu\nu}\, \mathrm{Tr}\, \mathbb{1} = 4 g_{\mu\nu}. \tag{3.19}$$

How about $n = 4$? We have, using (3.10),

$$\mathrm{Tr}(\gamma_\mu \gamma_\nu \gamma_\rho \gamma_\sigma) = \mathrm{Tr}\left[(2g_{\mu\nu} - \gamma_\nu \gamma_\mu)\gamma_\rho \gamma_\sigma\right] = 2g_{\mu\nu}\, \mathrm{Tr}(\gamma_\rho \gamma_\sigma) - \mathrm{Tr}(\gamma_\nu \gamma_\mu \gamma_\rho \gamma_\sigma). \tag{3.20}$$

Now, we can go on anticommuting the $\gamma_\mu$ with $\gamma_\rho$ and then with $\gamma_\sigma$. In this way, we end up with

$$\mathrm{Tr}(\gamma_\mu \gamma_\nu \gamma_\rho \gamma_\sigma) = 2g_{\mu\nu}\, \mathrm{Tr}(\gamma_\rho \gamma_\sigma) - 2g_{\mu\rho}\, \mathrm{Tr}(\gamma_\nu \gamma_\sigma) + 2g_{\mu\sigma}\, \mathrm{Tr}(\gamma_\nu \gamma_\rho) - \mathrm{Tr}(\gamma_\nu \gamma_\rho \gamma_\sigma \gamma_\mu). \tag{3.21}$$

However, in the last term on the right-hand side of (3.21) we can use the trace cyclicity and one thus gets, eventually,

$$\begin{aligned}
\mathrm{Tr}(\gamma_\mu \gamma_\nu \gamma_\rho \gamma_\sigma) &= \frac{1}{2}\left[2g_{\mu\nu}\, \mathrm{Tr}(\gamma_\rho \gamma_\sigma) - 2g_{\mu\rho}\, \mathrm{Tr}(\gamma_\nu \gamma_\sigma) + 2g_{\mu\sigma}\, \mathrm{Tr}(\gamma_\nu \gamma_\rho)\right] \\
&= 4(g_{\mu\nu}g_{\rho\sigma} - g_{\mu\rho}g_{\nu\sigma} + g_{\mu\sigma}g_{\nu\rho}),
\end{aligned} \tag{3.22}$$

where we have utilized the preceding result (3.19).

The above example makes it clear how to proceed further, i.e. for $n \geq 6$: one moves the first $\gamma$-matrix in the product step by step (employing the basic anticommutation relation (3.10)) to the last position, and then the trace cyclicity can be used. On the way, one encounters products with the number of $\gamma$-matrices less by two, so one can utilize the result for the preceding member of the whole hierarchy. Thus, it is quite clear that the results for traces in question are expressed as products of pertinent components of the metric tensor $g$; for $n = 2k$ these products consist of just $k$ factors. The resulting number $N$ of terms for the trace with $n = 2k$ grows rapidly with $n$; the recursive procedure outlined above shows clearly that $N(2k) = (2k - 1)N(2k - 2)$, which means that

$$N(2k) = (2k - 1)!! = \frac{(2k)!}{2^k \, k!} \tag{3.23}$$

(so, for $n = 6$ one gets 15 terms, for $n = 8$ there are 105 terms, etc.). One would certainly have a lot of fun computing such a trace for $n = 14$, which amounts to 135 135 terms (sic!), but rest assured that we will always get along with smaller numbers.[3] In any case, one might wonder what is it all good for; please, don't worry and be patient, you will see that the traces of products of $\gamma$-matrices will come in handy later (in QED, in particular). Perhaps one may refer to a well-known quotation (due to A. P. Chekhov (Čechov)) saying that "If in the first act (of a drama) there is a rifle hanging on the wall, then in the last act someone must fire it." (in fact, we will "fire the rifle" much earlier than in the last act of this lecture course).

There are many other special identities for $\gamma$-matrices that will be practically useful later on (they are collected in Appendix C), but now we are going to study some of their deeper structural properties that will be needed soon. The basic point of the analysis that follows is the observation that one can find an appropriate basis in the space of $4 \times 4$ matrices, made of products of $\gamma$-matrices. To this end, we will consider 16 matrices, denoted for convenience as $\Gamma_A$, $A = 1, \ldots, 16$, and defined as follows. First, $\Gamma_1 = \mathbb{1}$ (this can be obtained as e.g. $(\gamma_0)^2$); next, we take

$$\gamma_\mu, \quad \mu = 0, 1, 2, 3 \quad : \quad \Gamma_2, \Gamma_3, \Gamma_4, \Gamma_5, \tag{3.24}$$

and then one may form products of two, three and four $\gamma$-matrices. That's the end of the story — it is clear that products of five and more $\gamma$-matrices would not bring anything new, because of (3.11) (such expressions are reduced to products of less than five $\gamma$-matrices). Thus, let us denote

$$\gamma_0\gamma_1, \gamma_0\gamma_2, \gamma_0\gamma_3, \gamma_1\gamma_2, \gamma_1\gamma_3, \gamma_2\gamma_3 \quad : \quad \Gamma_6, \ldots, \Gamma_{11}. \tag{3.25}$$

Further, four independent products of three $\gamma$-matrices are equivalent to

$$\gamma_0\widetilde{\gamma}_5, \gamma_1\widetilde{\gamma}_5, \gamma_2\widetilde{\gamma}_5, \gamma_3\widetilde{\gamma}_5 \quad : \quad \Gamma_{12}, \Gamma_{13}, \Gamma_{14}, \Gamma_{15}, \tag{3.26}$$

where $\widetilde{\gamma}_5$ is defined as

$$\widetilde{\gamma}_5 = \gamma_0\gamma_1\gamma_2\gamma_3 \tag{3.27}$$

(we have chosen this provisional notation instead of (3.14) for simplicity). Finally, we set

$$\Gamma_{16} = \widetilde{\gamma}_5. \tag{3.28}$$

Using (3.10), it is easy to see that the square of any matrix $\Gamma_A$ is either $\mathbb{1}$ or $-\mathbb{1}$. In particular, one has

$$\begin{aligned}
\Gamma_A^2 &= \mathbb{1} && \text{for} \quad A = 1, 2, 6, 7, 8, 12, \\
\Gamma_A^2 &= -\mathbb{1} && \text{for} \quad A = 3, 4, 5, 9, 10, 11, 13, 14, 15, 16.
\end{aligned} \tag{3.29}$$

So, the set of $\Gamma_A$, $A = 1, \ldots, 16$, has the right number of terms to be a good candidate for a basis in the considered 16-dimensional space of $4 \times 4$ matrices. Before showing that the $\Gamma_A$ are indeed

---

[3]Note that discovering this remarkable number has been just serendipitous; obviously, an average QFT practitioner can hardly come across it in routine calculations.

linearly independent, we are going to present a few auxiliary statements (lemmas) describing some simple, but important, properties of the matrices $\Gamma_A$.

**L1 (commutation & anticommutation)**: *Any pair $\Gamma_A$, $\Gamma_B$ either commutes or anticommutes.*
This can be proved easily by employing the anticommutation relations (3.10) and (3.15).

**L2 (on traces)**: *Tr $\Gamma_A = 0$ for any $A > 1$.*
A part of this statement we have already proved before; in general, using (3.10) and (3.15) is sufficient.

**L3 (on the rearrangement)**: *When multiplying all $\Gamma_A$'s by a particular $\Gamma_B$ from left or right, one gets again the same set, up to signs and the order.*
One may prove such a statement simply "by inspection" (in principle, one should produce a pertinent multiplication table with $16 \times 16 = 256$ entries).

The above lemmas are now sufficient for establishing the fact that such $\Gamma_A$'s form a basis.

**L4 (on the linear independence)**: *The matrices $\Gamma_A$, $A = 1, \ldots, 16$, are linearly independent.*
Proof: Suppose that

$$a_1\Gamma_1 + a_2\Gamma_2 + \ldots + a_{16}\Gamma_{16} = 0 \tag{3.30}$$

for some coefficients $a_1, \ldots, a_{16}$. For convenience, let us denote the linear combination on the left-hand side of (3.30) simply as $L$. Obviously, Eq. (3.30) implies, for any $A = 1, \ldots, 16$,

$$\mathrm{Tr}(\Gamma_A L) = 0. \tag{3.31}$$

Now, using lemmas L2 and L3, along with the identities (3.28), (3.29), one gets $\mathrm{Tr}(\Gamma_A L) = \pm 4 a_A$, depending on whether $\Gamma_A^2$ is $\mathbb{1}$ or $-\mathbb{1}$. In any case, Eq. (3.31) thus implies $a_A = 0$ for any $A = 1, \ldots, 16$ and the statement L4 is thereby proved.

Now we are in a position to prove the following important statement:

**L5**: *Let $M$ be a matrix $4 \times 4$ that commutes with any $\gamma_\mu$, $\mu = 0, 1, 2, 3$. Then $M$ is a multiple of the unit matrix.*
Proof: According to the preceding lemma L4, the matrices $\Gamma_1, \ldots, \Gamma_{16}$ form a basis; thus, the matrix $M$ can be expressed as a linear combination

$$M = a_1\Gamma_1 + \ldots + a_{16}\Gamma_{16}. \tag{3.32}$$

The premise represents four conditions, namely $[M, \gamma_\mu] = 0$ for $\mu = 0, 1, 2, 3$. Let us start with $\mu = 0$. Using (3.32), the condition $[M, \gamma_0] = 0$ means

$$a_1\Gamma_1\gamma_0 + \ldots + a_{16}\Gamma_{16}\gamma_0 = a_1\gamma_0\Gamma_1 + \ldots + a_{16}\gamma_0\Gamma_{16}. \tag{3.33}$$

It is easy to find out that $\gamma_0$ commutes with $\Gamma_A$ for $A = 1, 2, 9, 10, 11, 13, 14, 15$ and anticommutes with $\Gamma_A$ for $A = 3, 4, 5, 6, 7, 8, 12, 16$. Thus, the terms involving the commuting $\Gamma_A$'s drop out of Eq. (3.33), while the anticommuting $\Gamma_A$'s survive, and Eq. (3.33) eventually becomes

$$a_3\Gamma_3 + a_4\Gamma_4 + a_5\Gamma_5 + a_6\Gamma_6 + a_7\Gamma_7 + a_8\Gamma_8 + a_{12}\Gamma_{12} + a_{16}\Gamma_{16} = 0. $$

Of course, according to the lemma L4 this amounts to

$$a_3 = a_4 = a_5 = a_6 = a_7 = a_8 = a_{12} = a_{16} = 0. \tag{3.34}$$

Consequently, after this first step, the expansion (3.32) is reduced to

$$M = a_1\Gamma_1 + a_2\Gamma_2 + a_9\Gamma_9 + a_{10}\Gamma_{10} + a_{11}\Gamma_{11} + a_{13}\Gamma_{13} + a_{14}\Gamma_{14} + a_{15}\Gamma_{15}. \tag{3.35}$$

One may now continue in this way, using for (3.35) the condition $[M, \gamma_1] = 0$. It reduces further the form of $M$, and then one goes on along the same line with $\gamma_2$ and $\gamma_3$. It turns out that eventually one is left with $M = a_1 \Gamma_1 = a_1 \cdot \mathbb{1}$ (since $\Gamma_1$ commutes with anything), and this is precisely what we wanted to prove. The reader is urged to check independently the steps involving the commutators $[M, \gamma_\mu]$ for $\mu = 1, 2, 3$.

The above series of statements concerning matrices $\Gamma_A$ culminates in a profound theorem, usually called the "fundamental theorem on $\gamma$-matrices". It can be formulated as follows.

**Theorem**: *Let $\gamma^\mu$ and $\gamma'^\mu$, $\mu = 0, 1, 2, 3$, be two sets of $4 \times 4$ matrices satisfying the relations $\{\gamma^\mu, \gamma^\nu\} = 2g^{\mu\nu}$, $\{\gamma'^\mu, \gamma'^\nu\} = 2g^{\mu\nu}$. Then there exists a non-singular matrix $S$, unique up to a multiplicative factor, such that*

$$\gamma'^\mu = S\gamma^\mu S^{-1} \tag{3.36}$$

*for any $\mu = 0, 1, 2, 3$. Further, if the $\gamma$-matrices satisfy the hermiticity conditions (3.12), the matrix $S$ can be chosen to be unitary.*

The proof of this remarkable statement is somewhat long and thus we will refrain from presenting it here. The interested reader can find the proof e.g. in the book [3].

Obviously, the importance of this theorem consists in the observation that all possible realizations of the Dirac $\gamma$-matrices are equivalent. Nevertheless, some particular representations may be more convenient than others in practical calculations. So, one might use a paraphrase of the familiar sentence from a famous book by George Orwell, namely: "All representations of $\gamma$-matrices are equal, but some of them are more equal than others".

We already know at least one explicit realization of $\gamma$-matrices; more precisely, we know the so-called standard representation of $\vec{\alpha}$ and $\beta$ (see (1.32)), and using (3.5) one then has

$$\gamma^0 = \begin{pmatrix} \mathbb{1} & 0 \\ 0 & -\mathbb{1} \end{pmatrix}, \qquad \gamma^j = \begin{pmatrix} 0 & \sigma_j \\ -\sigma_j & 0 \end{pmatrix}, \quad j = 1, 2, 3, \tag{3.37}$$

as the standard representation for $\gamma^\mu$. We have noticed before that within such a representation, the non-relativistic limit is characterized by a suppression of the lower two components of the wave function with respect to the upper ones. There are at least two more examples of $\gamma$-matrix representations that are worth mentioning here. One of them is the so-called **spinor** (or **chiral**) representation (the origin of these names will become clear later), which is

$$\gamma_S^0 = \begin{pmatrix} 0 & \mathbb{1} \\ \mathbb{1} & 0 \end{pmatrix}, \qquad \gamma_S^j = \begin{pmatrix} 0 & -\sigma_j \\ \sigma_j & 0 \end{pmatrix}, \quad j = 1, 2, 3. \tag{3.38}$$

Further, there is a remarkable representation, in which all $\gamma$-matrices are purely imaginary (this in turn means that in the Dirac equation only real coefficients are then involved). It is called the **Majorana representation** and the corresponding matrices $\gamma_M^\mu$ can be expressed with the help of the standard $\gamma$-matrices as

$$\gamma_M^0 = \gamma^0 \gamma^2, \quad \gamma_M^1 = -\gamma^1 \gamma^2, \quad \gamma_M^2 = -\gamma^2, \quad \gamma_M^3 = \gamma^2 \gamma^3. \tag{3.39}$$

One might also wonder whether there could be a representation involving purely real $\gamma$-matrices. The answer is no. The proof is quite tedious, but it could be a real challenge for a hard-working student. In any case, an enjoyable exercise would be finding the transformation matrices implementing the passage from the standard representation to the other two mentioned above.

After all those preparatory steps, we are ready to take up seriously the problem of relativistic invariance of the Dirac equation. This will be the main theme of the next chapter.

# Chapter 4

# Relativistic covariance of Dirac equation

---

The problem indicated in the title of this chapter can be formulated as follows. Let us consider the Dirac equation

$$i\gamma^\mu \frac{\partial \psi(x)}{\partial x^\mu} - m\psi(x) = 0 \tag{4.1}$$

in some coordinate frame, and a Lorentz transformation to another frame,

$$x'^\mu = \Lambda^\mu{}_\nu x^\nu . \tag{4.2}$$

The question is whether there is an appropriate linear transformation of the wave function, $\psi(x) \to \psi'(x')$, such that $\psi'(x')$ satisfies the equation

$$i\gamma^\mu \frac{\partial \psi'(x')}{\partial x'^\mu} - m\psi'(x') = 0 . \tag{4.3}$$

So, suppose that

$$\psi'(x') = S\psi(x) , \tag{4.4}$$

where $S$ is a $4 \times 4$ constant invertible matrix depending on $\Lambda$, i.e. $S = S(\Lambda)$. Our goal is to find an appropriate $S$ corresponding to a given $\Lambda$. To this end, one may use (4.4) to express $\psi(x)$ as

$$\psi(x) = S^{-1}\psi'(x') . \tag{4.5}$$

Inserting now (4.5) into (4.1), one has

$$i\gamma^\mu S^{-1} \frac{\partial \psi'(x')}{\partial x'^\lambda} \frac{\partial x'^\lambda}{\partial x^\mu} - m S^{-1}\psi'(x') = 0 . \tag{4.6}$$

However, from (4.2) it is clear that

$$\frac{\partial x'^\lambda}{\partial x^\mu} = \Lambda^\lambda{}_\mu . \tag{4.7}$$

Thus, using (4.7) and multiplying Eq. (4.6) by $S$ from the left, one gets

$$iS\gamma^\mu S^{-1} \Lambda^\lambda{}_\mu \frac{\partial \psi'(x')}{\partial x'^\lambda} - m\psi'(x') = 0 . \tag{4.8}$$

Obviously, if one wants to arrive at Eq. (4.3), the condition

$$\Lambda^\lambda{}_\mu S\gamma^\mu S^{-1} = \gamma^\lambda$$

is to be imposed. This can be recast in a more elegant form

$$\Lambda^\mu{}_\nu \gamma^\nu = S^{-1}\gamma^\mu S . \tag{4.9}$$

For a practical evaluation of $S = S(\Lambda)$, the exponential form of $\Lambda$ (see the formula (A.17) in Appendix A)

$$\Lambda = \exp\left(-\frac{i}{2}\omega^{\alpha\beta}I_{\alpha\beta}\right) \tag{4.10}$$

is instrumental. Let us recall that $\omega^{\alpha\beta} = -\omega^{\beta\alpha}$ are the six independent parameters of a general continuous Lorentz transformation and $I_{\alpha\beta} = -I_{\beta\alpha}$ are the corresponding generators. It is reasonable to write $S$ in analogy with (4.10) as

$$S = \exp\left(-\frac{i}{4}\omega^{\alpha\beta}\sigma_{\alpha\beta}\right), \tag{4.11}$$

where $\sigma_{\alpha\beta} = -\sigma_{\beta\alpha}$ is a set of unknown would-be generators (the factor $1/4$ in the exponent is introduced for later convenience). In this way, the solution of the problem is reduced to finding the set of the matrices $\sigma_{\alpha\beta}$. This means that it is sufficient to consider infinitesimal transformations. Let us denote the infinitesimal parameters in(4.10) and (4.11) as $\Delta\omega^{\alpha\beta}$. We know (see (A.19)) that the form of the generators in (4.10) then leads to

$$\Lambda^{\mu}{}_{\nu} = g^{\mu}{}_{\nu} + \Delta\omega^{\mu}{}_{\nu}, \tag{4.12}$$

or, equivalently,

$$\Lambda^{\mu\nu} = g^{\mu\nu} + \Delta\omega^{\mu\nu}.$$

For infinitesimal $S$ and $S^{-1}$ one may write

$$S = \mathbb{1} - \frac{i}{4}\sigma_{\mu\nu}\Delta\omega^{\mu\nu},$$
$$S^{-1} = \mathbb{1} + \frac{i}{4}\sigma_{\mu\nu}\Delta\omega^{\mu\nu}. \tag{4.13}$$

Using (4.12) and (4.13) in the condition (4.9) one has

$$\left(\mathbb{1} + \frac{i}{4}\sigma_{\alpha\beta}\Delta\omega^{\alpha\beta}\right)\gamma^{\mu}\left(\mathbb{1} - \frac{i}{4}\sigma_{\alpha\beta}\Delta\omega^{\alpha\beta}\right) = (g^{\mu\nu} + \Delta\omega^{\mu\nu})\gamma_{\nu}. \tag{4.14}$$

After some simple manipulations, (4.14) is recast as

$$-\frac{i}{4}\Delta\omega^{\alpha\beta}(\gamma^{\mu}\sigma_{\alpha\beta} - \sigma_{\alpha\beta}\gamma^{\mu}) = g^{\mu}{}_{\alpha}\Delta\omega^{\alpha\beta}\gamma_{\beta}. \tag{4.15}$$

Utilizing the antisymmetry of parameters $\Delta\omega^{\alpha\beta}$, one gets eventually the condition for the generators $\sigma_{\alpha\beta}$:

$$[\gamma_{\mu}, \sigma_{\alpha\beta}] = 2i(g_{\mu\alpha}\gamma_{\beta} - g_{\mu\beta}\gamma_{\alpha}). \tag{4.16}$$

For any fixed pair of indices $\alpha, \beta$ we thus have 64 equations for 16 unknowns (elements of the $4 \times 4$ matrix $\sigma_{\alpha\beta}$). So, at first sight one could say that $\sigma_{\alpha\beta}$ is overconstrained by the conditions (4.16). An uninspired way of solving Eq. (4.16) would consist in writing $\sigma_{\alpha\beta}$ as a linear combination of matrices $\Gamma_A$, $A = 1, \ldots, 16$, from the preceding chapter, and employ the commutation and anticommutation relations to fix the values of the relevant coefficients. This is possible, but rather tedious; one might call it a "poor man's way". Instead, let us try to guess some hint that would help us find a short cut to the desired solution. To this end, we are going to start e.g. with $\sigma_{01}$. The conditions (4.16) then give, for $\mu = 0, 1, 2, 3$,

$$[\gamma_0, \sigma_{01}] = 2i\gamma_1,$$
$$[\gamma_1, \sigma_{01}] = 2i\gamma_0,$$
$$[\gamma_2, \sigma_{01}] = 0,$$
$$[\gamma_3, \sigma_{01}] = 0. \tag{4.17}$$

Contemplating (4.17) one may guess that $\sigma_{01}$ must be proportional to $\gamma_0\gamma_1$; more precisely, a solution of (4.17) is, obviously,

$$\sigma_{01} = i\gamma_0\gamma_1 . \tag{4.18}$$

This is just the hint we need. One may, tentatively, generalize (4.18) to $\sigma_{\alpha\beta} = i\gamma_\alpha\gamma_\beta$ (of course, such an Ansatz is meaningful just for $\alpha \neq \beta$; for $\alpha = \beta$ the matrix $\sigma_{\alpha\beta}$ is trivial). Since we know that $\sigma_{\alpha\beta} = -\sigma_{\beta\alpha}$, this is equivalent to

$$\sigma_{\alpha\beta} = \frac{i}{2}[\gamma_\alpha, \gamma_\beta] . \tag{4.19}$$

It is not difficult to verify that the expression (4.19) satisfies indeed Eq. (4.16). For this purpose, one may employ the elementary algebraic identity

$$[A, BC] = \{A, B\}C - B\{A, C\} . \tag{4.20}$$

Then

$$[\gamma_\mu, \sigma_{\alpha\beta}] = [\gamma_\mu, i\gamma_\alpha\gamma_\beta] = 2ig_{\mu\alpha}\gamma_\beta - 2ig_{\mu\beta}\gamma_\alpha ,$$

and (4.16) is thereby proved.

So, we have guessed a particular solution of the conditions (4.16), but the question remains whether it is unique or not. To clarify this point, suppose there is another solution, denoted as $\sigma'_{\alpha\beta}$. Using (4.16) for $\sigma_{\alpha\beta}$ and $\sigma'_{\alpha\beta}$, one sees immediately that

$$[\sigma_{\alpha\beta} - \sigma'_{\alpha\beta}, \gamma_\mu] = 0$$

for any $\mu = 0, 1, 2, 3$. Then, according to the lemma L5 from preceding chapter, it must hold

$$\sigma'_{\alpha\beta} - \sigma_{\alpha\beta} = a \cdot \mathbb{1} ,$$

where $a$ is an arbitrary coefficient. Thus, (4.19) is in fact the general solution of Eq. (4.16), up to a multiple of unit matrix. Needles to say, such a trivial ambiguity has been clear from the very beginning; the non-trivial point here is that it is *the only* possible ambiguity. From now on, we will use (4.19) as the relevant formula for the generators $\sigma_{\alpha\beta}$.

An important remark is in order here. Our construction of the transformation matrix $S$ in (4.9) has been based on generators $\sigma_{\alpha\beta}$ that correspond to the Lorentz generators $I_{\alpha\beta}$. More precisely, the correspondence between (4.10) and (4.11) is

$$\frac{1}{2}\sigma_{\alpha\beta} \longleftrightarrow I_{\alpha\beta} . \tag{4.21}$$

As we know, the generators $I_{\alpha\beta}$ satisfy commutation relations characteristic of a Lie algebra. In particular (see (A.22)),

$$[I_{\mu\nu}, I_{\rho\sigma}] = i(g_{\mu\sigma}I_{\nu\rho} + g_{\nu\rho}I_{\mu\sigma} - g_{\mu\rho}I_{\nu\sigma} - g_{\nu\sigma}I_{\mu\rho}) . \tag{4.22}$$

One might suspect that the matrices $\frac{1}{2}\sigma_{\alpha\beta}$ satisfy the same commutation relation (in mathematical language, it would mean that the six matrices $\frac{1}{2}\sigma_{\alpha\beta}$ constitute a representation of the Lie algebra of the Lorentz group). It turns out that it is indeed the case. For the evaluation of the relevant commutators one may use the identity (an extension of (4.20))

$$[AB, CD] = A\{B, C\}D - AC\{B, D\} - C\{A, D\}B + \{A, C\}DB . \tag{4.23}$$

25

Then, one has

$$[\frac{1}{2}\sigma_{\mu\nu}, \frac{1}{2}\sigma_{\rho\sigma}] = \frac{i}{2} \cdot \frac{i}{2} [\gamma_\mu\gamma_\nu, \gamma_\rho\gamma_\sigma]$$

$$= -\frac{1}{4}(2g_{\nu\rho}\gamma_\mu\gamma_\sigma - 2g_{\nu\sigma}\gamma_\mu\gamma_\rho - 2g_{\mu\sigma}\gamma_\rho\gamma_\nu + 2g_{\mu\rho}\gamma_\sigma\gamma_\nu)$$

$$= -\frac{1}{2}\left[-g_{\nu\sigma}\left(\frac{1}{2}\{\gamma_\mu, \gamma_\rho\} + \frac{1}{2}[\gamma_\mu, \gamma_\rho]\right) + g_{\nu\rho}\left(\frac{1}{2}\{\gamma_\mu, \gamma_\sigma\} + \frac{1}{2}[\gamma_\mu, \gamma_\sigma]\right)\right.$$

$$\left. - g_{\mu\sigma}\left(\frac{1}{2}\{\gamma_\rho, \gamma_\nu\} + \frac{1}{2}[\gamma_\rho, \gamma_\nu]\right) + g_{\mu\rho}\left(\frac{1}{2}\{\gamma_\sigma, \gamma_\nu\} + \frac{1}{2}[\gamma_\sigma, \gamma_\nu]\right)\right]$$

$$= -\frac{1}{2}(g_{\overline{\nu\rho}}g_{\mu\sigma} + g_{\nu\rho}\frac{1}{i}\sigma_{\mu\sigma} - g_{\overline{\nu\sigma}}g_{\mu\rho} - g_{\nu\sigma}\frac{1}{i}\sigma_{\mu\rho}$$

$$- g_{\overline{\mu\sigma}}g_{\rho\nu} - g_{\mu\sigma}\frac{1}{i}\sigma_{\rho\nu} + g_{\overline{\mu\rho}}g_{\sigma\nu} + g_{\mu\rho}\frac{1}{i}\sigma_{\sigma\nu})$$

$$= i\left(g_{\mu\sigma}\frac{1}{2}\sigma_{\nu\rho} + g_{\nu\rho}\frac{1}{2}\sigma_{\mu\sigma} - g_{\mu\rho}\frac{1}{2}\sigma_{\nu\sigma} - g_{\nu\sigma}\frac{1}{2}\sigma_{\mu\rho}\right), \qquad (4.24)$$

and this is precisely the anticipated commutation relation for the generators $\sigma_{\alpha\beta}$ that matches Eq. (4.22).

Thus, in our straightforward way we have discovered a four-dimensional representation of the Lorentz group (or Lorentz algebra, if you want). It is rightly called the **bispinor** (or **Dirac spinor**) as we will see shortly. An elementary mathematical theory of representations of Lorentz group is described briefly in Appendix B.

It will be certainly instructive to present some explicit examples of the matrix $S$ implementing the transformation (4.4) of Dirac wave function. First, let us consider a spatial rotation around the third axis of coordinate system, i.e. in the plane (12). So, the relevant transformation (4.2) is given by

$$\begin{pmatrix} x^{0'} \\ x^{1'} \\ x^{2'} \\ x^{3'} \end{pmatrix} = \begin{pmatrix} 1 & 0 & 0 & 0 \\ 0 & \cos\varphi & \sin\varphi & 0 \\ 0 & -\sin\varphi & \cos\varphi & 0 \\ 0 & 0 & 0 & 1 \end{pmatrix} \begin{pmatrix} x^0 \\ x^1 \\ x^2 \\ x^3 \end{pmatrix}. \qquad (4.25)$$

The corresponding infinitesimal form thus amounts to

$$x^{1'} = x^1 + \delta\varphi\, x^2,$$
$$x^{2'} = -\delta\varphi\, x^1 + x^2. \qquad (4.26)$$

Using the notation (4.12), this means

$$(\Delta\omega)^1{}_2 = \delta\varphi, \qquad (\Delta\omega)^2{}_1 = -\delta\varphi. \qquad (4.27)$$

It is easy to see that raising the indices in (4.27) leads to

$$(\Delta\omega)^{12} = -\delta\varphi, \qquad (\Delta\omega)^{21} = \delta\varphi. \qquad (4.28)$$

Then, according to the formula (4.13), one has

$$S = \mathbb{1} - 2 \cdot \frac{i}{4}(\Delta\omega)^{12}\sigma_{12} = \mathbb{1} + \frac{i}{2}\delta\varphi\sigma_{12} \qquad (4.29)$$

for the infinitesimal transformation, i.e. for the representation of the finite rotation (4.25) one may write

$$S(\varphi) = \exp\left(\frac{i}{2}\varphi\sigma_{12}\right). \qquad (4.30)$$

26

Now, in the standard representation of $\gamma$-matrices one gets

$$\sigma_{12} = i\gamma_1\gamma_2 = i \begin{pmatrix} 0 & \sigma_1 \\ -\sigma_1 & 0 \end{pmatrix} \begin{pmatrix} 0 & \sigma_2 \\ -\sigma_2 & 0 \end{pmatrix} = \begin{pmatrix} \sigma_3 & 0 \\ 0 & \sigma_3 \end{pmatrix} = \Sigma_3 .$$

Thus, the result (4.30) can be recast as

$$S(\varphi) = \exp\left(\frac{i}{2}\varphi\Sigma_3\right). \tag{4.31}$$

Expanding the exponential (4.31) in Taylor series, one gets, taking into account that $(\Sigma_3)^2 = \mathbb{1}$,

$$S(\varphi) = \cos\frac{\varphi}{2} \cdot \mathbb{1} + i \sin\frac{\varphi}{2}\Sigma_3 . \tag{4.32}$$

This means, in particular,

$$S(2\pi) = -\mathbb{1} . \tag{4.33}$$

Thus, the full rotation with $\varphi = 360°$ changes the sign of a wave function in question. This is a typical property of spinors (well-known already from the non-relativistic description of a spin-1/2 particle in terms of a two-component wave function). For this reason, it is natural to call the four-component Dirac wave function the bispinor (in Appendix B, one may find a more precise explanation of this concept).

Next, let us consider the case of a Lorentz boost; in particular, the example we take up is the uniform motion along the coordinate axis 1 with a velocity $v$. The corresponding transformation of spacetime coordinates reads

$$\begin{pmatrix} x^{0'} \\ x^{1'} \\ x^{2'} \\ x^{3'} \end{pmatrix} = \begin{pmatrix} \text{ch}\,\varphi & -\text{sh}\,\varphi & 0 & 0 \\ -\text{sh}\,\varphi & \text{ch}\,\varphi & 0 & 0 \\ 0 & 0 & 1 & 0 \\ 0 & 0 & 0 & 1 \end{pmatrix} \begin{pmatrix} x^0 \\ x^1 \\ x^2 \\ x^3 \end{pmatrix}, \tag{4.34}$$

where

$$\text{ch}\,\varphi = \frac{1}{\sqrt{1-v^2}}, \qquad \text{sh}\,\varphi = \frac{v}{\sqrt{1-v^2}}.$$

For the infinitesimal transformation we then have

$$\begin{aligned} x^{0'} &= x^0 - \delta\varphi\, x^1 , \\ x^{1'} &= -\delta\varphi\, x^0 + x^1 . \end{aligned} \tag{4.35}$$

Using the notation (4.12), this means

$$(\Delta\omega)^0{}_1 = -\delta\varphi, \qquad (\Delta\omega)^1{}_0 = -\delta\varphi .$$

The corresponding infinitesimal transformation of the Dirac wave function is thus, according to (4.13),

$$S = \mathbb{1} - 2 \cdot \frac{i}{4}(\Delta\omega)^{01}\sigma_{01} = \mathbb{1} - \frac{i}{2}\delta\varphi\sigma_{01} ,$$

so that the relevant finite transformation may be written as

$$S(\varphi) = \exp\left(-\frac{i}{2}\varphi\sigma_{01}\right). \tag{4.36}$$

27

In the standard representation of $\gamma$-matrices one gets

$$\sigma_{01} = i\gamma_0\gamma_1 = i \begin{pmatrix} 1 & 0 \\ 0 & -1 \end{pmatrix} \begin{pmatrix} 0 & -\sigma_1 \\ \sigma_1 & 0 \end{pmatrix} = -i \begin{pmatrix} 0 & \sigma_1 \\ \sigma_1 & 0 \end{pmatrix}.$$

Obviously, $(\sigma_{01})^2 = -1$, and the exponential (4.36) can thus be worked out as

$$S(\varphi) = \operatorname{ch}\frac{\varphi}{2} \cdot 1 - i\operatorname{sh}\frac{\varphi}{2}\sigma_{01}. \tag{4.37}$$

In this way, we have established Lorentz covariance of the Dirac equation. Up to now we have considered continuous Lorentz transformations (spatial rotations and boosts) described, in general, by six parameters. The case of discrete symmetries will be discussed in the next chapter.

To extend our technical tools for "diracology", let us now mention a simple but very useful formula that relates $S^{-1}$ and $S^\dagger$. It holds

$$S^{-1} = \gamma_0 S^\dagger \gamma_0. \tag{4.38}$$

The proof of this identity is quite easy. Let us write, for convenience,

$$S = \exp(-i\Omega), \tag{4.39}$$

with

$$\Omega = \frac{1}{4}\omega^{\alpha\beta}\sigma_{\alpha\beta}.$$

Expanding the exponential (4.39) in Taylor series, one has

$$S = 1 + \frac{1}{1!}(-i\Omega) + \frac{1}{2!}(-i\Omega)^2 + \dots \tag{4.40}$$

Using the relation $\gamma_\mu^\dagger = \gamma_0\gamma_\mu\gamma_0$ (see (3.13)), it is easy to arrive at the identity

$$\Omega^\dagger = \gamma_0\Omega\gamma_0. \tag{4.41}$$

This, of course, means that such a rule is valid for any power of $\Omega$, i.e.

$$(\Omega^n)^\dagger = \gamma_0\Omega^n\gamma_0. \tag{4.42}$$

So, taking into account (4.42), from (4.40) one gets immediately

$$S^\dagger = \gamma_0 \left[ 1 + \frac{1}{1!}i\Omega + \frac{1}{2!}(i\Omega)^2 + \dots \right] \gamma_0. \tag{4.43}$$

The expression in square brackets is just $\exp(i\Omega) = S^{-1}$. Thus, we have $S^\dagger = \gamma_0 S^{-1}\gamma_0$ and the identity (4.38) is thereby proved.

One more remark is in order here. For spatial rotations, $\Omega$ is clearly Hermitian, so that the corresponding $S$ in (4.39) is unitary. For boosts, $\Omega$ is anti-Hermitian, and $S$ is thus Hermitian. This corresponds precisely to the known properties of the matrices $\Lambda$ of Lorentz transformations.

With the basic elements of the covariant formalism at hand, let us now return briefly to the continuity equation (2.4) for the probability density and probability current. Let us see how it can be recovered from the covariant form of Dirac equation. We have

$$i\gamma^\mu\partial_\mu\psi = m\psi, \tag{4.44}$$

and its Hermitian conjugation becomes, upon multiplication by $\gamma_0$ from the right,

$$-i\partial_\mu\psi^\dagger\gamma_0\gamma^\mu = m\psi^\dagger\gamma_0 \,. \tag{4.45}$$

One can see that in (4.45) the expression

$$\bar{\psi} = \psi^\dagger\gamma_0 \tag{4.46}$$

emerges naturally. It is called the **Dirac conjugation** and we will use it frequently from now on. So, (4.45) is recast as

$$-i\partial_\mu\bar{\psi}\gamma^\mu = m\bar{\psi} \,. \tag{4.47}$$

Now, multiplying equations (4.44) and (4.47) by $\bar{\psi}$ and $\psi$ from left and right, respectively, the difference of the resulting expressions gives immediately

$$\partial_\mu(\bar{\psi}\gamma^\mu\psi) = 0 \,. \tag{4.48}$$

Taking into account the relations between matrices $\gamma^\mu$ and $\vec{\alpha}, \beta$, it is obvious that (4.48) coincides with (2.4). Thus, we have recovered the continuity equation in the covariant form

$$\partial_\mu j^\mu = 0 \tag{4.49}$$

involving a **four-current** $j^\mu$,

$$j^\mu = \bar{\psi}\gamma^\mu\psi \,. \tag{4.50}$$

Our notation suggests that $j^\mu$ might be a four-vector under Lorentz transformations. It is indeed so. Denoting

$$j^{\mu'}(x') = \bar{\psi}'(x')\gamma^\mu\psi'(x') \,, \tag{4.51}$$

and using the transformation law (4.4), as well as the definition (4.46), one has

$$\begin{aligned} j^{\mu'}(x') &= \psi^{\dagger'}(x')\gamma_0\gamma^\mu S\psi(x) \\ &= \psi^\dagger(x)S^\dagger\gamma_0\gamma^\mu S\psi(x) \\ &= \bar{\psi}(x)\gamma_0 S^\dagger\gamma_0\gamma^\mu S\psi(x) \,. \end{aligned} \tag{4.52}$$

So, utilizing the identity (4.38), the last expression becomes

$$j^{\mu'}(x') = \bar{\psi}(x)S^{-1}\gamma^\mu S\psi(x) \,,$$

and, taking into account (4.9), one gets finally

$$j^{\mu'}(x') = \Lambda^\mu{}_\nu j_\nu(x) \,, \tag{4.53}$$

which is the anticipated result.

Thus, we have seen that one can construct a four-vector as a bilinear form made of bispinors (in this sense, a bispinor is a "square root of a four-vector"). We will see later on that the example described above can be generalized in a systematic way; such constructions are particularly useful within the framework of field theory.

# Chapter 5

# *C*, *P* and *T*

In this chapter, we are going to examine discrete symmetries of the Dirac equation. In the title, they are shown in alphabetical order, but now we will perform a cyclic permutation and start with *P*. So, what is *P*? It is **spatial inversion**, or **space reflection**, if you want. Usually, it is also called the **parity transformation** (hence *P*). Such a transformation simply means

$$(x^0, \vec{x}) \longrightarrow (x^0, -\vec{x}) \,. \tag{5.1}$$

It is certainly a Lorentz transformation, since it preserves the spacetime interval $x^2 = (x^0)^2 - \vec{x}^2$. The corresponding transformation matrix is, obviously,

$$\Lambda_P = \begin{pmatrix} 1 & 0 & 0 & 0 \\ 0 & -1 & 0 & 0 \\ 0 & 0 & -1 & 0 \\ 0 & 0 & 0 & -1 \end{pmatrix} \,. \tag{5.2}$$

Using the standard terminology (see Appendix A), spatial inversion (5.1) belongs to the set of orthochronous transformations, with $\det \Lambda_P = -1$.

Now, the question is what could be the corresponding transformation of the Dirac wave function satisfying Eq. (3.7). The relation (4.9) is quite general; it is not restricted to the continuous proper Lorentz transformations discussed in detail in the preceding chapter. So, using (4.9) and taking into account the simple structure of the matrix $\Lambda_P$, one gets immediately the conditions for the relevant matrix $S_P$:

$$\begin{aligned} S_P^{-1} \gamma^0 S_P &= \gamma^0 \,, \\ S_P^{-1} \gamma^k S_P &= -\gamma^k \,, \qquad k = 1, 2, 3 \,. \end{aligned} \tag{5.3}$$

Thus, the sought matrix $S_P$ should commute with $\gamma^0$ and anticommute with any $\gamma^k$, $k = 1, 2, 3$. A solution is then clear:

$$S_P = a \cdot \gamma^0 \,, \tag{5.4}$$

where $a$ is an arbitrary constant factor. It remains to be clarified whether (5.4) is the general solution or not. So, suppose there are two matrices $R$ and $S$ satisfying (5.3). It then means

$$\begin{aligned} S^{-1} \gamma^0 S &= R^{-1} \gamma^0 R \,, \\ S^{-1} \gamma^k S &= R^{-1} \gamma^k R \,, \qquad k = 1, 2, 3 \,. \end{aligned} \tag{5.5}$$

In other words, one has

$$S^{-1} \gamma^\mu S = R^{-1} \gamma^\mu R \tag{5.6}$$

for any $\mu = 0, 1, 2, 3$. Eq. (5.6) can be recast as

$$RS^{-1}\gamma^{\mu}(RS^{-1})^{-1} = \gamma^{\mu},$$

and this means that $RS^{-1}$ commutes with any $\gamma^{\mu}$, $\mu = 0, 1, 2, 3$. According to the lemma L5 from Chapter 3 this implies that

$$RS^{-1} = b \cdot \mathbb{1},\tag{5.7}$$

where $b$ is an arbitrary factor. Thus,

$$R = b \cdot S.\tag{5.8}$$

In view of the relation (5.8) one may conclude that (5.4) is indeed the general solution of conditions (5.3). Conventionally, we will use $S_P = \gamma_0$ henceforth.

Thus, we have established the covariance of Dirac equation under the parity transformation: if $\psi(x)$ is a solution in a given reference frame, then for $x' = (x^0, -\vec{x})$ the function

$$\psi'(x') = \gamma_0\psi(x)\tag{5.9}$$

is the corresponding solution in the primed system. Note that this is tantamount to the statement that if $\psi(x)$ is a solution of the Dirac equation, then

$$\psi_P(x) = \gamma_0\psi(x^0, -\vec{x})\tag{5.10}$$

is its solution as well.

One may notice that we have achieved such a result quite easily and the parity symmetry seems to be almost automatic in the present case. In fact, as we will see later on, it is not difficult to find an example of a relativistic equation that does exhibit parity violation.

With the knowledge of the parity transformation at hand, we may now extend our previous considerations concerning bilinear forms made of Dirac spinors (cf. the discussion following the formula (4.50)). Although such a technical progress is not of immediate importance for our study of Dirac equation, it will be useful later on, within the framework of field theory (so, it will be another "rifle hanging on the wall" à la A. P. Chekhov). In any case, at the moment it may serve as a refreshing exercise for a loyal reader.

For simplicity, let us start with the expression $\bar{\psi}\psi$. Using the relation (4.38), it is easy to see that such a form is a scalar under proper Lorentz transformations. Moreover, from (5.9) it is obvious that it is invariant under spatial inversion as well. So, in this sense, $\bar{\psi}\psi$ is a true **scalar**. Next, let us consider the combination $\bar{\psi}\gamma_5\psi$. For $x' = \Lambda x$ one gets, in general,

$$\bar{\psi}'(x')\gamma_5\psi'(x') = \psi^{\dagger}(x)S^{\dagger}\gamma_0\gamma_5 S\psi(x) = \bar{\psi}(x)\gamma_0 S^{\dagger}\gamma_0\gamma_5 S\psi(x) = \bar{\psi}(x)S^{-1}\gamma_5 S\psi(x).\tag{5.11}$$

For a proper Lorentz transformation, the generators of $S$ are made of products of two $\gamma$-matrices, and therefore they commute with $\gamma_5$; this in turn means that $[S, \gamma_5] = 0$. Thus, we see that $\bar{\psi}\gamma_5\psi$ is invariant in such a case. On the other hand, for the spatial inversion one has $S = \gamma_0$, so that $S^{-1}\gamma_5 S = -\gamma_5$ and one thus gets

$$\bar{\psi}'(x')\gamma_5\psi'(x') = -\bar{\psi}(x)\gamma_5\psi(x).\tag{5.12}$$

So, we end up with the conclusion that $\bar{\psi}\gamma_5\psi$ is a **pseudoscalar**.

In a similar way, one may compare the behaviour of $\bar{\psi}\gamma^{\mu}\psi$ and $\bar{\psi}\gamma^{\mu}\gamma_5\psi$. We have already observed that $j^{\mu} = \bar{\psi}\gamma^{\mu}\psi$ is a four-vector (cf. (4.53)) under proper Lorentz transformations. Obviously, for spatial inversion the relation (4.53) holds equally well; this means that one has, schematically,

$$(j^0, \vec{j}) \xrightarrow{P} (j^0, -\vec{j}),\tag{5.13}$$

31

i.e. $j^\mu$ is a true four-vector. For $j_5^\mu = \bar{\psi}\gamma^\mu\gamma_5\psi$ one gets, after a simple manipulation,

$$j_5^\mu(x') = \psi^\dagger(x)S^\dagger\gamma_0\gamma^\mu\gamma_5 S\psi(x) = \bar{\psi}(x)S^{-1}\gamma^\mu\gamma_5 S\psi(x). \tag{5.14}$$

For proper Lorentz transformations one has $[S, \gamma_5] = 0$ and (5.14) then amounts to

$$j_5^{\mu\prime}(x') = \Lambda^\mu{}_\nu j_5^\nu(x).$$

For spatial inversion, $\gamma_5 S = -S\gamma_5$ and one thus gets

$$(j_5^0, \vec{j}_5) \xrightarrow{P} (-j_5^0, \vec{j}_5). \tag{5.15}$$

This means that $j_5^\mu$ is a **pseudovector** (**axial vector**). Thus, in the above examples, the matrix $\gamma_5$ is responsible for the prefix "pseudo-" in the notation of the quantities in question.

Let us now proceed to the item $T$ of our list. It denotes **time reversal** (or **time inversion**, if you want). Before examining a pertinent transformation for Dirac equation, let us return briefly to the non-relativistic Schrödinger equation mentioned in the first two chapters; it may provide us with an inspiring hint. From (1.15), (1.16) it is clear that the replacement $t \to -t$ changes the sign of the time derivative, and the complex conjugation of the wave function does the same. Thus, one may observe that the free-particle Schrödinger equation is invariant under time reversal, in the sense that if $\psi(t, \vec{x})$ is a solution, then $\psi^*(-t, \vec{x})$ is a solution as well. So, an important point is that the transformation $t \to -t$ is to be accompanied by the complex conjugation. This has a clear and desirable physical effect. Upon such a transformation, the energy is not changed, while the momentum changes its sign; to see this explicitly, please recall the form of a plane wave, involving the familiar factor $\exp[-i(Et - \vec{p} \cdot \vec{x})]$.

So, let us come back to our staple food, the Dirac equation. The considered transformation of spacetime coordinates is now described by means of the matrix

$$\Lambda = \Lambda_T = \begin{pmatrix} -1 & 0 & 0 & 0 \\ 0 & 1 & 0 & 0 \\ 0 & 0 & 1 & 0 \\ 0 & 0 & 0 & 1 \end{pmatrix}. \tag{5.16}$$

Motivated by the preceding considerations, we may try an Ansatz for the corresponding transformation of the Dirac wave function, defined as

$$\psi'(x') = B\psi^*(x), \tag{5.17}$$

where $x' = (-x^0, \vec{x})$ and $B$ is an invertible constant $4 \times 4$ matrix. By the way, in contrast to the preceding case of spatial inversion, (5.17) represents an **antilinear transformation**, due to the involvement of the complex conjugation — this is a well-known characteristic of the time reversal in quantum theory. So, our task is now to find the matrix $B$ (if it exists). To this end, we start with the complex conjugation of the Dirac equation, i.e.

$$-i\gamma^{\mu*}\partial_\mu\psi^* - m\psi^* = 0. \tag{5.18}$$

Using (5.17), we express $\psi^*$ as

$$\psi^*(x) = B^{-1}\psi'(x'), \tag{5.19}$$

and plugging this into Eq. (5.18) one gets

$$-i\gamma^{\mu*}B^{-1}\frac{\partial}{\partial x'^\lambda}\psi'(x')\frac{\partial x'^\lambda}{\partial x^\mu} - mB^{-1}\psi'(x') = 0. \tag{5.20}$$

32

Now, taking into account that

$$\frac{\partial x'^{\lambda}}{\partial x^{\mu}} = \Lambda_T{}^{\lambda}{}_{\mu} \,,$$

and multiplying Eq. (5.20) by $B$ from the left, one obtains

$$-iB\gamma^{\mu *}B^{-1}\Lambda_T{}^{\lambda}{}_{\mu}\frac{\partial \psi'(x')}{\partial x'^{\lambda}} - m\psi'(x') = 0 \,. \tag{5.21}$$

The condition of the covariance of Dirac equation under time reversal thus reads

$$B^{-1}\gamma^{\mu}B = -\Lambda_T{}^{\mu}{}_{\nu}\gamma^{\nu *} \,. \tag{5.22}$$

To work out the general relation (5.22) explicitly, one should realize that the complex conjugation of $\gamma$-matrices depends on their particular representation. In the standard representation, $\gamma^{\mu *} = \gamma^{\mu}$ for $\mu = 0, 1, 3$ and $\gamma^{\mu *} = -\gamma^{\mu}$ for $\mu = 2$. Thus, in this "household representation" the conditions (5.22) read

$$\begin{aligned}
B^{-1}\gamma^0 B &= \gamma^0 \,, \\
B^{-1}\gamma^1 B &= -\gamma^1 \,, \\
B^{-1}\gamma^2 B &= \gamma^2 \,, \\
B^{-1}\gamma^3 B &= -\gamma^3 \,.
\end{aligned} \tag{5.23}$$

It is easy to guess that a solution of (5.23) is

$$B = a \cdot \gamma^1 \gamma^3 \,, \tag{5.24}$$

where $a$ is an arbitrary constant factor. One can also show that such a solution is unique; to prove this, one may proceed in the same way as before, in the case of the parity transformation. If we want to make (5.17) antiunitary operator, the factor $a$ in (5.24) should be chosen so that $|a| = 1$. A conventional choice used frequently in the literature is $a = i$. Sticking to such a convention, our result can be written as

$$\psi'(x') = i\gamma^1 \gamma^3 \psi^*(x) \,, \tag{5.25}$$

with $x' = (-x^0, \vec{x})$.

Let us remind the reader that the concept of antiunitary vs. unitary operators is mostly due to E. P. Wigner, who is the author of a fundamental theorem on symmetries in quantum theory, which bears his name.[4]

Finally, let us take up the item $C$ in our list. It is the so-called **charge conjugation**, and in distinction to the preceding two discrete symmetries, $C$ is not related to any spacetime transformation. Rather, it is an **internal symmetry** and its substantial ingredient is the complex conjugation. An obvious motivation for investigating such a symmetry is the existence of free-particle solutions with positive and negative energy; one may thus naturally contemplate the possibility of a transformation turning one type of a solution into another.

We are going to start with the Ansatz

$$\psi'(x) = A\psi^*(x) \,, \tag{5.26}$$

---

[4]Biographical remark: Eugene Paul Wigner (1902–1995) (originally Wigner Jenő Pál) was an eminent theorist in the field of quantum theory. Born in Hungary, he emigrated to U. S. in 1930s and received Nobel Prize in 1963 together with Maria Goeppert Mayer for the work on symmetries in nuclear and particle physics. He was a brother–in–law of Paul Dirac.

where $A$ is an invertible constant matrix. Let us stress once again that spacetime coordinates are left unchanged. Expressing then $\psi^*(x)$ as

$$\psi^*(x) = A^{-1}\psi'(x),\tag{5.27}$$

and inserting (5.27) into the complex conjugate of the Dirac equation (5.18), one gets first

$$-i\gamma^{\mu*}A^{-1}\partial_\mu\psi'(x) - mA^{-1}\psi'(x) = 0,$$

and, subsequently,

$$-iA\gamma^{\mu*}A^{-1}\partial_\mu\psi'(x) - m\psi'(x) = 0.\tag{5.28}$$

The condition of the invariance under the considered transformation thus reads

$$-A\gamma^{\mu*}A^{-1} = \gamma^\mu,$$

or, equivalently,

$$-\gamma^{\mu*} = A^{-1}\gamma^\mu A.\tag{5.29}$$

As we have already noted before, there is no universal pattern of complex conjugation for $\gamma$-matrices, so one has to resort to an explicit representation. Thus, for the good old standard representation the set of conditions (5.29) is worked out as

$$\begin{aligned}
A^{-1}\gamma^0 A &= -\gamma^0,\\
A^{-1}\gamma^1 A &= -\gamma^1,\\
A^{-1}\gamma^2 A &= \gamma^2,\\
A^{-1}\gamma^3 A &= -\gamma^3.
\end{aligned}\tag{5.30}$$

A solution of (5.30) is immediately obvious: it is

$$A = a \cdot \gamma^2,\tag{5.31}$$

where $a$ is an arbitrary constant. Again, one can prove easily that (5.31) is in fact the general solution of the conditions (5.30). It is an old hat by now, relying on the lemma L5 from Chapter 3. Conventionally, one may set $a = i$; then $A$ is unitary matrix (and Hermitian as well). Thus, we have established a remarkable internal symmetry of Dirac equation, described by the antiunitary transformation[5]

$$\psi'(x) = i\gamma^2\psi^*(x).\tag{5.32}$$

Instead of (5.32), it is more practical to express the charge conjugation transformation in terms of the Dirac conjugation $\bar\psi$ (which, as we know, appears naturally e.g. in bilinear covariant forms discussed above). To this end, $\psi^*$ is expressed as

$$\psi^* = \left(\psi^\dagger\right)^{\mathsf{T}} = \left(\bar\psi\gamma^0\right)^{\mathsf{T}}.\tag{5.33}$$

Within the standard representation, (5.33) becomes

$$\psi^* = \gamma^0\bar\psi^{\mathsf{T}},$$

---

[5]Let us emphasize that (5.32) holds within the standard representation of $\gamma$-matrices. A watchful reader may have noticed that for the Majorana representation mentioned in Chapter 3 one would get $A = a \cdot \mathbb{1}$, i.e. conventionally $\psi'(x) = \psi^*(x)$.

and (5.32) may be recast as

$$\psi' = C\bar{\psi}^{\mathrm{T}},\tag{5.34}$$

where

$$C = i\gamma^2\gamma^0.\tag{5.35}$$

Let us also rewrite (5.29) as a corresponding condition for the matrix $C$. First, one has

$$-\gamma^{\mu*} = (C\gamma^0)^{-1}\gamma^\mu C\gamma^0.$$

Using the simple properties of $\gamma^0$ in the standard representation, one gets finally, after some simple manipulations,

$$-\gamma^{\mu\mathrm{T}} = C^{-1}\gamma^\mu C.\tag{5.36}$$

A remark is in order here. We have arrived at (5.34) with $C$ satisfying (5.36) within the standard representation of $\gamma$-matrices. In fact, these relations are quite general; what is specific for the standard representation, is the result (5.35) for $C$. So, let us show how an alternative derivation of (5.36) can proceed. First, an equation for $\bar{\psi}$ is obtained easily; it reads

$$-i\partial_\mu\bar{\psi}\gamma^\mu - m\bar{\psi} = 0.\tag{5.37}$$

Next, using the Ansatz (5.34), $\bar{\psi}^{\mathrm{T}}$ is expressed as

$$\bar{\psi}^{\mathrm{T}} = C^{-1}\psi'.\tag{5.38}$$

Eq. (5.37) is recast as

$$-i\gamma^{\mu\mathrm{T}}\partial_\mu\left(\bar{\psi}^{\mathrm{T}}\right) - m\bar{\psi}^{\mathrm{T}} = 0,$$

and inserting there (5.38), one gets readily

$$-iC\gamma^{\mu\mathrm{T}}C^{-1}\partial_\mu\psi' - m\psi' = 0.$$

The requirement of invariance then reads

$$-C\gamma^{\mu\mathrm{T}}C^{-1} = \gamma^\mu,$$

i.e.

$$-\gamma^{\mu\mathrm{T}} = C^{-1}\gamma^\mu C,$$

so that (5.36) is recovered. Note that it is also quite easy to obtain the formula (5.35) directly from (5.36). Indeed, using the familiar properties of $\gamma$-matrices, one can see that in the standard representation $\gamma^{\mu\mathrm{T}} = \gamma^\mu$ for $\mu = 0, 2$ and $\gamma^{\mu\mathrm{T}} = -\gamma^\mu$ for $\mu = 1, 3$. The relation (5.36) thus means that $C$ should commute with $\gamma^1$, $\gamma^3$ and anticommute with $\gamma^0$, $\gamma^2$. In this way, one is led immediately to the solution (5.35) (up to an arbitrary coefficient). Let us also notice that the form (5.35) reveals the following simple properties of the matrix $C$:

$$C^{-1} = C^\dagger = C^{\mathrm{T}} = -C.\tag{5.39}$$

It is interesting that the relation $C^{\mathrm{T}} = -C$ (i.e. $C$ is an antisymmetric matrix) is quite general, i.e. it holds in any representation of $\gamma$-matrices. To see this, let us consider two different representations, $\gamma^\mu$ and $\widetilde{\gamma}^\mu$. According to the fundamental theorem on the $\gamma$-matrices, formulated in Chapter 3, there is a similarity transformation between the two sets,

$$\widetilde{\gamma}^\mu = U\gamma^\mu U^{-1}.\tag{5.40}$$

Then

$$\widetilde{\gamma}^{\mu\,\mathsf{T}} = (U^{-1})^{\mathsf{T}}\gamma^{\mu\,\mathsf{T}}U^{\mathsf{T}} = (U^{\mathsf{T}})^{-1}(-C^{-1}\gamma^{\mu}C)U^{\mathsf{T}}$$
$$= -(U^{\mathsf{T}})^{-1}C^{-1}U^{-1}\widetilde{\gamma}^{\mu}UCU^{\mathsf{T}} = -(UCU^{\mathsf{T}})^{-1}\widetilde{\gamma}^{\mu}UCU^{\mathsf{T}}. \tag{5.41}$$

Thus, if one denotes the charge conjugation matrix for the representation $\widetilde{\gamma}^{\mu}$ as $\widetilde{C}$, from (5.41) one has

$$\widetilde{C} = UCU^{\mathsf{T}}. \tag{5.42}$$

Now, we already know that in the standard representation $C^{\mathsf{T}} = -C$. From the relation (5.42) one gets

$$\widetilde{C}^{\mathsf{T}} = UC^{\mathsf{T}}U^{\mathsf{T}},$$

and thus we see that the antisymmetry of $C$ in the standard representation implies the antisymmetry of $\widetilde{C}$ in any other representation.

For an illustration, let us show the explicit form of the matrix $C$ in the standard representation:

$$C = i\gamma^{2}\gamma^{0} = \begin{pmatrix} 0 & 0 & 0 & -1 \\ 0 & 0 & 1 & 0 \\ 0 & -1 & 0 & 0 \\ 1 & 0 & 0 & 0 \end{pmatrix}. \tag{5.43}$$

Let us add that another universal property of the charge conjugation is the unitarity of the matrix $C$. Indeed, according to (5.39), this holds in the standard representation and the passage to any other set of $\gamma$-matrices is implemented by means of the similarity transformation (5.40), where one may set $U^{-1} = U^{\dagger}$ (assuming that the Dirac matrices in question possess standard properties under Hermitian conjugation). Employing then the relation (5.42), one can see easily that $C^{-1} = C^{\dagger}$ is valid for any relevant $\gamma$-matrix representation.

We may now resume, for a moment, our earlier theme of the "fun with $\gamma$-matrices"; we are going to add a remarkable item to the collection of $\gamma$-matrix identities. It turns out that the trace of a product of $\gamma$-matrices is not changed when the order of matrices is reversed, i.e.

$$\mathrm{Tr}(\gamma_{\alpha}\gamma_{\beta}\cdots\gamma_{\tau}\gamma_{\omega}) = \mathrm{Tr}(\gamma_{\omega}\gamma_{\tau}\cdots\gamma_{\beta}\gamma_{\alpha}). \tag{5.44}$$

The proof of the "palindromic" relation (5.44) is quite easy and relies on the identity (5.36). First, if the number of $\gamma$-matrices under the trace is odd, (5.44) holds trivially. For an even number $n$ of $\gamma$-matrices one has

$$\mathrm{Tr}(\gamma_{\alpha}\gamma_{\beta}\cdots\gamma_{\tau}\gamma_{\omega}) = \mathrm{Tr}(\gamma_{\alpha}\gamma_{\beta}\cdots\gamma_{\tau}\gamma_{\omega})^{\mathsf{T}} = \mathrm{Tr}(\gamma_{\omega}^{\mathsf{T}}\gamma_{\tau}^{\mathsf{T}}\cdots\gamma_{\beta}^{\mathsf{T}}\gamma_{\alpha}^{\mathsf{T}}),$$

and then the relation (5.36) can be employed. Thus one gets

$$\mathrm{Tr}(\gamma_{\alpha}\gamma_{\beta}\cdots\gamma_{\tau}\gamma_{\omega}) = (-1)^{n}\,\mathrm{Tr}(C^{-1}\gamma_{\omega}CC^{-1}\gamma_{\tau}C\cdots C^{-1}\gamma_{\alpha}C),$$

and using trace cyclicity (as well as $(-1)^{n} = 1$) one arrives at (5.44).

After this relatively long exposition concerning $C$ one might naturally ask why it is called "charge conjugation", when there was no charge at play. Well, one natural answer comes from the form of Dirac equation for the charged particle in an external electromagnetic field. Denoting the corresponding four-potential as $A_{\mu}$, the "minimal" interaction is incorporated in the equation

$$\left[i\gamma^{\mu}(\partial_{\mu} - ieA_{\mu}) - m\right]\psi = 0. \tag{5.45}$$

36

Since the transformation

$$\psi_C = C\overline{\psi}^\mathsf{T}$$

involves the complex conjugation, it is easy to see that $\psi_C$ satisfies Eq. (5.45) with $e$ replaced by $-e$. Moreover, we have motivated our discussion of the charge conjugation by the desire to find a symmetry transformation connecting the solutions with positive and negative energy. We will see later (within the framework of field theory) that the negative energy solutions are closely related to antiparticles (carrying the charge opposite to particles). So, this is another, quite standard, justification of the label "charge conjugation".

We have devoted this chapter to the discrete symmetries $C$, $P$, $T$ and now one might wonder whether this is all, i.e. whether one would not be able to find another independent symmetry of such a type. For instance, one might try full spacetime inversion $x^0 \rightarrow -x^0$, $\vec{x} \rightarrow -\vec{x}$ without complex conjugation of the wave function. In such a case, one has to solve the conditions $\Lambda^\mu_{\ \nu}\gamma^\nu = S^{-1}\gamma^\mu S$ with $\Lambda = -\mathbb{1}$. It means that the matrix $S$ should anticommute with any $\gamma^\mu$, $\mu = 0, 1, 2, 3$; of course, the answer is then obvious, $S = \gamma_5$. It is easy to realize that such a transformation is in fact equivalent to the product $CPT$. If one tries other possibilities, seemingly not covered by $C$, $P$ and $T$, one always arrives at a combination of some of them. In this sense, our set of discrete symmetries is complete.

# Chapter 6

# Plane-wave solutions of Dirac equation: $u$ and $v$

The subject indicated in the title has been already touched earlier, in Chapter 2, but here we are going to discuss it in detail; the solutions we have in mind are plane waves, describing particle states with a definite energy and momentum. One might wonder why should we pay so much attention to plane waves, which may be viewed as a description of an almost trivial physical situation. Well, a knowledgeable reader may guess that the plane-wave solutions will come in handy later on, when considering problems of particle scattering within perturbation theory: in this context one should recall the famous Born approximation for the potential scattering. Of course, the plane waves for a Dirac particle are more complicated than what one may remember from the introductory quantum mechanics course, because of spin degrees of freedom; such technical aspects may add more flavour to the subject that otherwise might seem rather boring at first sight. So much for the motivation and apologies, and now let us proceed to calculations.

For convenience, let us reiterate here the familiar equation of our interest:

$$i\gamma^\mu \frac{\partial \psi}{\partial x^\mu} - m\psi = 0 \,. \tag{6.1}$$

Following our earlier treatment of the Klein–Gordon equation (cf. (1.12)), it is straightforward to figure out an appropriate form of the Dirac plane waves in question. One may write

$$\psi_{(+)}(x) = u(p)\, e^{-ipx} \,,$$
$$\psi_{(-)}(x) = v(p)\, e^{ipx} \,, \tag{6.2}$$

where, according to the preliminary discussion in Chapter 2 (cf. (2.23)), we envisage the existence of solutions with positive and negative energy. The coefficients $u(p)$ and $v(p)$ are four-component bispinor amplitudes of the considered plane waves and their properties will be the main subject of our subsequent study. Substituting the expressions (6.2) into Eq. (6.1), one gets readily

$$(p_\mu \gamma^\mu - m)u(p) = 0 \,,$$
$$(p_\mu \gamma^\mu + m)v(p) = 0 \,. \tag{6.3}$$

The linear combination of $\gamma$-matrices appearing in (6.3) is denoted as

$$p_\mu \gamma^\mu = \not{p} \,. \tag{6.4}$$

The symbol $\not{p}$ is to be pronounced as "$p$ slash" (this somewhat crazy notation is due to R. Feynman). Thus, linear algebraic equations (6.3) are recast as

$$(\not{p} - m)u(p) = 0 ,$$
$$(\not{p} + m)v(p) = 0 .$$

(6.5)

Multiplying the first equation (6.5) by the matrix $\not{p} + m$, or the second equation by $\not{p} - m$, one gets

$$(\not{p}\not{p} - m^2)u(p) = 0 ,$$

(6.6)

and the same for $v(p)$. However, it is easy to see that $\not{p}\not{p} = p^2 = p_0^2 - \vec{p}^2$. Indeed, one has

$$\not{p}\not{p} = p_\mu \gamma^\mu p_\nu \gamma^\nu = \frac{1}{2} p_\mu p_\nu \{\gamma^\mu, \gamma^\nu\} = \frac{1}{2} p_\mu p_\nu \cdot 2g^{\mu\nu} = p_\mu p^\mu = p^2 .$$

(6.7)

Thus, we get the condition $p^2 = m^2$, i.e. $p_0^2 = \vec{p}^2 + m^2$, which is certainly good (though expected) news. Conventionally, one may choose $p_0 > 0$ (without loss of generality); then one may find readily the values of the energy for $\psi_{(+)}$ and $\psi_{(-)}$ in (6.2). One has

$$i \frac{\partial}{\partial x_0} \psi_{(+)}(x) = p_0 \psi_{(+)}(x) ,$$
$$i \frac{\partial}{\partial x_0} \psi_{(-)}(x) = -p_0 \psi_{(-)}(x) ,$$

(6.8)

so that one may conclude that $\psi_{(+)}$ and $\psi_{(-)}$ correspond to a positive and negative energy, respectively.

Our main goal is to find an explicit form of the amplitudes $u(p)$ and $v(p)$. For this purpose, it is useful to realize that $\psi_{(-)}$ can be related to $\psi_{(+)}$ via charge conjugation; it means that one may define $v(p)$ as

$$v(p) = u_C(p) = C\bar{u}^\mathsf{T}(p) ,$$

(6.9)

where $C$ is determined by (5.36). So, we will concentrate now on $u(p)$. Let us start with $\vec{p} = 0$, i.e. the particle at rest. One then has $p_0 = m$; denoting the corresponding $u(p)$ as $u(m, \vec{0})$, Eq. (6.5) is reduced to

$$(\gamma_0 - \mathbb{1})u(m, \vec{0}) = 0$$

(6.10)

(similarly, for $v(m, \vec{0})$ one gets (6.10) with $\mathbb{1}$ replaced by $-\mathbb{1}$). From now on we will employ the standard representation of $\gamma$-matrices. From (6.10) one then gets two linearly independent solutions, e.g.

$$u^{(1)}(m, \vec{0}) = \begin{pmatrix} 1 \\ 0 \\ 0 \\ 0 \end{pmatrix} , \qquad u^{(2)}(m, \vec{0}) = \begin{pmatrix} 0 \\ 1 \\ 0 \\ 0 \end{pmatrix}$$

(6.11)

(a watchful reader has certainly observed that we have just reproduced our earlier result (2.23)). The form of $u(p)$ for $\vec{p} \neq 0$ could be obtained from $u(m, \vec{0})$ by means of an appropriate Lorentz boost, but there is an elegant short cut to the desired solution, utilizing a simple trick. Clearly, if one acts with $\not{p} + m$ on $u(m, \vec{0})$, the result certainly satisfies the equation $(\not{p} - m)u = 0$, because of $(\not{p} - m)(\not{p} + m) = 0$ for $p^2 = m^2$, and for $p = (m, \vec{0})$ it is proportional to $u(m, \vec{0})$. So, one may employ an Ansatz

$$u^{(r)}(p) = N(\not{p} + m)u^{(r)}(m, \vec{0}) , \qquad r = 1, 2 ,$$

(6.12)

39

where $N$ is an appropriate normalization factor. For the standard representation of $\gamma$-matrices one has

$$\not{p} + m = \begin{pmatrix} p_0 + m & -\vec{\sigma} \cdot \vec{p} \\ \vec{\sigma} \cdot \vec{p} & -p_0 + m \end{pmatrix},$$ (6.13)

and $u^{(r)}(m, \vec{0})$ can be written as

$$u^{(r)}(m, \vec{0}) = \begin{pmatrix} \varphi^{(r)} \\ 0 \end{pmatrix},$$ (6.14)

with

$$\varphi^{(1)} = \begin{pmatrix} 1 \\ 0 \end{pmatrix}, \qquad \varphi^{(2)} = \begin{pmatrix} 0 \\ 1 \end{pmatrix}.$$

Using (6.12), (6.13) and (6.14), one thus gets

$$u^{(r)}(p) = N(E + m) \begin{pmatrix} \varphi^{(r)} \\ \dfrac{\vec{\sigma} \cdot \vec{p}}{E + m} \varphi^{(r)} \end{pmatrix},$$ (6.15)

where we have denoted $E = p_0$. Notice that the formula (6.15) demonstrates clearly the suppression of the lower components for $|\vec{p}| \ll m$, i.e. in the non-relativistic approximation (as we have stressed before, it is a characteristic feature of the standard representation of $\gamma$-matrices).

Let us now fix the value of the factor $N$. It is reasonable to formulate the normalization condition in terms of the quantity $\bar{u}u$, since it is a scalar. The convention we are going to use is

$$\bar{u}(p)u(p) = 2m.$$ (6.16)

Note that it is not universally used in the literature; some authors prefer the normalization defined by $\bar{u}u = 1$. We will see later why the choice (6.16) is convenient. So, using (6.12) and denoting, for brevity, $u(m, \vec{0})$ as $u(0)$, one gets first

$$\begin{aligned}
\bar{u}(p)u(p) &= u^\dagger(p)\gamma_0 u(p) = N^2 u^\dagger(0)(\not{p} + m)^\dagger \gamma_0(\not{p} + m)u(0) \\
&= N^2 u^\dagger(0)\gamma_0(\not{p} + m)^2 u(0) = N^2 u^\dagger(0)\gamma_0(p^2 + 2m\not{p} + m^2)u(0) \\
&= 2m N^2 u^\dagger(0)\gamma_0(\not{p} + m)u(0)
\end{aligned}$$

(for simplicity, we take $N$ to be real). Then, employing (6.13), the last expression is recast as

$$\begin{aligned}
\bar{u}(p)u(p) &= 2m N^2 (\varphi^\dagger, 0) \begin{pmatrix} p_0 + m & -\vec{\sigma} \cdot \vec{p} \\ -\vec{\sigma} \cdot \vec{p} & p_0 - m \end{pmatrix} \begin{pmatrix} \varphi \\ 0 \end{pmatrix} \\
&= 2m N^2 (p_0 + m)\varphi^\dagger \varphi = 2m N^2 (E + m).
\end{aligned}$$ (6.17)

The condition (6.16) thus gives

$$N = (E + m)^{-1/2},$$ (6.18)

and the final form of (6.15) becomes

$$u^{(r)} = \sqrt{E + m} \begin{pmatrix} \varphi^{(r)} \\ \dfrac{\vec{\sigma} \cdot \vec{p}}{E + m} \varphi^{(r)} \end{pmatrix}.$$ (6.19)

Next, how about $v(p)$ defined by (6.9)? For brevity, we will write simply $v$, $u$ instead of $v(p)$, $u(p)$ in what follows. Using (6.9), one has

$$\bar{v}v = \bar{u}_C u_C = u_C^\dagger \gamma_0 u_C = (\bar{u}^\mathsf{T})^\dagger C^\dagger \gamma_0 C \bar{u}^\mathsf{T}.$$ (6.20)

Now it is useful to take into account the trivial fact that the expression (6.20) is a number, i.e. the matrix $1 \times 1$. This, of course, means that it is equal to its transpose. Thus, we may write

$$\bar{v}v = \left[ (\bar{u}^{\mathsf{T}})^{\dagger} C^{\dagger} \gamma_0 C \, \bar{u}^{\mathsf{T}} \right]^{\mathsf{T}} = \bar{u} C^{\mathsf{T}} \gamma_0^{\mathsf{T}} (C^{-1})^{\mathsf{T}} \bar{u}^{\dagger} \,, \tag{6.21}$$

where we have employed the relation $C^{\dagger} = C^{-1}$. Using the known properties of $C$ and $\gamma_0$ in the standard representation, from (6.21) one thus gets finally

$$\bar{v}v = \bar{u} C \gamma_0 C^{-1} \gamma_0 u = -\bar{u}u \,. \tag{6.22}$$

The last result is quite remarkable: it means that the normalization condition for $v(p)$,

$$\bar{v}(p)v(p) = -2m \,, \tag{6.23}$$

is a necessary consequence of our choice (6.16) for $u(p)$.

As we have already noted at the beginning of this chapter, some knowledge of the properties of the plane-wave solutions of Dirac equation will be important later on, in computation of scattering amplitudes and cross sections of various physical processes. Well, the formula (6.19) perhaps does not look quite encouraging from the point of view of the envisaged algebraic manipulations, but there is a good news for a skeptical reader. In fact, we will never need an explicit form of $u(p)$ or $v(p)$ like (6.19) or (6.9); instead, when working out the physically relevant quantities, one usually needs just some specific bilinear combinations of these functions, which are well suited for creating efficient computational algorithms (by the way, the knowledge of traces of $\gamma$-matrix products then becomes essential). An example of the above-mentioned bilinear combination is the sum over two spin states

$$\sum_{r=1}^{2} u^{(r)}(p) \bar{u}^{(r)}(p) = \not{p} + m \tag{6.24}$$

(such an identity is usually called the "completeness relation"). As a warm-up exercise, let us prove Eq. (6.24) by means of a straightforward calculation. So, using (6.12) one has

$$u^{(r)}(p) \bar{u}^{(r)}(p) = N^2 (\not{p} + m) u^{(r)}(0) u^{(r)\,\dagger}(0) (\not{p} + m)^{\dagger} \gamma_0$$
$$= N^2 (\not{p} + m) u^{(r)}(0) \bar{u}^{(r)}(0) (\not{p} + m) \,, \tag{6.25}$$

where we have utilized the familiar identity $\not{p}^{\dagger} = \gamma_0 \not{p} \gamma_0$. From (6.11) it is easy to see that

$$\sum_{r=1}^{2} u^{(r)}(0) \bar{u}^{(r)}(0) = \begin{pmatrix} 1 & 0 \\ 0 & 0 \end{pmatrix} = \frac{1}{2} (1 + \gamma_0) \,. \tag{6.26}$$

Then one has

$$\sum_{r=1}^{2} u^{(r)}(p) \bar{u}^{(r)}(p) = N^2 (\not{p} + m) \frac{1 + \gamma_0}{2} (\not{p} + m) \,. \tag{6.27}$$

Working out the product of $\gamma$-matrices in (6.27) is straightforward. One gets, after some obvious manipulations,

$$X \equiv (\not{p} + m)(1 + \gamma_0)(\not{p} + m)$$
$$= 2m^2 + 2m\not{p} + \not{p}\gamma_0\not{p} + m\not{p}\gamma_0 + m\gamma_0\not{p} + m^2\gamma_0 \,. \tag{6.28}$$

41

Employing the basic anticommutation relation for $\gamma$-matrices, one has $\not{p}\gamma_0 + \gamma_0\not{p} = 2p_0$, so that (6.28) can be further simplified as

$$X = 2m^2 + 2m\not{p} + \not{p}(2p_0 - \not{p}\gamma_0) + 2mp_0 + m^2\gamma_0$$
$$= 2m^2 + 2m\not{p} + 2p_0\not{p} + 2mp_0 = 2(p_0 + m)(\not{p} + m) \,. \tag{6.29}$$

Thus, substituting (6.29) into (6.27) and using the value (6.18) for $N$, one recovers indeed the formula (6.24). The corresponding result for the bispinors $v^{(r)}(p)$ reads

$$\sum_{r=1}^{2} v^{(r)}(p)\bar{v}^{(r)}(p) = \not{p} - m \,, \tag{6.30}$$

and it can be proved by using (6.24) and the charge conjugation transformation. Indeed, from (6.9) one has

$$\bar{v} = v^\dagger\gamma_0 = \left(\bar{u}^\mathsf{T}\right)^\dagger C^\dagger\gamma_0 \,,$$

so that

$$v^{(r)}\bar{v}^{(r)} = C\bar{u}^{(r)\,\mathsf{T}}\left(\bar{u}^{(r)\,\mathsf{T}}\right)^\dagger C^\dagger\gamma_0 = C\bar{u}^{(r)\,\mathsf{T}}\left(\left(u^{(r)\,\dagger}\gamma_0\right)^\dagger\right)^\mathsf{T} C^\dagger\gamma_0$$

$$= C\bar{u}^{(r)\,\mathsf{T}}\left(\gamma_0 u^{(r)}\right)^\mathsf{T} C^\dagger\gamma_0 = C\bar{u}^{(r)\,\mathsf{T}} u^{(r)\,\mathsf{T}}\gamma_0 C^{-1}\gamma_0$$

$$= C\left(u^{(r)}\bar{u}^{(r)}\right)^\mathsf{T}\gamma_0 C^{-1}\gamma_0 \,. \tag{6.31}$$

Thus, using (6.24), the sum in question becomes

$$\sum_{r=1}^{2} v^{(r)}(p)\bar{v}^{(r)}(p) = C(\not{p} + m)^\mathsf{T}\gamma_0 C^{-1}\gamma_0 = C(-C^{-1}\not{p}C + m)\gamma_0 C^{-1}\gamma_0$$

$$= (-\not{p}C + mC)\gamma_0 C^{-1}\gamma_0 = -\not{p}C\gamma_0 C^{-1}\gamma_0 + mC\gamma_0 C^{-1}\gamma_0$$

$$= \not{p} - m \,, \tag{6.32}$$

and (6.30) is thereby proved.

At this point, one can also see what is the advantage of using the normalizations (6.16) and (6.23). For the convention $\bar{u}u = 1$ and $\bar{v}v = -1$ one would get in (6.24) the result $(\not{p}+m)/2m$ and (6.23) would become $(\not{p} - m)/2m$. In some situations, the limit $m \to 0$ may be of interest; within our normalization convention this can be implemented directly at the level of the formulae (6.23), (6.24), while in the alternative scheme more care is needed.

# Chapter 7

# Description of spin states of Dirac particle

The solutions of the Dirac equation, described in the preceding chapter (see (6.11), (6.12)) have a straightforward physical interpretation. They correspond to a particle that has (apart from definite values of the energy and momentum) a definite projection of the spin to the 3rd coordinate axis in its rest system. In particular, for $r = 1$ such a projection (corresponding to $\frac{1}{2}\Sigma_3$) is $+1/2$ and for $r = 2$ its value is $-1/2$. Of course, one would like to have a more direct explicit characterization of the spin states in the reference frame, where the particle is moving. However, it turns out that one cannot simply rely on the eigenvalues of the matrix $\vec{s} \cdot \vec{\Sigma}$ with $\vec{s}$ being the unit vector for a fixed space direction. The point is that the plane-wave amplitude $u(p)$ is an eigenvector of the matrix $\not{p}$ (with the eigenvalue $m$, see (6.5)), but $\not{p}$ does not, in general, commute with $\vec{s} \cdot \vec{\Sigma}$. We will return to this issue in more technical terms later on.

So, how should one proceed in order to achieve the desired goal? Obviously, a right strategy could be to start in the rest system, perform an appropriate Lorentz boost and find out how the condition for the spin projection gets transformed. To this end, let us consider two solutions of the equation $(\not{p} - m)u = 0$ for $p = p^{(0)} = (m, \vec{0})$ that are eigenvectors of the matrix $\vec{s} \cdot \vec{\Sigma}$ with $|\vec{s}| = 1$. We have in mind the standard representation of the $\gamma$-matrices, so that $\vec{\Sigma}$ has the usual form

$$\vec{\Sigma} = \begin{pmatrix} \vec{\sigma} & 0 \\ 0 & \vec{\sigma} \end{pmatrix}. \tag{7.1}$$

Using the familiar properties of the Pauli matrices, it is clear that $(\vec{s} \cdot \vec{\Sigma})^2 = |\vec{s}|^2 = 1$; thus, the eigenvalues of $\vec{s} \cdot \vec{\Sigma}$ are $\pm 1$. Note that the spin projection matrix is in fact $\vec{s} \cdot \vec{S}$, with $\vec{S} = \frac{1}{2}\vec{\Sigma}$, but for convenience we are going to work with $\vec{s} \cdot \vec{\Sigma}$ in the sequel. The two linearly independent eigenvectors of $\vec{s} \cdot \vec{\Sigma}$ in the particle rest system can be written as

$$\vec{s} \cdot \vec{\Sigma}\, u(0, \vec{s}) = u(0, \vec{s})\,, \tag{7.2a}$$

$$-\vec{s} \cdot \vec{\Sigma}\, u(0, -\vec{s}) = u(0, -\vec{s})\,, \tag{7.2b}$$

which is a form most appropriate for our purpose. Let us now start processing Eq. (7.2a) along the lines indicated above. A key ingredient of the calculational procedure is the remarkable identity

$$\vec{\Sigma} = \gamma_5 \vec{\alpha}\,. \tag{7.3}$$

It can be verified easily within the standard representation, since here we have

$$\gamma_5 = i\gamma^0\gamma^1\gamma^2\gamma^3 = \begin{pmatrix} 0 & \mathbb{1} \\ \mathbb{1} & 0 \end{pmatrix} \tag{7.4}$$

43

(the reader is urged to verify this result independently) and the matrices $\vec{\alpha}$ are given by (1.32) (in fact, the formula (7.3) is valid in any representation of $\gamma$-matrices; for technical details see Appendix C). So, using (7.3) and the relation $\vec{\alpha} = \gamma_0 \vec{\gamma}$, Eq. (7.2a) is recast as

$$\gamma_5 \gamma_0 \vec{s} \cdot \vec{\gamma} u(0, \vec{s}) = u(0, \vec{s}) . \tag{7.5}$$

Next, as we know, in the rest system it holds

$$\gamma_0 u(0, \vec{s}) = u(0, \vec{s})$$

(cf. (6.10)). Thus, (7.5) becomes

$$-\gamma_5 \vec{s} \cdot \vec{\gamma} u(0, \vec{s}) = u(0, \vec{s}) . \tag{7.6}$$

The matrix product on the left-hand side of Eq. (7.6) can be formally rewritten as $\gamma_5 \slashed{s}^{(0)}$, if one adopts a natural notation

$$s^{(0)} = (0, \vec{s}) . \tag{7.7}$$

So, an intermediate result of the first step of our calculation reads

$$\gamma_5 \slashed{s}^{(0)} u(0, \vec{s}) = u(0, \vec{s}) . \tag{7.8}$$

Next, let us consider the Lorentz boost. The coordinate system, in which the particle has a momentum $\vec{p}$ (velocity $\vec{v}$) is moving with velocity $-\vec{v}$ relatively to the particle rest frame. Thus, the reference system of our interest is connected with the rest frame by means of the Lorentz transformation described by a matrix denoted briefly as $\Lambda(-\vec{v})$. Dirac spinors are then transformed by means of the corresponding matrix $S(-\vec{v})$ (cf. (4.9)). Denoting for brevity $u(0, \vec{s})$ simply as $u(0)$, we thus have $u(p) = S(-\vec{v})u(0)$, i.e.

$$u(0) = S^{-1}(-\vec{v})u(p) . \tag{7.9}$$

Of course, one should keep in mind that

$$S^{-1}(-\vec{v}) = S(\vec{v}) \leftrightarrow \Lambda(\vec{v}) . \tag{7.10}$$

So, substituting (7.9) into (7.8), one has

$$\gamma_5 \slashed{s}^{(0)} S^{-1}(-\vec{v})u(p) = S^{-1}(-\vec{v})u(p) ,$$

i.e.

$$\gamma_5 S(-\vec{v}) \slashed{s}^{(0)} S^{-1}(-\vec{v})u(p) = u(p) , \tag{7.11}$$

where we have utilized the fact that $S(-\vec{v})$ commutes with $\gamma_5$. Working out the left-hand side of Eq. (7.11), one gets

$$S(-\vec{v}) \slashed{s}^{(0)} S^{-1}(-\vec{v}) = S^{-1}(\vec{v}) s_\mu^{(0)} \gamma^\mu S(\vec{v}) = s_\mu^{(0)} \Lambda^\mu{}_\nu(\vec{v}) \gamma^\nu . \tag{7.12}$$

In this way, one arrives at the combination

$$\Lambda^\mu{}_\nu(\vec{v}) s_\mu^{(0)} \equiv s_\nu . \tag{7.13}$$

This relation can be written in the matrix form as

$$s = \Lambda^{\mathsf{T}}(\vec{v}) s^{(0)} . \tag{7.14}$$

44

Using the notation introduced in Appendix A, one may employ the known result for $\Lambda^T$, namely (see (A.7) or (A.8))

$$\Lambda^T = \left(\overline{\Lambda}\right)^{-1}.$$

So, (7.14) can be recast as

$$s = \left(\overline{\Lambda}(\vec{v})\right)^{-1} s^{(0)} = \overline{\Lambda}(-\vec{v}) s^{(0)}. \tag{7.15}$$

Having in mind the definition of the matrix $\overline{\Lambda}$, Eq. (7.15) can be written in terms of components as

$$s_\mu = \Lambda_\mu{}^\rho(-\vec{v}) s^{(0)}_\rho \tag{7.16}$$

or, equivalently,

$$s^\mu = \Lambda^\mu{}_\nu(-\vec{v}) s^{(0)\,\nu}. \tag{7.17}$$

Then, using (7.13) and (7.16), the expression (7.12) becomes $\not{s}$ and the relation (7.11) can be thus rewritten as

$$\gamma_5 \not{s} u(p) = u(p). \tag{7.18}$$

The result (7.18) is quite remarkable. We have found out that the original condition (7.2a) for the particle at rest is equivalent to the relation (7.18) for the particle in motion. The four-component quantity $s$ is obtained from $s^{(0)}$, defined by (7.7), by means of the same Lorentz transformation that turns the rest four-momentum $p^{(0)} = (m, \vec{0})$ into $p = (E, \vec{p})$. This, of course, means that $s = s(p)$ is a four-vector. Moreover, since obviously $p^{(0)} \cdot s^{(0)} = 0$, it also holds

$$p \cdot s(p) = 0. \tag{7.19}$$

The four-vector $s$ has another obvious property; it is space-like, and

$$s^2 = -1. \tag{7.20}$$

This is a direct consequence of $(s^{(0)})^2 = -1$ and Lorentz invariance. Now one may check immediately the consistency of the condition (7.18). One would like to see that $\gamma_5 \not{s}$ commutes with $\not{p}$. It is quite easy; one has

$$[\not{p}, \gamma_5 \not{s}] = \not{p} \gamma_5 \not{s} - \gamma_5 \not{s} \not{p} = -\gamma_5 \{\not{p}, \not{s}\} = -2\gamma_5 p \cdot s = 0.$$

Moreover,

$$(\gamma_5 \not{s})^2 = \mathbb{1}. \tag{7.21}$$

Indeed, taking into account (7.20), one has

$$(\gamma_5 \not{s})^2 = \gamma_5 \not{s} \gamma_5 \not{s} = -\not{s}\not{s} = 1.$$

So, owing to (7.21), the eigenvalues of $\gamma_5 \not{s}$ are $\pm 1$.

The condition (7.18) represents a covariant characterization of the spin state of the Dirac particle, so it is quite natural to call $s$ the **spin four-vector**; an alternative term that is used is the **polarization** four-vector. From now on, we will use the notation $u(p, s)$ for the solution of the equation $(\not{p} - m)u = 0$ satisfying (7.18):

$$\gamma_5 \not{s} u(p, s) = u(p, s). \tag{7.22}$$

Next, the question is what happens with $u(0, -\vec{s})$ satisfying (7.2b). The answer is quite clear: Following the same procedure as above, one gets a spin four-vector equal to $-s$, since in this

45

case we would start with $u(0, -\vec{s})$ in the rest system, and the ensuing Lorentz transformation is linear. Thus, we end up with a solution $u(p, -s)$ of $(\not{p} - m)u = 0$, satisfying

$$-\gamma_5 \not{s} u(p, -s) = u(p, -s). \tag{7.23}$$

Just to be sure: Please note that while $s$ and $-s$ are linearly dependent, $u(p, s)$ and $u(p, -s)$ are linearly independent.

As regards the solutions of the equation $(\not{p} + m)v = 0$, we may rely on the approach outlined in Chapter 6, i.e. define $v(p, s)$ by means of the charge conjugation of $u(p, s)$:

$$v(p, s) = u_C(p, s) = C\bar{u}^{\mathsf{T}}(p, s). \tag{7.24}$$

Then it is not difficult to show that

$$\gamma_5 \not{s} v(p, s) = v(p, s). \tag{7.25}$$

Indeed, starting with (7.22) one gets first, after some simple manipulations,

$$-\bar{u}(p, s)\not{s}\gamma_5 = \bar{u}(p, s). \tag{7.26}$$

The transpose of (7.26) becomes

$$C^{-1}\gamma_5 \not{s} C\bar{u}^{\mathsf{T}}(p, s) = \bar{u}^{\mathsf{T}}(p, s), \tag{7.27}$$

so that finally one has

$$\gamma_5 \not{s} C\bar{u}^{\mathsf{T}}(p, s) = C\bar{u}^{\mathsf{T}}(p, s), \tag{7.28}$$

and (7.25) is thereby proved. Note that when deriving (7.27) we have used the relation $\gamma_5^{\mathsf{T}} = C^{-1}\gamma_5 C$ (the reader is encouraged to verify this identity independently). Clearly, in the same way one may obtain $v(p, -s)$ as $u_C(p, -s)$ that satisfies

$$-\gamma_5 \not{s} v(p, -s) = v(p, -s). \tag{7.29}$$

At this point, a conceptual remark is in order. Our road to the covariant treatment of the spin states of a free Dirac particle relied on an elementary description of the states in question in the rest system, and the ensuing Lorentz transformation. In fact, one might now forget about any reference to the rest system and adopt a more abstract approach, defining e.g. $u(p, s)$ directly as a common eigenvector of $\not{p}$ and $\gamma_5 \not{s}$, with $s$ being a space-like four-vector orthogonal to $p$ and normalized as $s^2 = -1$. We will come back to this viewpoint in more detail later on.

Hopefully, the preceding discussion contained some exciting moments (in particular, the appearance of somewhat mysterious $\gamma_5$ in the spin description might be surprising at first sight). Now we have to add some more technicalities to our description of the functions $u(p, \pm s)$ and $v(p, \pm s)$ that you may find rather boring, but they will turn out to be very important later (another "rifle hanging on the wall"... ). The technique we are going to develop here usually appears in textbooks under the heading "projection operators for energy and spin". So, what are the projection operators in question? Let us start with $u(p, s)$. As we know, it satisfies the equation

$$(\not{p} - m)u(p, s) = 0.$$

This can be rewritten, somewhat artificially, as

$$\frac{m + \not{p}}{2m} u(p, s) = u(p, s). \tag{7.30}$$

46

Using it, we are on the right track, since the matrix

$$\Lambda_+ = \frac{m + \not{p}}{2m} \tag{7.31}$$

is a projection operator, in the sense that $(\Lambda_+)^2 = \Lambda_+$ (i.e. it is an idempotent matrix). Indeed,

$$(\Lambda_+)^2 = \left(\frac{m + \not{p}}{2m}\right)^2 = \frac{m^2 + 2m\not{p} + \not{p}\not{p}}{4m^2} = \frac{m^2 + 2m\not{p} + p^2}{4m^2} = \frac{m + \not{p}}{2m}.$$

The notation chosen in (7.31) refers to the fact that $u(p, s)$ is the amplitude of a plane wave with positive energy. Similarly, the relation (7.22) can be recast as

$$\frac{1 + \gamma_5 \not{s}}{2} u(p, s) = u(p, s), \tag{7.32}$$

and the matrix

$$\Sigma(s) = \frac{1 + \gamma_5 \not{s}}{2} \tag{7.33}$$

is another projection operator, since

$$\left(\Sigma(s)\right)^2 = \frac{1}{4}(1 + 2\gamma_5 \not{s} + \gamma_5 \not{s} \gamma_5 \not{s}) = \frac{1}{4}(1 + 2\gamma_5 \not{s} - s^2) = \Sigma(s),$$

where we have used (7.20) in the last step.

Furthermore, since we know that $\not{p}$ and $\gamma_5 \not{s}$ commute, $\Lambda_+$ and $\Sigma(s)$ commute as well. This in turn means that the product $\Lambda_+ \Sigma(s)$ is also a projection operator. Obviously, the same can be done for $u(p, -s)$, replacing $\Sigma(s)$ with $\Sigma(-s)$.

For a $v(p, s)$, which is a solution of the equation

$$(\not{p} + m)v(p, s) = 0, \tag{7.34}$$

one may proceed analogously. Eq. (7.34) may be recast as

$$\frac{m - \not{p}}{2m} v(p, s) = v(p, s),$$

and it is easily seen that

$$\Lambda_- = \frac{m - \not{p}}{2m} \tag{7.35}$$

is a projection operator. As for the spin operators, we get the same as before.

Our findings can be summarized as follows. One may construct four matrices

$$\begin{aligned}
P_1 &= \Lambda_+ \Sigma(s) &&= \frac{m + \not{p}}{2m} \frac{1 + \gamma_5 \not{s}}{2}, \\
P_2 &= \Lambda_+ \Sigma(-s) &&= \frac{m + \not{p}}{2m} \frac{1 - \gamma_5 \not{s}}{2}, \\
P_3 &= \Lambda_- \Sigma(s) &&= \frac{m - \not{p}}{2m} \frac{1 + \gamma_5 \not{s}}{2}, \\
P_4 &= \Lambda_- \Sigma(-s) &&= \frac{m - \not{p}}{2m} \frac{1 - \gamma_5 \not{s}}{2},
\end{aligned} \tag{7.36}$$

that have the basic property of projection operators, i.e. satisfy

$$P_j^2 = P_j, \qquad j = 1, \ldots, 4. \tag{7.37}$$

47

These projection operators are in one-to-one correspondence with functions $u(p, \pm s)$, $v(p, \pm s)$, in such a way that

$$P_1 u(p, s) = u(p, s), \qquad P_2 u(p, -s) = u(p, -s),$$
$$P_3 v(p, s) = v(p, s), \qquad P_4 v(p, -s) = v(p, -s). \tag{7.38}$$

It is not difficult to find out that the matrices $P_j$, $j = 1, \ldots, 4$ have the following properties:

$$\sum_{j=1}^{4} P_j = \mathbb{1},$$
$$P_i \cdot P_j = 0 \qquad \text{for} \quad i \neq j, \tag{7.39}$$
$$P_j^\dagger = \gamma_0 P_j \gamma_0.$$

Proving (7.39) is left to the reader as a simple exercise.

The relations (7.38) indicate that the projection operators $P_j$ can eventually be expressed in terms of bilinear combinations of the functions $u(p, \pm s)$, $v(p, \pm s)$. It is indeed so, as we will see shortly. First, let us introduce a simpler notation, adapted directly to the formalism of projection operators, namely

$$w_1 = u(p, s), \qquad w_2 = u(p, -s), \qquad w_3 = v(p, s), \qquad w_4 = v(p, -s). \tag{7.40}$$

Thus, the identities (7.38) are recast as

$$P_j w_j = w_j, \qquad j = 1, \ldots, 4, \tag{7.41}$$

and from this one gets readily

$$\overline{w}_j P_j = \overline{w}_j. \tag{7.42}$$

These relations imply

$$\overline{w}_j w_k = 0 \qquad \text{for} \quad j \neq k \tag{7.43}$$

as an immediate consequence of the second identity in (7.39). Let us also recall that the normalization conditions read

$$\overline{w}_j w_j = 2m \qquad \text{for} \quad j = 1, 2,$$
$$\overline{w}_j w_j = -2m \qquad \text{for} \quad j = 3, 4. \tag{7.44}$$

Thus, we have a basis in the 4-dimensional complex space $\mathbb{C}^4$, made of $w_1, \ldots, w_4$. Now, let $\phi$ be an arbitrary element of such a space; it can be expanded as

$$\phi = C_1 w_1 + \ldots + C_4 w_4. \tag{7.45}$$

Using (7.43) and (7.44), the coefficients $C_j$ can be expressed as

$$C_1 = \frac{1}{2m} \overline{w}_1 \phi, \qquad\qquad C_2 = \frac{1}{2m} \overline{w}_2 \phi,$$
$$C_3 = -\frac{1}{2m} \overline{w}_3 \phi, \qquad\qquad C_4 = -\frac{1}{2m} \overline{w}_4 \phi. \tag{7.46}$$

On the other hand, owing to the properties of the projection operators, one gets readily

$$P_1 \phi = C_1 w_1, \qquad \ldots, \qquad P_4 \phi = C_4 w_4. \tag{7.47}$$

Thus, (7.47) along with (7.46) give

$$P_1 \phi = \frac{1}{2m} \overline{w}_1 \phi w_1, \quad \ldots, \quad P_4 \phi = -\frac{1}{2m} \overline{w}_4 \phi w_4. \tag{7.48}$$

Now the results for $P_1, \ldots, P_4$ are quite clear, but in order to have a foolproof argument, let us write (7.48) in terms of components:

$$(P_1)_{ab} \phi_b = \frac{1}{2m} \overline{w}_{1b} \phi_b w_{1a} = \frac{1}{2m} w_{1a} \overline{w}_{1b} \phi_b, \tag{7.49}$$

etc. (the indices $a$, $b$ take on values $1, \ldots, 4$). From (7.48) or (7.49) one may thus conclude that

$$P_1 = \frac{1}{2m} w_1 \overline{w}_1, \qquad\qquad P_2 = \frac{1}{2m} w_2 \overline{w}_2,$$
$$P_3 = -\frac{1}{2m} w_3 \overline{w}_3, \qquad\qquad P_4 = -\frac{1}{2m} w_4 \overline{w}_4. \tag{7.50}$$

Taking into account (7.36), one then gets finally

$$
\begin{aligned}
u(p,s)\overline{u}(p,s) &= (\not{p} + m)\frac{1 + \gamma_5 \not{s}}{2}, \\
u(p,-s)\overline{u}(p,-s) &= (\not{p} + m)\frac{1 - \gamma_5 \not{s}}{2}, \\
v(p,s)\overline{v}(p,s) &= (\not{p} - m)\frac{1 + \gamma_5 \not{s}}{2}, \\
v(p,-s)\overline{v}(p,-s) &= (\not{p} - m)\frac{1 - \gamma_5 \not{s}}{2}.
\end{aligned}
\tag{7.51}
$$

These formulae are highly important in computations of observable quantities, like decay rates and scattering cross sections, for physical processes; we will enjoy such applications later on. Recall that we have already mentioned such a technical point in the preceding chapter, in connection with the so-called completeness relations (see (6.24) and (6.30)). The formulae (7.51) represent, in a sense, an "anatomy" of the completeness relations, since they refer to particular spin (polarization) states and by summing over $\pm s$ one recovers the completeness relations. Practically, this means that the formulae (7.51) will be instrumental in calculations concerning processes that involve polarized particles.

# Chapter 8

# Helicity and chirality

In the preceding chapter we have noted that, in general, $\not{p}$ does not commute with $\vec{s} \cdot \vec{\Sigma}$, where $\vec{s}$ represents a fixed space direction. Let us now work out the commutator in question explicitly. One has, using the identity (7.3), as well as other properties of $\gamma$-matrices

$$[\not{p}, \vec{s} \cdot \vec{\Sigma}] = [p_0 \gamma_0 - \vec{p} \cdot \vec{\gamma}, \vec{s} \cdot \vec{\Sigma}] = [p_0 \gamma_0 - \vec{p} \cdot \vec{\gamma}, \gamma_5 \gamma_0 \vec{s} \cdot \vec{\gamma}] = -[\vec{p} \cdot \vec{\gamma}, \gamma_5 \gamma_0 \vec{s} \cdot \vec{\gamma}]$$
$$= -\gamma_5 \gamma_0 [\vec{p} \cdot \vec{\gamma}, \vec{s} \cdot \vec{\gamma}] = -\gamma_5 \gamma_0 p^j s^k [\gamma^j, \gamma^k] = 2i\gamma_5 \gamma_0 p^j s^k \sigma^{jk} = 2i\gamma_5 \gamma_0 p^j s^k \varepsilon^{jkl} \Sigma^l \,,$$

$$(8.1)$$

where we have employed the general representation of the spin matrices $\vec{\Sigma}$ in the last step (see the formula (C.36) in Appendix C). Thus, the relation (8.1) may be finally recast as

$$[\not{p}, \vec{s} \cdot \vec{\Sigma}] = 2i\gamma_5 \gamma_0 (\vec{p} \times \vec{s}) \cdot \vec{\Sigma} \,. \qquad (8.2)$$

This result means that the considered commutator vanishes if and only if $\vec{p} \times \vec{s} = 0$. Obviously, there are just two exceptional situations, in which this can happen; either $\vec{p} = 0$ (the familiar case of the rest system), or $\vec{p} \neq 0$ and $\vec{s}$ parallel to $\vec{p}$.

Let us now focus on the second case. Thus, we are going to consider the projection of the particle spin onto the direction of momentum (again, instead of the true spin matrices $\vec{S} = \frac{1}{2}\vec{\Sigma}$ we will work with $\vec{\Sigma}$, for convenience). Such a quantity is called, traditionally, the **helicity**. The etymology of this technical term may be understood as follows. Let us visualize, naïvely, a particle in the above-mentioned spin state as e.g. a small disc rotating in the plane perpendicular to the direction of motion (defined by the unit vector $\vec{n} = \vec{p}/|\vec{p}|$). Then, a positive value of the projection of its angular momentum (spin) onto the direction $\vec{n}$ reminds one of the motion of a right-handed screw; similarly, the negative value of the considered projection would correspond to a left-handed screw. Geometrically, the motion of a screw is associated with the shape (curve) called helix (which, of course, is encountered in many other situations, including everyday life). For etymological completeness, let us add that the name originates from the Greek word $\varepsilon \lambda \iota \xi$ (meaning twisted). So, the Dirac particle with a definite momentum and positive helicity is called right-handed, and a state with negative helicity corresponds to a left-handed particle.

The straightforward definition of the helicity introduced above is quite transparent. Nevertheless, it is certainly desirable to have also an equivalent description in terms of an appropriate spin four-vector; in other words, instead of the scalar product $\vec{n} \cdot \vec{\Sigma}$ one would like to have a covariant matrix form that may be employed within the framework of the technique developed in the preceding chapter. So, we want to find a four-vector $s = s(p)$ with the standard general properties

$$s \cdot p = 0 \,, \qquad s^2 = -1 \,, \qquad (8.3)$$

50

such that $\gamma_5 \not{s}$ is equivalent to $\vec{n} \cdot \vec{\Sigma}$ when acting on the solution of the equation $(\not{p} - m)u = 0$. Note that in this case we do not intend to rely on a starting point in the rest system and the ensuing Lorentz transformation; rather we would like to implement the viewpoint advocated in the remark following the relation (7.29) in the preceding chapter, i.e. find an appropriate $s$ directly. For this purpose, we will start with an "educated guess": having in mind that the only relevant space direction is here $\vec{n} = \vec{p}/|\vec{p}|$, we will try to construct the four-vector $s$ so that its space part be parallel to $\vec{n}$. More precisely, for a right-handed state, we will employ the Ansatz

$$s_R(p) = \left( s_R^0, \lambda \frac{\vec{p}}{|\vec{p}|} \right),$$ (8.4)

with $\lambda > 0$; the parameters $s_R^0$ and $\lambda$ are to be determined by means of the conditions (8.3). So, let us work it out. The requirement $p \cdot s_R(p) = 0$ yields, obviously,

$$0 = E s_R^0 - \lambda \frac{\vec{p}}{|\vec{p}|} \cdot \vec{p} = E s_R^0 - \lambda |\vec{p}|,$$

so that

$$s_R^0 = \lambda \frac{|\vec{p}|}{E} = \lambda \beta,$$ (8.5)

where $\beta$ is the particle velocity. Further, the normalization condition leads to

$$-1 = (s_R(p))^2 = \lambda^2 \beta^2 - \lambda^2,$$

and thus

$$\lambda^2 = \frac{1}{1 - \beta^2} = \frac{1}{1 - \frac{|\vec{p}|^2}{E^2}} = \frac{E^2}{m^2}.$$ (8.6)

Taking into account (8.5) and (8.6), the result for $s_R(p)$ reads, finally,

$$s_R(p) = \left( \frac{|\vec{p}|}{m}, \frac{E}{m} \frac{\vec{p}}{|\vec{p}|} \right).$$ (8.7)

Now, let us check whether we are indeed on the right track, i.e. whether the spin four-vector (8.7) has the desired relation to the helicity. Fortunately, it is so. It holds: Let $u$ be a solution of the equation $(\not{p} - m)u = 0$. Then

$$\gamma_5 \not{s}_R(p)u = \frac{\vec{\Sigma} \cdot \vec{p}}{|\vec{p}|} u.$$ (8.8)

The easiest way of proving Eq. (8.8) is to work out both sides independently, until one gets identical expressions. Let us start with the left-hand side. Using the formula (8.7) and taking into account that $\not{p} - m = E\gamma_0 - \vec{p} \cdot \vec{\gamma} - m$, one has

$$\gamma_5 \not{s}_R(p)u = \gamma_5 \left( \frac{|\vec{p}|}{m} \gamma_0 - \frac{E}{m} \frac{1}{|\vec{p}|} \vec{p} \cdot \vec{\gamma} \right) u$$

$$= \gamma_5 \left[ \left( \frac{|\vec{p}|}{m} - \frac{E^2}{m|\vec{p}|} \right) \gamma_0 + \frac{E}{|\vec{p}|} \right] u = \frac{1}{|\vec{p}|} \gamma_5 (E - m\gamma_0) u.$$ (8.9)

For the right-hand side of (8.8) one gets, using the identity (7.3),

$$\frac{\vec{\Sigma} \cdot \vec{p}}{|\vec{p}|} u = \frac{1}{|\vec{p}|} \gamma_5 \vec{\alpha} \cdot \vec{p} u = \frac{1}{|\vec{p}|} \gamma_5 \gamma_0 \vec{\gamma} \cdot \vec{p} u = \frac{1}{|\vec{p}|} \gamma_5 \gamma_0 (E\gamma_0 - m) u = \frac{1}{|\vec{p}|} \gamma_5 (E - m\gamma_0) u.$$ (8.10)

51

So, our calculation turned out well; the coincidence of (8.9) and (8.10) proves the validity of (8.8).

If we now denote the plane-wave amplitude $u(p)$ describing a right-handed particle (i.e. carrying positive helicity) as $u_R(p)$, the identity (8.8) means that

$$\gamma_5 \slashed{s}_R(p) u_R(p) = u_R(p) \,. \tag{8.11}$$

Recalling the results of the preceding chapter, one may then also write (cf. (7.51))

$$u_R(p) \bar{u}_R(p) = (\slashed{p} + m) \frac{1 + \gamma_5 \slashed{s}_R}{2} \,. \tag{8.12}$$

Further, considering a free-particle state with negative helicity, the relevant spin four-vector is

$$s_L = -s_R \,. \tag{8.13}$$

For the corresponding plane-wave amplitude $u_L(p)$ one has

$$\gamma_5 \slashed{s}_L(p) u_L(p) = u_L(p) \,, \tag{8.14}$$

and thus also

$$u_L(p) \bar{u}_L(p) = (\slashed{p} + m) \frac{1 + \gamma_5 \slashed{s}_L}{2} \,. \tag{8.15}$$

Now, it is natural to ask what one gets for a pertinent $v(p, s)$ corresponding to a definite helicity. The answer is quite straightforward and is based on an identity analogous to (8.8). It holds: Let $v$ be a solution of the equation $(\slashed{p} + m)v = 0$. Then

$$\gamma_5 \slashed{s}_R v = \frac{\vec{\Sigma} \cdot (-\vec{p})}{|\vec{p}|} v \,, \tag{8.16}$$

where $s_R = s_R(p)$ is given by (8.7).

The proof of (8.16) is practically the same as in the case of the identity (8.8).

A remark is in order here. Recall that the function $v(p)$ is the amplitude of the plane wave

$$\psi_{(-)}(x) = v(p) e^{ipx} \,. \tag{8.17}$$

In our conventional notation $p = (E, \vec{p})$, and the solution (8.17) carries negative energy $-E$ and the momentum $-\vec{p}$. So, the scalar product on the right-hand side of Eq. (8.16) corresponds indeed to the straightforward definition of the helicity. In analogy with (8.11) we may therefore write

$$\gamma_5 \slashed{s}_R v_R(p) = v_R(p) \,, \tag{8.18}$$

where $v_R(p)$ corresponds to positive particle helicity (right-handed state). Thus, we may also write

$$v_R(p) \bar{v}_R(p) = (\slashed{p} - m) \frac{1 + \gamma_5 \slashed{s}_R}{2} \,. \tag{8.19}$$

Similarly, we write

$$\gamma_5 \slashed{s}_L v_L(p) = v_L(p) \tag{8.20}$$

for left-handed states, with $s_L = -s_R$ according to (8.13), and

$$v_L(p) \bar{v}_L(p) = (\slashed{p} - m) \frac{1 + \gamma_5 \slashed{s}_L}{2} \,. \tag{8.21}$$

52

There is another important issue arising in connection with the discussion of helicity states, namely the problem of the limit of zero mass (shortly, the "massless limit"). Of course, we know that at present there is no serious candidate for a spin-1/2 massless particle, but despite this, the massless limit is interesting for practical, technical, reasons. The point is that in high-energy limit, $E \gg m$, the energy of a relativistic particle $E = \sqrt{\vec{p}^2 + m^2}$ is very close to $|\vec{p}|$; in other words, in such a situation one is indeed close to the limit $m \to 0$. Now, from (8.7) it is clear that the limit of $s_R(p)$ for $m \to 0$ does not exist. On the other hand, the limit of $u(p)$ or $v(p)$ does exist, since the equation $\not{p}u = 0$ certainly has a solution. Thus, we expect that the identities (8.12) or (8.19) have a well-defined limit for $m \to 0$. To verify this, let us work out the limit of the right-hand side of the relation (8.12) explicitly. The starting point is the expression

$$\frac{1}{2}(\not{p} + m)(1 + \gamma_5 \not{s}_R) = \frac{1}{2}(\not{p} + m - \gamma_5 \not{p} \not{s}_R + m\gamma_5 \not{s}_R). \tag{8.22}$$

Taking into account the formula (8.7), the last term in parentheses obviously has a finite limit for $m \to 0$, since the coefficient $m$ cancels the $1/m$ factors appearing in $s_R(p)$. In particular,

$$\lim_{m \to 0} m\gamma_5 \not{s}_R(p) = \lim_{m \to 0} m\gamma_5 \left( \frac{|\vec{p}|}{m}\gamma_0 - \frac{E}{m}\frac{\vec{p}}{|\vec{p}|} \cdot \vec{\gamma} \right) = \gamma_5 \not{p}, \tag{8.23}$$

since $\lim_{m \to 0} E/|\vec{p}| = 1$. The third term in parentheses is more complicated, since at first sight there is no clear suppression of the $1/m$ factor in $s_R(p)$. So, let us evaluate the relevant limit in detail. One has

$$\lim_{m \to 0} \not{p} \not{s}_R = \lim_{m \to 0} (E\gamma_0 - \vec{p} \cdot \vec{\gamma}) \left( \frac{|\vec{p}|}{m}\gamma_0 - \frac{E}{m|\vec{p}|}\vec{p} \cdot \vec{\gamma} \right)$$

$$= \lim_{m \to 0} \left[ \frac{E|\vec{p}|}{m} + \frac{E^2}{m|\vec{p}|}\vec{p} \cdot \vec{\gamma}\gamma_0 - \frac{|\vec{p}|}{m}\vec{p} \cdot \vec{\gamma}\gamma_0 + \frac{E}{m|\vec{p}|}(\vec{p} \cdot \vec{\gamma})(\vec{p} \cdot \vec{\gamma}) \right]. \tag{8.24}$$

For the last term in square brackets one gets readily

$$(\vec{p} \cdot \vec{\gamma})(\vec{p} \cdot \vec{\gamma}) = \frac{1}{2}p^j p^k \{\gamma^j, \gamma^k\} = \frac{1}{2}p^j p^k (-2\delta^{jk}) = -|\vec{p}|^2. \tag{8.25}$$

The first term then gets cancelled and the expression (8.24) becomes

$$\lim_{m \to 0} \frac{E^2 - |\vec{p}|^2}{m|\vec{p}|}\vec{p} \cdot \vec{\gamma}\gamma_0 = \lim_{m \to 0} \frac{m}{|\vec{p}|}\vec{p} \cdot \vec{\gamma}\gamma_0 = 0. \tag{8.26}$$

Thus, surprisingly, the term that looked potentially troublesome, vanishes in the massless limit! Putting all above results together, the limit of the expression (8.22) for $m \to 0$ becomes

$$\frac{1}{2}(\not{p} + \gamma_5 \not{p}) = \not{p}\frac{1 - \gamma_5}{2}. \tag{8.27}$$

Let us summarize the results of our calculation. We have found that for $m = 0$ one may write

$$u_R(p)\bar{u}_R(p) = \not{p}\frac{1 - \gamma_5}{2}. \tag{8.28}$$

In the same way, we would get

$$u_L(p)\bar{u}_L(p) = \not{p}\frac{1 + \gamma_5}{2}. \tag{8.29}$$

53

Concerning the plane-wave amplitudes $v(p)$ for $m \to 0$, one obtains, clearly (cf. (8.19)),

$$v_R(p)\bar{v}_R(p) = \not{p}\frac{1 + \gamma_5}{2}, \tag{8.30}$$

and finally

$$v_L(p)\bar{v}_L(p) = \not{p}\frac{1 - \gamma_5}{2}. \tag{8.31}$$

Let us now return, for a moment, to the remarkable finding (8.26), i.e.

$$\lim_{m \to 0} \not{p}\not{s}_R(p) = 0. \tag{8.32}$$

To get a better insight into this result, it is useful to examine the behaviour of $s_R(p)$ in the high-energy limit (which, technically, is tantamount to $m \to 0$). It turns out that it behaves asymptotically like the four-momentum $p$, up to a remainder that is suppressed as $1/E$. More precisely, it holds, for $E \gg m$:

$$s_R^\mu(p) = \frac{p^\mu}{m} + \Delta^\mu(p), \tag{8.33}$$

where $\Delta^\mu(p)$ is of the order $\mathcal{O}(m/E)$, i.e. it vanishes for $E \to \infty$, or, equivalently, for $m \to 0$. The proof of such a statement is not difficult. Using the formula (8.7), one gets, after a trivial manipulation,

$$\Delta^\mu(p) = s_R^\mu(p) - \frac{p^\mu}{m} = \frac{E - |\vec{p}|}{m}\left(-1, \frac{\vec{p}}{|\vec{p}|}\right). \tag{8.34}$$

The factor $(E - |\vec{p}|)/m$ can be recast as

$$\frac{E - |\vec{p}|}{m} = \frac{(E - |\vec{p}|)(E + |\vec{p}|)}{m(E + |\vec{p}|)} = \frac{m}{E + |\vec{p}|}. \tag{8.35}$$

For $E \gg m$, $|\vec{p}|$ is of the same order of magnitude as $E$, and the proof is thereby completed. Let us add that from (8.33) it is easy to see that $p \cdot \Delta(p) = -m$. In any case, our conclusion is that $\Delta^\mu(p) \to 0$ for $m \to 0$. Thus, one has

$$\not{p}\not{s}_R(p) = \not{p}\left(\frac{\not{p}}{m} + \not{\Delta}(p)\right) = \frac{p^2}{m} + \not{p}\not{\Delta}(p) = m + \not{p}\not{\Delta}(p).$$

According to our preceding discussion, the second term in the last expression vanishes for $m \to 0$, and thus we recover the result (8.32).

Let us add that the relation (8.33) is, technically, quite remarkable in its own right; we will encounter such a type of algebraic identity later on, in a completely different physical situation.

Our preceding description of the limiting case $m = 0$ can be reformulated in a different way, which in fact is even simpler and more transparent. One may return to the original natural definition of the helicity in terms of the scalar product

$$\hat{h} = \frac{\vec{\Sigma} \cdot \vec{p}}{|\vec{p}|} \tag{8.36}$$

standing on the right-hand side of Eq. (8.8). Clearly, such a definition is directly applicable even for a massless particle. Let us consider first the plane-wave amplitude $u(p)$. For $m = 0$ it satisfies the equation $(E\gamma_0 - \vec{p} \cdot \vec{\gamma})u(p) = 0$ with $E = |\vec{p}|$. For our purpose, it is useful to write it as

$$\vec{\alpha} \cdot \vec{p}\,u = |\vec{p}|u. \tag{8.37}$$

54

Let us now consider a solution of Eq. (8.37) with definite helicity, i.e

$$\frac{\vec{\Sigma} \cdot \vec{p}}{|\vec{p}|} u = h \, u \, , \tag{8.38}$$

where, of course, $h$ can be $+1$ or $-1$. Employing the identity (7.3) and Eq. (8.37), one gets from (8.38)

$$\gamma_5 u = h \, u \, . \tag{8.39}$$

Thus, it turns out that for a massless particle described by means of a plane-wave amplitude the helicity is equal to an eigenvalue of the matrix $\gamma_5$ (as a consistency check one should realize that $\gamma_5$ has eigenvalues $\pm 1$, since $\gamma_5^2 = 1$). Such an eigenvalue has a special name; traditionally it is called **chirality**. So, the right-handed $u$, i.e. $u_R$, has the chirality $+1$ and the left-handed $u_L$ carries the chirality $-1$. This way of distinguishing between right and left explains the etymological origin of the term chirality: in Greek, χειρ means the hand, so that the chirality could also be dubbed the "handedness". Note that this term was introduced into relativistic quantum physics by Satosi Watanabe in 1957 (see ref. [34]).

Similarly, if one considers a plane-wave amplitude $v(p)$ (corresponding to a state with negative energy $-E = -|\vec{p}|$ and momentum $-\vec{p}$), the helicity operator is

$$\widehat{h} = \frac{\vec{\Sigma} \cdot (-\vec{p})}{|\vec{p}|} \, , \tag{8.40}$$

and the $v(p)$ satisfies for $m = 0$ the equation

$$\vec{\alpha} \cdot \vec{p} \, v = |\vec{p}| v \, . \tag{8.41}$$

Then, in the same manner as before, one gets, for the solution with a definite helicity $h$,

$$\gamma_5 v = -h \, v \, . \tag{8.42}$$

Thus, for a plane-wave amplitude $v$, the helicity is equal to minus chirality in the massless case. Our results can be summarized briefly as the following simple rule:

$$
\begin{aligned}
&\text{For } u(p) \text{ with } m = 0 \, , \quad \text{helicity} = \text{chirality} \, , \\
&\text{For } v(p) \text{ with } m = 0 \, , \quad \text{helicity} = -\text{chirality} \, .
\end{aligned}
\tag{8.43}
$$

In more detail, this means that for $m = 0$ one has

$$
\begin{aligned}
\gamma_5 u_R &= u_R \, , \\
\gamma_5 u_L &= -u_L \, , \\
\gamma_5 v_R &= -v_R \, , \\
\gamma_5 v_L &= v_L \, .
\end{aligned}
\tag{8.44}
$$

Using the relations (8.44), one may recover easily our previous results for $u_R \bar{u}_R$, etc. (see (8.28), (8.29), (8.30) and (8.31)). To this end, let us first recast (8.44) in terms of appropriate projection operators. One gets, after some simple manipulations,

$$\gamma_5 u_R = u_R \quad \Rightarrow \quad \frac{1 + \gamma_5}{2} u_R = u_R \, , \tag{8.45}$$

$$\gamma_5 u_L = -u_L \quad \Rightarrow \quad \frac{1 - \gamma_5}{2} u_L = u_L \, , \tag{8.46}$$

and similarly for $v_R$ and $v_L$. Clearly, the matrices $(1 \pm \gamma_5)/2$ are mutually orthogonal projection operators, so that one may write e.g.

$$\frac{1 + \gamma_5}{2} u_R = u_R \,,$$
$$\frac{1 + \gamma_5}{2} u_L = 0 \,. \tag{8.47}$$

It is easy to see that the Dirac conjugation of (8.47) amounts to

$$\bar{u}_R \frac{1 - \gamma_5}{2} = \bar{u}_R \,,$$
$$\bar{u}_L \frac{1 - \gamma_5}{2} = 0 \,. \tag{8.48}$$

Then, multiplying (8.48) by $u_R$ and $u_L$, respectively, and summing the resulting relations, one gets

$$(u_R \bar{u}_R + u_L \bar{u}_L) \frac{1 - \gamma_5}{2} = u_R \bar{u}_R \,. \tag{8.49}$$

However, the sum $u_R \bar{u}_R + u_L \bar{u}_L$ is equal to $\not{p}$, since it is the massless limit of the completeness relation (polarization sum) for the functions $u(p, s)$ (cf. (7.51) and (6.24)). Thus, we end up with the result

$$u_R(p) \bar{u}_R(p) = \not{p} \frac{1 - \gamma_5}{2} \tag{8.50}$$

for $m = 0$. The other relations shown above can be proved in an analogous manner. The simple trick exemplified by the relations (8.47) is worth remembering as an ideal starting point for a quick derivation of the important formulae like (8.50) (i.e. a derivation written on the back of an envelope, or in the sand of a beach on a desert island).

We have seen that the concept of chirality is quite useful from the technical point of view, since it obviously leads to a considerable simplification of the treatment of the helicity states in high-energy limit. In fact, such a concept has much broader impact as it is very important also in field theory models; in particular, it plays a key role in the formulation of the standard model of electroweak interactions.

# Chapter 9

# Weyl equation

In the preceding chapter, we have examined the massless limit of solutions of the Dirac equation and we have uncovered a remarkable rule for the description of right-handed and left-handed states in terms of the chirality. Now, as a natural continuation of the study of the $m \to 0$ limit, one might try to go back "ad fontes" and take $m = 0$ from the very beginning. Well, we have already noted that at present there is no truly massless spin-$\frac{1}{2}$ particle; nevertheless, for various reasons, such a case retains some purely theoretical interest, and thus it certainly makes sense to devote one chapter to this issue.

Let us start with re-examining the guesswork presented in the introductory Chapter 1, which has led to the Dirac equation. If $m = 0$, one may use a simple Ansatz

$$i \frac{\partial \psi}{\partial t} = -i \vec{\alpha} \cdot \vec{\nabla} \psi \tag{9.1}$$

for the relativistic equation in question (cf. (1.23) with $\hbar = c = 1$). In the same way as before, the matrices $\alpha^j$, $j = 1, 2, 3$, should satisfy the anticommutation relations

$$\{\alpha^j, \alpha^k\} = 2\delta^{jk} . \tag{9.2}$$

The matrix coefficient $\beta$ is absent now, and this makes a significant difference. The point is that the relation (9.2) can be satisfied by means of $2 \times 2$ matrices; obviously, the Pauli matrices $\sigma^j$ represent a possible choice. More precisely, there are two inequivalent options for the triplet of $\alpha^j$ in (9.2), namely

$$1) \quad \alpha^j = \sigma^j , \tag{9.3}$$

$$2) \quad \alpha^j = -\sigma^j . \tag{9.4}$$

The sets (9.3) and (9.4) are indeed inequivalent, since there is no invertible matrix $S$ such that

$$-\sigma^j = S\sigma^j S^{-1} . \tag{9.5}$$

We have come across this fact in Chapter 1: such a matrix $S$ would have to anticommute with all $\sigma^j$, $j = 1, 2, 3$, and there is no such thing within the space of $2 \times 2$ matrices (let us recall that it was the problem of finding a matrix $\beta$ anticommuting with all $\alpha^j$, which has led us to $4 \times 4$ matrices in the Dirac equation with $m \neq 0$).

Of course, instead of (9.3) or (9.4), one could use any matrices related to these by means of a similarity transformation, i.e. $\pm U\sigma^j U^{-1}$; thus, there is an infinite number of triplets of $2 \times 2$ matrices $\alpha^j$ satisfying (9.2), but they fall into two distinct classes, generated either by (9.3) or

(9.4). Taking (9.3) or (9.4) as the familiar representatives of these classes, one has two possible types of relativistic equations in question, namely

$$i\frac{\partial\psi}{\partial t} = -i\vec{\sigma}\cdot\vec{\nabla}\psi\,, \tag{9.6}$$

$$i\frac{\partial\psi}{\partial t} = i\vec{\sigma}\cdot\vec{\nabla}\psi\,. \tag{9.7}$$

Equations (9.6) and (9.7) are called **Weyl equations** (they are due to Hermann Weyl, who introduced them in 1929, shortly after the formulation of the Dirac equation).

As in the case of the Dirac equation, we should check the relativistic covariance of the equations (9.6), (9.7). To this end, let us first rewrite them in a "covariant form". It means that Eq. (9.6) can be recast as

$$\sigma^\mu\partial_\mu\psi = 0\,, \tag{9.8}$$

and (9.7) as

$$\widetilde{\sigma}^\mu\partial_\mu\psi = 0\,, \tag{9.9}$$

where

$$\sigma^\mu = (\mathbb{1},\vec{\sigma})\,, \qquad \widetilde{\sigma}^\mu = (\mathbb{1},-\vec{\sigma})\,. \tag{9.10}$$

In what follows, let us concentrate e.g. on Eq. (9.8); Eq. (9.9) can be treated in an analogous manner. So, let $x' = \Lambda x$ be a Lorentz transformation of spacetime coordinates; the corresponding transformation of the wave function $\psi$ can be written as

$$\psi'(x') = S\psi(x)\,, \tag{9.11}$$

where $S = S(\Lambda)$ is a constant invertible $2\times 2$ matrix. Substituting $\psi(x) = S^{-1}\psi'(x')$ in Eq. (9.8), one gets, after a simple manipulation,

$$0 = \sigma^\mu S^{-1}\frac{\partial\psi'}{\partial x'^\lambda}\frac{\partial x'^\lambda}{\partial x^\mu} = \Lambda^\lambda{}_\mu\sigma^\mu S^{-1}\frac{\partial\psi'}{\partial x'^\lambda}\,. \tag{9.12}$$

In order to obtain from (9.12) the Weyl equation (9.8) in primed coordinates, the most general condition to be imposed is, obviously,

$$\Lambda^\lambda{}_\mu\sigma^\mu S^{-1} = R\sigma^\lambda\,, \tag{9.13}$$

where $R$ is another non-singular matrix; the point is that if (9.13) is satisfied, then multiplying Eq. (9.12) by $R^{-1}$, the form of the equation (9.8) is restored. Thus, the condition that should determine the transformation matrix $S$ can be written as

$$\Lambda^\mu{}_\nu\sigma^\nu = R\sigma^\mu S \tag{9.14}$$

(simultaneous determination of the auxiliary matrix $R$ is a part of the procedure). Notice that the relation (9.14) differs from Eq. (4.9) that we have employed earlier for the Dirac equation. It is so because here the mass term is absent; recovering such a term in the Dirac equation gives in fact an additional requirement to be imposed on the transformation matrix $S$.

For the proper (continuous) Lorentz transformations, the procedure of finding the corresponding matrices $R$ and $S$ is quite lengthy, so in current literature the solution for the matrix $S$ is usually shown immediately, relying on the theory of representations of the Lorentz group (see Appendix B). Using the standard notation, the relevant representation here is $(0,\frac{1}{2})$, i.e. it is one of the two inequivalent 2-dimensional spinor representations. Similarly, for the second Weyl equation (9.9) the pertinent representation of the Lorentz group is $(\frac{1}{2}, 0)$. Nevertheless, a direct solution of the matrix equation (9.14) should be a challenge for any enthusiast, so let us now take it up. Denoting the six independent parameters of an infinitesimal Lorentz transformation as $\Delta\omega^{\alpha\beta}$ ($\Delta\omega^{\beta\alpha} = -\Delta\omega^{\alpha\beta}$), one may write

$$S = \mathbb{1} - \frac{i}{4}\Delta\omega^{\alpha\beta}W_{\alpha\beta}, \tag{9.15}$$

$$R = \mathbb{1} - \frac{i}{4}\Delta\omega^{\alpha\beta}V_{\alpha\beta} \tag{9.16}$$

(of course, $W$ stands here for Weyl); our goal is to find the generators $W_{\alpha\beta}$ and $V_{\alpha\beta}$. Before starting the calculation, a remark is in order. The solution of Eq. (9.14) for the matrices $R$ and $S$ has an obvious ambiguity: if a pair $R$, $S$ is a solution, then $R'$, $S'$ such that $R' = \lambda R$, $S' = \lambda^{-1}S$ represent a solution as well. We will eliminate such an essentially trivial ambiguity by restricting ourselves to the generators that are traceless (i.e. we discard possible multiples of the unit matrix). So, substituting the expressions (9.15), (9.16) into Eq. (9.14) and using the familiar form of the infinitesimal $\Lambda$, one obtains the conditions for the generators $W_{\alpha\beta}$, $V_{\alpha\beta}$ (in much the same way as in the case of the Dirac equation in Chapter 4). These read

$$2i(g_{\mu\alpha}\sigma_\beta - g_{\mu\beta}\sigma_\alpha) = V_{\alpha\beta}\sigma_\mu + \sigma_\mu W_{\alpha\beta}. \tag{9.17}$$

Thus, for a fixed pair $(\alpha, \beta)$ we have 16 equations for 8 unknowns (elements of the matrices $V_{\alpha\beta}$ and $W_{\alpha\beta}$). The procedure of solving the equations (9.17) is straightforward though somewhat tedious, so we will describe only some salient features of the whole calculation.

As a first step, take $(\alpha, \beta) = (j, k)$ with $j, k = 1, 2, 3$. For $\mu = 0$ in Eq. (9.17) one gets readily (recall that $\sigma^0 = \mathbb{1}$)

$$V_{jk} = -W_{jk}. \tag{9.18}$$

For $\mu = 1, 2, 3$ one then gets, e.g. for $W_{12}$,

$$[\sigma_1, W_{12}] = -2i\sigma_2,$$
$$[\sigma_2, W_{12}] = 2i\sigma_1, \tag{9.19}$$
$$[\sigma_3, W_{12}] = 0.$$

Notice that the aforementioned ambiguity is obviously manifested here (to any solution $W_{12}$ one may add an arbitrary multiple of the unit matrix). The only traceless solution of the equations (9.19) is $W_{12} = \sigma_3$. In a similar way, one obtains $W_{13} = -\sigma_2$ and $W_{23} = \sigma_1$. These results can be summarized in a compact form as

$$W_{jk} = \varepsilon_{jkl}\sigma_l. \tag{9.20}$$

Next, consider the combination $(\alpha, \beta) = (0, j)$, $j = 1, 2, 3$. Taking $\mu = 0$ in Eq. (9.17), one gets

$$V_{0j} + W_{0j} = 2i\sigma_j. \tag{9.21}$$

The remaining conditions (9.17), i.e. those with $\mu = 1, 2, 3$, then yield, after some simple manipulations, e.g. for $W_{01}$,

$$[\sigma_1, W_{01}] = 0,$$
$$[\sigma_2, W_{01}] = 2\sigma_3, \tag{9.22}$$
$$[\sigma_3, W_{01}] = -2\sigma_2.$$

The only traceless solution of these equations is $W_{01} = i\sigma_1$. Analogously, one can get $W_{02} = i\sigma_2$ and $W_{03} = i\sigma_3$.

Putting all these results together, we conclude that the only solution of the conditions (9.17) such that $\text{Tr}\, W_{\alpha\beta} = 0$, $\text{Tr}\, V_{\alpha\beta} = 0$ is given by

$$W_{jk} = \varepsilon_{jkl}\sigma_l\,, \qquad W_{0j} = i\sigma_j\,, \tag{9.23}$$

and

$$V_{jk} = -W_{jk}\,, \qquad V_{0j} = W_{0j}\,. \tag{9.24}$$

Using this in (9.15) and (9.16), and taking into account that $W_{jk}$ is Hermitian, while $W_{0j}$ is anti-Hermitian, one can see that $R = S^\dagger$ (but $S^\dagger$ is not, in general, equal to $S^{-1}$).

It is to be expected that the generators $W_{\alpha\beta}$ correspond to a representation of the Lorentz algebra; in particular, the matrices $w_{\alpha\beta} = \frac{1}{2}W_{\alpha\beta}$ should satisfy the commutation relations of the form (A.22). It is indeed so, though the explicit calculation is somewhat lengthy (another challenge for a diligent reader). In any case, the matrices $w_{\alpha\beta}$ are seen to coincide with the generators of the representation $(\frac{1}{2}, 0)$ described in the Appendix B.

Let us now turn to discrete symmetries. First, we are going to examine the spatial inversion (the parity operation). In such a case, the matrix of Lorentz transformation is $\text{diag}(+1, -1, -1, -1)$. Using this in the relation (9.14), one gets first, for $\mu = 0$, $\mathbb{1} = RS$, i.e. $R = S^{-1}$. Then, for $\mu = j = 1, 2, 3$ one has the requirement

$$-\sigma^j = S^{-1}\sigma^j S\,,$$

but we already know that such a $2 \times 2$ matrix does not exist. Thus, we see that the Weyl equation (9.6) is not invariant under space inversion. Of course, the same is true for the second Weyl equation (9.7) as well. This is the example promised earlier, in Chapter 5 — a simple relativistic equation involving parity violation. In fact, it is not difficult to realize that the spatial inversion converts Eq. (9.6) into (9.7) and vice versa. Explicitly, this means that if $\psi(x)$ is a solution of Eq. (9.6), then

$$\psi_P(x) = \psi(x_0, -\vec{x}) \tag{9.25}$$

satisfies Eq. (9.7).

Next, let us consider the charge conjugation, i.e. an operation of internal symmetry involving complex conjugation. The corresponding transformation of the wave function can be written as

$$\psi_C(x) = C\psi^*(x)\,. \tag{9.26}$$

The complex conjugation of Eq. (9.6) reads

$$\partial_0\psi^* + \sigma_j^*\partial_j\psi^* = 0\,,$$

so that for $\psi_C$ one gets

$$\partial_0\psi_C + C\sigma_j^*C^{-1}\partial_j\psi_C = 0\,.$$

So, in order to restore Eq. (9.6) for $\psi_C$, one would like to impose the conditions

$$\sigma_j = C\sigma_j^*C^{-1}\,, \qquad j = 1, 2, 3\,. \tag{9.27}$$

Taking into account familiar properties of the Pauli matrices, the requirement (9.27) means

$$\begin{aligned}
\sigma_1 &= C\sigma_1 C^{-1}\,, \\
-\sigma_2 &= C\sigma_2 C^{-1}\,, \\
\sigma_3 &= C\sigma_3 C^{-1}\,.
\end{aligned} \tag{9.28}$$

60

In other words, the matrix $C$ should commute with $\sigma_1$ and $\sigma_3$, and anticommute with $\sigma_2$. It turns out that such a matrix does not exist. Proving this is a refreshing algebraic exercise and the reader is encouraged to do it. (Hint: Use the most general form of $C$ written as a linear combination of the unit matrix and Pauli matrices.) Thus, the Weyl equation (9.6) is not invariant under charge conjugation (of course, the same is true for Eq. (9.7)).

On the other hand, while there is no similarity transformation between $\sigma_j^*$ and $\sigma_j$, such a transformation between $\sigma_j^*$ and $-\sigma_j$ does exist. Indeed, it holds

$$\sigma_2 \sigma_j^* \sigma_2^{-1} = -\sigma_j \tag{9.29}$$

(of course, $\sigma_2^{-1} = \sigma_2$). This in turn means that the transformation (that may be called charge conjugation)

$$\psi_C(x) = \sigma_2 \psi^*(x) \tag{9.30}$$

converts a solution of Eq. (9.6) into a solution of (9.7) (and vice versa). Thus, the combination of the spatial inversion and charge conjugation is a symmetry of the Weyl equation, i.e. if $\psi(x)$ is a solution of Eq. (9.6), then

$$\psi_{CP}(x) = \sigma_2 \psi^*(x_0, -\vec{x}) \tag{9.31}$$

is a solution as well (and the same is true for Eq. (9.7)).

Further, let us examine the possible invariance of the Weyl equation under time reversal. In analogy with our earlier discussion of the Dirac equation, an appropriate Ansatz for the corresponding transformation is

$$\psi'(x') = B\psi^*(x), \tag{9.32}$$

where $x' = (-x_0, \vec{x})$. An observant reader will have no problem with finding out that the relevant condition for the matrix $B$ reads

$$\Lambda^\mu{}_\nu \sigma^{\nu*} = -B^{-1} \sigma^\mu B, \tag{9.33}$$

where the matrix of the Lorentz transformation is now $\Lambda = \mathrm{diag}(-1, +1, +1, +1)$. This means that $B$ should commute with $\sigma_2$ and anticommute with $\sigma_1$ and $\sigma_3$. An obvious choice is therefore $B = \sigma_2$. So, if $\psi(x)$ is a solution of Eq. (9.6), then

$$\psi_T(x) = \sigma_2 \psi^*(-x_0, \vec{x}) \tag{9.34}$$

is a solution as well, and the $T$-invariance of the Weyl equation is thereby established.

Note that this last discrete symmetry does not come as a surprise, as it is perfectly consistent with the famous $CPT$ theorem (an interested reader can find the relevant information e.g. in [6] or [21]): once we have established the invariance under $CP$ (see (9.31)), the $T$-invariance must hold as well, since the $CPT$ symmetry is guaranteed in the considered case. One can verify it explicitly: it is easy to see that the Weyl equation is invariant under the transformation

$$\psi'(x') = \psi(x), \tag{9.35}$$

where $x' = (-x^0, -\vec{x})$, and this is just the $CPT$.

As the last issue to be discussed in this chapter, we are going to describe briefly the salient features of the plane-wave solutions of the Weyl equations. We will consider the equation of the first type (9.6) as an instructive example; the relevant results can be extended in a straightforward manner to Eq. (9.7). The solutions in question can be written in a form analogous to what we know from our previous study of the Dirac equation. Thus, we have

$$\psi_{(+)}(x) = u(p)\, e^{-ipx} \tag{9.36}$$

and

$$\psi_{(-)}(x) = v(p)\, e^{ipx}\,, \tag{9.37}$$

where $p = (p_0, \vec{p})$ with $p_0 = |\vec{p}|$. As the notation employed in (9.36), (9.37) indicates, $\psi_{(+)}$ corresponds to the positive energy $E = |\vec{p}|$ and momentum $\vec{p}$, while $\psi_{(-)}$ carries the negative energy $E = -|\vec{p}|$ and momentum $-\vec{p}$. Substituting the Ansatz (9.36) into Eq. (9.6), one gets readily

$$(\vec{\sigma} \cdot \vec{p})u = |\vec{p}|u\,. \tag{9.38}$$

This result is quite remarkable, since it means that

$$\frac{\vec{\sigma} \cdot \vec{p}}{|\vec{p}|}u = u\,, \tag{9.39}$$

i.e. the solution with positive energy has automatically positive helicity. Similarly, for the negative energy solution (9.37) one obtains

$$\frac{\vec{\sigma} \cdot (-\vec{p})}{|\vec{p}|}v = -v\,, \tag{9.40}$$

which means that $\psi_{(-)}$ has negative helicity. From the above exposition it is also clear that for the second Weyl equation (9.7) the correlation between the energy and helicity is reversed: a plane-wave solution of Eq. (9.7) with positive energy has negative helicity and a negative energy plane wave is endowed with positive helicity.

The fact that for a given energy and momentum one can have just a single value of the helicity is a clear physical manifestation of the non-invariance of the Weyl equations under space reflection; we have in mind a natural intuitive picture of the parity violation as the absence of a "mirror symmetry" between right-handed and left-handed states.

A historical remark is perhaps in order here. The two-component Weyl equation, formulated almost simultaneously with the four-component Dirac equation, was rejected soon after its birth, just because of the inherent parity violation (the invariance under spatial inversion was taken for granted then, as an automatic symmetry of the fundamental laws of nature). The situation has changed in 1957 with the advent of the experimentally confirmed parity violation in weak interactions. The Weyl equation has been rehabilitated and accepted then as a proper description of the neutrino (that was believed for a long time to be strictly massless); thus, the "two-component Weyl neutrino" was taken to be responsible for the left-right asymmetry of weak interactions. Such an idea inspired indeed a very successful theory of weak forces, but now we know that the neutrinos do have tiny masses; so, a true source of the parity violation must lie deeper. In any case, the Weyl equation(s) still represent a useful theoretical tool in building models of fundamental physics, where one may start with massless spin-$\frac{1}{2}$ particles that acquire masses by means of some additional mechanism. From the purely mathematical point of view, the two-dimensional spinor representations of the Lorentz group, implemented in the Weyl equations, are the basic building blocks for the construction of any higher-dimensional representation; this certainly underlines the fundamental nature of the original Weyl's work.

# Chapter 10

# Wave packets. Zitterbewegung

Up to now we have discussed mostly the plane waves, as the free-particle solutions of the Dirac equation. Such states are rather boring from the physical point of view, but, as we have already stressed earlier, their detailed description is important with regard to the perturbative calculations of the scattering and decay processes to be pursued later on. In fact, there is one point that has not been clarified yet, namely the relevance of the negative energy states. Naïvely, one might be tempted to ignore them completely, but it would not be consistent as they form, along with the positive energy states, a complete system of solutions of the quantum mechanical Dirac equation.

Apart from this general aspect, which is common to all models of relativistic quantum mechanics, the presence of negative energy solutions has an intriguing and rather surprising manifestation when one considers the time evolution of a wave packet, i.e. a superposition of the plane waves (that may represent a spatially localized state); here we are alluding to the strange German word in the title. So, let us now pack the plane waves thoroughly, the German dictionary will be consulted in the end.

We will start with a packet of the positive energy plane waves, i.e. the superposition we have in mind may be written as

$$\Psi_{(+)}(x) = \int \frac{d^3 p}{(2\pi)^{3/2}(2E)^{1/2}} \sum_s b(p,s)u(p,s)\, e^{-ipx}\,, \tag{10.1}$$

where $p = (E, \vec{p})$ with $E = p_0 = (\vec{p}^2 + m^2)^{1/2}$. The reader should not feel uneasy about the unfamiliar factor $(2\pi)^{-3/2}(2E)^{-1/2}$ (to be denoted for brevity as $N(p)$ in what follows); it will prove to be an appropriate contribution to the normalization of the expansion coefficients $b(p,s)$. The sum over the spin label $s$ is to be understood as including the two possible states corresponding e.g. to polarizations $\pm s$. First of all, let us compute the normalization integral; since we know that the product $\psi^\dagger \psi$ in general represents the probability density, the relevant integral is

$$I = \int d^3 x\, \Psi_{(+)}^\dagger(x)\Psi_{(+)}(x)$$

$$= \int d^3 x \iint d^3 p\, d^3 q\, N(p)N(q) \sum_{s,s'} b(p,s)b^*(q,s')u^\dagger(q,s')u(p,s)\, e^{iqx-ipx}\,. \tag{10.2}$$

The integration in (10.2) gets simplified readily. First, one has, obviously

$$\int d^3 x\, e^{iqx-ipx} = e^{i(q_0-p_0)x_0} \int d^3 x\, e^{-i(\vec{q}-\vec{p})\vec{x}} = (2\pi)^3 \delta^{(3)}(\vec{p}-\vec{q})\, e^{i(q_0-p_0)x_0}\,. \tag{10.3}$$

However, since $p_0 = \sqrt{\vec{p}^2 + m^2}$, $q_0 = \sqrt{\vec{q}^2 + m^2}$, the values of $p_0$ and $q_0$ become effectively equal, due to the presence of $\delta^{(3)}(\vec{p} - \vec{q})$. Further, taking into account that the dependence on $p$ and $q$ in (10.2) means in fact a dependence on $\vec{p}$ and $\vec{q}$, the integral then becomes

$$ I = \int d^3p \, (2\pi)^3 N^2(p) \sum_{s,s'} b(p,s) b^*(p,s') u^\dagger(p,s') u(p,s) . \tag{10.4} $$

The scalar product $u^\dagger u = \bar{u}\gamma_0 u$ in (10.4) can be worked out by means of the Gordon identity (cf. Appendix C). In general, it holds

$$ \bar{u}(p)\gamma^\mu u(p') = \frac{1}{2m} \bar{u}(p) \left[ (p + p')^\mu + i\sigma^{\mu\nu}(p - p')_\nu \right] u(p') . \tag{10.5} $$

Then one gets

$$ u^\dagger(p,s') u(p,s) = \bar{u}(p,s')\gamma^0 u(p,s) = \frac{1}{2m}\bar{u}(p,s') u(p,s) 2p_0 . \tag{10.6} $$

According to the results summarized in Chapter 7, one has $\bar{u}(p,s') u(p,s) = 2m\delta_{s,s'}$ and the expression (10.6) thus becomes

$$ u^\dagger(p,s') u(p,s) = 2E\delta_{ss'} . \tag{10.7} $$

One then gets, finally,

$$ I = \int d^3p \, (2\pi)^3 2EN^2(p) \sum_s |b(p,s)|^2 = \int d^3p \sum_s |b(p,s)|^2 \tag{10.8} $$

(so, now it is also clear why the choice of the normalization factor $N(p)$ in (10.1) is convenient). Thus, the expansion coefficients $b(p,s)$ may be normalized according to

$$ \int d^3p \sum_s |b(p,s)|^2 = 1 . \tag{10.9} $$

One may now employ the wave packet (10.1), normalized properly according to (10.9), to compute expectation values of various physical quantities. The quantity that we will choose as an example is the velocity. Admittedly, at first sight it does not look as the most attractive and urgent case, but we will see that in fact it does lead to interesting and rather unexpected results.

Before starting the calculation, we must make a brief digression concerning the quantum mechanical picture of the quantity we have in mind. The components of velocity are defined in terms of the time derivative of the corresponding coordinates, so one should invoke the Heisenberg picture (representation), in which the observables are time-dependent, in general. Let us recall that an observable $A(t)$ in such a picture satisfies the equation

$$ \frac{d}{dt}A(t) = i[H, A(t)] , \tag{10.10} $$

where $H$ is the relevant Hamiltonian. So, let us take $A(t)$ to be the coordinate (vector) $\vec{x}(t)$. Of course, in the non-relativistic case with the Hamiltonian $H = \vec{p}^2/2m$, one gets from (10.10) readily $\vec{v} = \vec{p}/m$, if the canonical commutation relations between the momentum and coordinate components are utilized. For the Dirac Hamiltonian the situation is different. One has

$$ \frac{dx^k(t)}{dt} = i[H, x^k(t)] = i[H, e^{iHt} x^k e^{-iHt}] = i \, e^{iHt}[H, x^k] e^{-iHt} , \tag{10.11} $$

64

where now $H = \alpha^j p^j + \beta m$. The commutator in (10.11) is worked out easily; one gets

$$[H, x^k] = [\alpha^j p^j + \beta m, x^k] = -i\alpha^k , \qquad (10.12)$$

as $[p^j, x^k] = -i\delta^{jk}$. Thus, for the operator of velocity one has finally

$$v^k(t) = \frac{\mathrm{d}x^k(t)}{\mathrm{d}t} = e^{iHt} \alpha^k e^{-iHt} . \qquad (10.13)$$

This result is in fact rather strange. The operator on the right-hand side obviously has just the eigenvalues $\pm 1$ (since for $\alpha^k$ it is so, and (10.13) represents a similarity transformation); so it does not comply with the common idea about what the velocity of a relativistic particle should be (in contrast to the non-relativistic case, which is quite transparent). On the other hand, the familiar relativistic relation for the particle velocity can be written simply in terms of the energy and momentum as

$$\vec{v} = \frac{\vec{p}}{E} . \qquad (10.14)$$

Thus, one might try to transform our result (10.13) in an appropriate way to the momentum representation, and see whether the form (10.14) emerges there. Practically, it means that we should evaluate the expectation value of the velocity operator in the state described by the wave packet (10.1), and write it in a form involving the probability density $|b(p, s)|^2$. For this purpose, it is convenient to employ the Schrödinger picture of time evolution, because the wave function (10.1) is already represented so; thus, one has to compute the expectation value of the time-independent $\vec{\alpha}$ (cf. (10.13)) in the state given by (10.1) (having in mind the normalization condition (10.9)). We will return to the Heisenberg picture later on.

The quantity we are going to evaluate is given by

$$\langle v^j \rangle = \int \mathrm{d}^3 x \, \Psi^{\dagger}_{(+)}(x) \alpha^j \Psi_{(+)}(x)$$

$$= \int \mathrm{d}^3 x \sum_{s,s'} \iint \mathrm{d}^3 p \, \mathrm{d}^3 q \, N(p) N(q) b^*(p, s) b(q, s') \bar{u}(p, s) \gamma^j u(q, s') \, e^{ipx - iqx} . \qquad (10.15)$$

Proceeding in the same way as before (cf. (10.2) through (10.4)), one gets first

$$\langle v^j \rangle = \int \mathrm{d}^3 p \, (2\pi)^3 N^2(p) \sum_{s,s'} b^*(p, s) b(p, s') \bar{u}(p, s) \gamma^j u(p, s') . \qquad (10.16)$$

Next, using the Gordon identity (10.5), one has

$$\bar{u}(p, s) \gamma^j u(p, s') = \frac{1}{2m} \bar{u}(p, s) u(p, s') 2 p_j = 2 p^j \delta_{ss'} . \qquad (10.17)$$

Substituting this and the value of $N(p)$ into (10.16), one obtains finally

$$\langle v^j \rangle = \int \mathrm{d}^3 p \sum_s |b(p, s)|^2 \frac{p^j}{E} . \qquad (10.18)$$

So, we have arrived at the envisaged result; the familiar relativistic velocity appears here in the right place, and this is certainly reassuring.

However, this cannot be the whole story. As we know, the complete system of solutions of the Dirac equation comprises also the plane waves with amplitudes $v(p, s)$, i.e. the negative energy states. Thus, the most general wave packet should be written in the form

$$\Psi(x) = \int \mathrm{d}^3 p \, N(p) \sum_s \left[ b(p, s) u(p, s) \, e^{-ipx} + d^*(p, s) v(p, s) \, e^{ipx} \right] . \qquad (10.19)$$

Note that we have used here a conventional notation $d^*(p, s)$ instead of $d(p, s)$ (which would seem more natural) with regard to later applications of such a formula within the field theory; please be patient and you will see. As before, we will consider first the normalization integral

$$I = \int d^3x\, \Psi^\dagger(x)\Psi(x)\,.$$

Using the form (10.19), one gets for $I$ the following long expression:

$$I = \int d^3x \iint d^3p\, d^3q \sum_{s,s'} N(p)N(q)$$

$$\times \left[ b(p, s)b^*(q, s')\bar{u}(q, s')\gamma_0 u(p, s)\, e^{-ipx+iqx} + d^*(p, s)d(q, s')\bar{v}(q, s')\gamma_0 v(p, s)\, e^{ipx-iqx} \right.$$

$$\left. + b^*(q, s')d^*(p, s)\bar{u}(q, s')\gamma_0 v(p, s)\, e^{iqx+ipx} + d(q, s')b(p, s)\bar{v}(q, s')\gamma_0 u(p, s)\, e^{-iqx-ipx} \right].$$

$$(10.20)$$

The delta functions that emerge, as usual, upon the spatial integration result in the identification $\vec{q} = \vec{p}$ in the first two terms in the square brackets, and $\vec{q} = -\vec{p}$ in the third and fourth term. Further, the time-dependent factor $\exp(2ip_0x_0)$ then survives in the third term and $\exp(-2ip_0x_0)$ in the fourth term. For the remaining algebra one has to employ all variants of the Gordon identity (see Appendix C). As a result, the expression in the square brackets is simplified considerably: for the first term one gets $|b(p, s)|^2 2E\delta_{ss'}$ and for the second term $|d(p, s)|^2 2E\delta_{ss'}$, while the third and the fourth terms vanish. The reader is encouraged to verify these statements with the help of the identities (C.33).

Now, putting all this together, including also the ubiquitous overall factor $(2\pi)^3$ accompanying the delta functions, one has finally

$$\int d^3x\, \Psi^\dagger(x)\Psi(x) = \int d^3p \sum_s \left( |b(p, s)|^2 + |d(p, s)|^2 \right). \qquad (10.21)$$

Up to now, everything seems to be OK. So, let us compute the expectation value of the velocity, in order to generalize our previous result. As before, it is given by

$$\langle v^k \rangle = \int d^3x\, \Psi^\dagger(x)\alpha^k\Psi(x) = \int d^3x\, \overline{\Psi}(x)\gamma^k\Psi(x)\,. \qquad (10.22)$$

The calculation is straightforward and proceeds along the same lines as before. In comparison with the previous case of the restricted wave packet there is more algebraic work involving the Gordon identities and the result is rather long. It reads

$$\langle v^k \rangle = \int d^3p \sum_s \left( |b(p, s)|^2 + |d(p, s)|^2 \right) \frac{p^k}{E}$$

$$+ \int d^3p \sum_{s,s'} \frac{i}{2m} \left( b^*(\widetilde{p}, s')d^*(p, s)\bar{u}(\widetilde{p}, s')\sigma^{k0}v(p, s)\, e^{2iEx_0} \right.$$

$$\left. - d(\widetilde{p}, s')b(p, s)\bar{v}(\widetilde{p}, s')\sigma^{k0}u(p, s)\, e^{-2iEx_0} \right)$$

$$+ \int d^3p \sum_{s,s'} \frac{1}{2m} \left( d(\widetilde{p}, s')b(p, s)\bar{v}(\widetilde{p}, s')u(p, s)\, e^{-2iEx_0} \right.$$

$$\left. - b^*(\widetilde{p}, s')d^*(p, s)\bar{u}(\widetilde{p}, s')v(p, s)\, e^{2iEx_0} \right) \frac{p^k}{E}, \qquad (10.23)$$

where we have denoted $\widetilde{p} = (p_0, -\vec{p})$. Note that in several books one may find the expression (10.23) without the last two lines (see e.g. [1, 6]); at one place there is even an explicit statement that such terms vanish (see [5]). It is not true, since, in general, $\bar{v}(\widetilde{p}, s')u(p, s) \neq 0$ and similarly for $\bar{u}(\widetilde{p}, s')v(p, s)$. In any case, the result (10.23) is quite intriguing. Apart from the "normal" term in the first line, there are "anomalous" time-dependent oscillatory terms involving an interference of positive and negative energy waves, which are *a priori* unexpected and their physical interpretation is not clear (the point of the conundrum is that there is no external force to cause the oscillations). Obviously, a minimum frequency of such an oscillatory motion is at least $2m$ (i.e. $2mc^2/\hbar$ in ordinary units), which for the electron amounts to about $10^{21}$ s$^{-1}$. This effect was observed theoretically for the first time in 1930 by E. Schrödinger, who called it, in German, die Zitterbewegung (it means "jittery", or "quivering" motion).

Such a phenomenon may cast some doubt on the interpretation of the quantity in question as a true velocity expectation value, and this in turn is related to the interpretation of the space coordinates as the true quantum mechanical observables; it is indeed so that the operator of multiplication by the space coordinate is not satisfactory in this context (see e.g. [8, 24]). In general, the issue of a correct definition of a position operator is one of the fundamental consistency problems of relativistic quantum mechanics.

On the other hand, even quite recently there have been serious attempts to prove or disprove experimentally the reality of the "Zitterbewegung" (not directly for electrons in the vacuum, but for some more appropriate materials) and some results seem to be positive (see e.g. [38]). Anyway, one should say, with due caution, that this enigmatic issue is still not settled at present. An illuminating discussion of Zitterbewegung can be found also in the book [4].

The Schrödinger picture of time evolution makes it clear that the phenomenon of Zitterbewegung is intimately related to the presence of negative energy waves in the packet. The anomalous terms contributing to the particle velocity can be also represented in a compact form in the Heisenberg picture. Below we describe it briefly, following the treatment presented in the book [11].

One may start with Eq. (10.13). For convenience, we will denote the expression on its right-hand side as $\vec{\alpha}(t)$. It satisfies the differential equation (Heisenberg equation of motion)

$$\dot{\vec{\alpha}}(t) = i[H, \vec{\alpha}(t)],\tag{10.24}$$

where we use, for the sake of brevity, the Newton's dot symbol for the time derivative. Obviously, the commutator can be recast as

$$[H, \vec{\alpha}(t)] = e^{iHt}[H, \vec{\alpha}]\,e^{-iHt} = e^{iHt}\left(\{H, \vec{\alpha}\} - 2\vec{\alpha}H\right)e^{-iHt}$$
$$= e^{iHt}\left(2\vec{p} - 2\vec{\alpha}H\right)e^{-iHt} = 2\vec{p} - 2\vec{\alpha}(t)H,$$

where we have utilized the familiar anticommutation relations for $\vec{\alpha}$, as well as the fact that $H$ commutes with $\vec{p}$. Thus, Eq. (10.24) becomes

$$\dot{\vec{\alpha}}(t) = 2i\vec{p} - 2i\vec{\alpha}(t)H.\tag{10.25}$$

Now, one may differentiate Eq. (10.25) with respect to time, taking into account that $\vec{p}$ and $H$ are constants of motion; one gets

$$\ddot{\vec{\alpha}}(t) = -2i\dot{\vec{\alpha}}(t)H,\tag{10.26}$$

and this can be readily integrated so that

$$\dot{\vec{\alpha}}(t) = \dot{\vec{\alpha}}(0)\,e^{-2iHt}.\tag{10.27}$$

However, from (10.25) one has

$$\dot{\vec{\alpha}}(0) = 2i\vec{p} - 2i\vec{\alpha}H = -2i(\vec{\alpha} - \vec{p}H^{-1})H, \tag{10.28}$$

and substituting this into (10.27), the expression for $\dot{\vec{\alpha}}(t)$ becomes

$$\dot{\vec{\alpha}}(t) = -2i(\vec{\alpha} - \vec{p}H^{-1})H\, e^{-2iHt}. \tag{10.29}$$

Further, one may use it in (10.25) and thus we end up with a relation for $\vec{\alpha}(t)$, namely

$$-2i(\vec{\alpha} - \vec{p}H^{-1})H\, e^{-2iHt} = 2i\vec{p} - 2i\vec{\alpha}(t)H,$$

from which one obtains, after a simple manipulation,

$$\vec{\alpha}(t) = \frac{\vec{p}}{H} + \left(\vec{\alpha} - \frac{\vec{p}}{H}\right)e^{-2iHt} \tag{10.30}$$

(where, of course, $1/H$ means $H^{-1}$). A technical remark is in order here. We have reproduced here the calculation presented in [11], since it is a very efficient and elegant way of obtaining the result (10.30). An alternative procedure would be to start with (10.13) and employ the well-known general formula

$$e^{A}B\,e^{-A} = B + \frac{1}{1!}[A, B] + \frac{1}{2!}[A, [A, B]] + \dots \tag{10.31}$$

along with the commutator

$$[H, \vec{\alpha}] = 2\vec{p} - 2\vec{\alpha}H$$

shown above. Taking into account that $[H, \vec{p}] = 0$ and proceeding with due care, the result (10.30) can be recovered (as usual, a diligent reader is encouraged to perform this calculation).

The first term on the right-hand side of (10.30) is the "normal" one, since it corresponds to the standard relation for velocity of a relativistic particle (10.14). The second term is "anomalous", as it is time-dependent (oscillatory) and is clearly responsible for the Zitterbewegung that we have uncovered previously. Finally, it is also instructive to get an expression for the time evolution of the coordinate. We have

$$\dot{\vec{x}}(t) = \vec{\alpha}(t) = \frac{\vec{p}}{H} + \left(\vec{\alpha} - \frac{\vec{p}}{H}\right)e^{-2iHt},$$

and integrating it we obtain

$$\vec{x}(t) = \vec{a} + \frac{\vec{p}}{H}t + \frac{i}{2}\left(\vec{\alpha} - \frac{\vec{p}}{H}\right)H^{-1}\, e^{-2iHt}, \tag{10.32}$$

where $\vec{a}$ is a constant. The last term on the right-hand side of (10.32) is another representation of the quivering motion of a free particle, the mysterious Zitterbewegung. One may also notice that the amplitude of such oscillations is at most of the order $1/m$, the Compton wavelength of the particle, since for a definite momentum and energy $1/E \leq 1/m$ and, of course, $|\vec{p}/E| \leq 1$.

We have seen that the (inevitable) presence of the negative energy solutions leads to an unexpected and counter-intuitive behaviour of a wave packet representing, in general, a localized state of a Dirac particle. In this context, one may naturally ask when the admixture of the negative energy waves can be substantial, or when it is negligible. Technically, it is the problem of the relative magnitude of the expansion coefficients $b(p, s)$ and $d(p, s)$.

68

As an explicit illustration, let us consider an example of a wave packet that for $t = 0$ has the Gaussian form

$$\Psi(0, \vec{x}) = \frac{1}{(\pi a^2)^{3/4}} w \, \exp\left(-\frac{\vec{x}^2}{2a^2}\right),$$ (10.33)

where $w$ is e.g.

$$w = \begin{pmatrix} 1 \\ 0 \\ 0 \\ 0 \end{pmatrix}.$$ (10.34)

Note that $\Psi(0, \vec{x})$ given by (10.33) is normalized to unity. It represents an initial condition for a time-dependent solution of the Dirac equation that can be generally written in the form (10.19). Setting $x_0 = 0$ in (10.19) one has

$$\Psi(0, \vec{x}) = \int d^3 p \, N(p) \sum_s \left[ b(p, s) u(p, s) \, e^{i \vec{p} \vec{x}} + d^*(p, s) v(p, s) \, e^{-i \vec{p} \vec{x}} \right].$$ (10.35)

Evaluation of the expansion coefficients in (10.35) amounts to the implementation of inverse Fourier transformation; practically, it means that one multiplies (10.35) by $e^{-i \vec{q} \vec{x}}$ and integrates over $\vec{x}$. The calculation is not difficult and the result is

$$\left(\frac{a^2}{\pi}\right)^{3/4} e^{-\frac{1}{2} a^2 \vec{q}^2} w = \frac{1}{\sqrt{2E(q)}} \sum_s \left[ b(q, s) u(q, s) + d^*(\widetilde{q}, s) v(\widetilde{q}, s) \right],$$ (10.36)

where we have denoted, as before, $\widetilde{q} = (q_0, -\vec{q})$ and, of course, $E(q) = (\vec{q}^2 + m^2)^{1/2}$. In the sequel, we will change the notation (just for convenience), writing $p$ instead of $q$. The coefficients $b(p, s)$ and $d^*(p, s)$ can now be determined with the help of the familiar orthogonality relations for $u$ and $v$. For instance, multiplying Eq. (10.36) by $u^\dagger(p, s')$ from left and using the Gordon identities, one gets

$$u^\dagger(p, s') u(p, s) = 2E \delta_{ss'}$$ (10.37)

and

$$u^\dagger(p, s') v(\widetilde{p}, s) = 0.$$ (10.38)

Similarly, the multiplication by $v^\dagger(\widetilde{p}, s')$ yields

$$v^\dagger(\widetilde{p}, s') u(p, s) = 0$$ (10.39)

and

$$v^\dagger(\widetilde{p}, s') v(\widetilde{p}, s) = 2E \delta_{ss'}.$$ (10.40)

Final results for the expansion coefficients then read

$$b(p, s) = \frac{1}{(2E)^{1/2}} \left(\frac{a^2}{\pi}\right)^{3/4} e^{-\frac{1}{2} a^2 \vec{p}^2} u^\dagger(p, s) w,$$

$$d^*(\widetilde{p}, s) = \frac{1}{(2E)^{1/2}} \left(\frac{a^2}{\pi}\right)^{3/4} e^{-\frac{1}{2} a^2 \vec{p}^2} v^\dagger(\widetilde{p}, s) w.$$ (10.41)

Now, to get explicit values of the functions (10.41), one may use the form of $u(p, s)$ and $v(p, s)$ discussed in Chapter 5. It is

$$u(p, s) = \sqrt{E + m} \begin{pmatrix} \varphi^{(r)} \\ \dfrac{\vec{\sigma} \cdot \vec{p}}{E + m} \varphi^{(r)} \end{pmatrix},$$

(10.42)

$$v(p, s) = \sqrt{E + m} \begin{pmatrix} \dfrac{\vec{\sigma} \cdot \vec{p}}{E + m} \varphi^{(r)} \\ \varphi^{(r)} \end{pmatrix},$$

where $\varphi^{(r)}$, with $r = 1, 2$, is e.g. $\begin{pmatrix} 1 \\ 0 \end{pmatrix}$ or $\begin{pmatrix} 0 \\ 1 \end{pmatrix}$. Choosing $\varphi^{(r)} = \begin{pmatrix} 1 \\ 0 \end{pmatrix}$, one gets non-trivial values of the expressions (10.41):

$$d^*(p, s) = d(p, s) = \left( \frac{E + m}{2E} \right)^{1/2} \left( \frac{a^2}{\pi} \right)^{3/4} e^{-\frac{1}{2}a^2 \vec{p}^2} \frac{p_3}{E + m},$$

(10.43)

$$b(p, s) = \left( \frac{E + m}{2E} \right)^{1/2} \left( \frac{a^2}{\pi} \right)^{3/4} e^{-\frac{1}{2}a^2 \vec{p}^2}.$$

Thus, the ratio of typical non-zero expansion coefficients is quite simple, namely

$$\left| \frac{d^*(p, s)}{b(p, s)} \right| = \frac{|p_3|}{E + m} \lesssim \frac{|\vec{p}|}{E + m}.$$

(10.44)

Now we are in a position to find out when the contribution of negative energy waves may be substantial. In any case, the exponential factor $\exp\left( -\frac{1}{2}a^2\vec{p}^2 \right)$ has non-negligible value (both for $b$ and $d$) if $|\vec{p}|a \lesssim 1$, i.e.

$$|\vec{p}| \lesssim \frac{1}{a}.$$

(10.45)

As regards the magnitude of the constant $a$ (the width of the wave packet), there are two possibilities. First,

$$a \gg \frac{1}{m} \quad \text{i.e.} \quad \frac{1}{a} \ll m.$$

(10.46)

In such a situation, in the region (10.45) of the non-negligible values of $\exp\left( -\frac{1}{2}a^2\vec{p}^2 \right)$ one has $|\vec{p}| \lesssim 1/a \ll m$, i.e. $|\vec{p}| \ll m$, and according to (10.44) one then has $|d/b| \ll 1$, i.e. the contribution of the negative energy waves is suppressed. Second,

$$a \lesssim \frac{1}{m} \quad \text{i.e.} \quad \frac{1}{a} \gtrsim m.$$

(10.47)

In this case, for $|\vec{p}| \simeq 1/a$ one has $|\vec{p}| \gtrsim m$ and the ratio $d/b$ is not suppressed for such values of $|\vec{p}|$.

Note that the condition (10.46) corresponds to a particle localized in a region much larger than its Compton wavelength. Our example thus demonstrates that a substantial contribution of negative energy states may be expected for a wave packet localized in a region with linear dimension of the order of Compton wavelength or less.

A physical system with the size corresponding to (10.46) is e.g. the hydrogen atom, where the electron is localized in a region with the linear dimension characterized by the Bohr radius; in our system of units it is

$$a_{\text{Bohr}} = \frac{1}{m\alpha},\qquad(10.48)$$

where $\alpha$ is the fine-structure constant, $\alpha \doteq 1/137$ (thus, the size of the hydrogen atom is about two orders of magnitude larger than the Compton wavelength of the electron). Of course, our preceding considerations concerning the wave packets correspond to localized free particles, but one might hope that even for the electron bound in an external field one could get results that are not heavily contaminated by the puzzling effects of relativistic quantum mechanics mentioned above. Such an expectation is strengthened by the fact that the hydrogen atom is described quite successfully even within the non-relativistic quantum mechanics.

Needless to say, these intuitive arguments turn out to be true. The treatment of the hydrogen atom within the framework of relativistic quantum mechanics has become one of the early triumphs of the Dirac equation. The relevant calculations were carried out in 1928 independently by Walter Gordon and Charles Darwin (the grandson of the famous Charles Darwin) and, among other things, explained very naturally the so-called fine structure of the hydrogen energy levels. Although the calculation is really impressive (but also long and tedious), it will not be reproduced here, since we would like to proceed faster on our way to quantum field theory. The interested reader can find the details e.g. in [1], [6] or [7].

# Chapter 11

# Klein paradox

As the title of this chapter indicates, we are going to discuss another intriguing phenomenon that emerges within the framework of relativistic quantum mechanics. This time, the effect in question concerns the theoretical description of a simple scattering process involving a one-dimensional potential step. The reader may remember that solving a problem like this is a traditional mundane topic of one of the first technical exercises in an introductory course of the non-relativistic quantum mechanics. There one can demonstrate some effects that are impossible classically (such as quantum tunnelling), but otherwise well understood and interpreted in terms of the behaviour of the particle wave function. In contrast to that, here we will see that in the relativistic case one encounters some surprising counterintuitive new phenomena that may reveal limits of applicability of relativistic quantum mechanics as a single-particle theory.

So, let us formulate the problem we have in mind. We are going to consider a Dirac particle moving in a static external field, described in terms of the corresponding potential energy that has the form

$$
\begin{aligned}
V(\vec{x}) &= V(z)\,, \\
V(z) &= 0 \quad \text{for } z < 0\,, \\
V(z) &= V \quad \text{for } z \geq 0\,,
\end{aligned}
\tag{11.1}
$$

where $V$ is a constant, $V > 0$. Thus, the potential energy defining our model depends non-trivially just on the 3rd coordinate $z$ and has the idealized shape of a "potential step" with sharp boundary. Suppose that the particle is in a stationary state with a definite energy $E > 0$. The time-independent part of the wave function then satisfies the equation

$$
\left(-i\vec{\alpha} \cdot \vec{\nabla} + \beta m + V(\vec{x})\right)\psi(\vec{x}) = E\psi(\vec{x})\,.
\tag{11.2}
$$

In view of the form of our $V(\vec{x})$ we may examine the solution of Eq. (11.2) separately in the two distinct regions indicated in (11.1); for convenience, the half-spaces with $z < 0$ and $z \geq 0$ will be denoted as the regions I and II, respectively. Thus, we have two simple equations to be solved, namely

$$
\left(-i\vec{\alpha} \cdot \vec{\nabla} + \beta m\right)\psi_{\mathrm{I}}(\vec{x}) = E\psi_{\mathrm{I}}(\vec{x})
\tag{11.3}
$$

in the region I, and

$$
\left(-i\vec{\alpha} \cdot \vec{\nabla} + \beta m\right)\psi_{\mathrm{II}}(\vec{x}) = (E - V)\psi_{\mathrm{II}}(\vec{x})
\tag{11.4}
$$

in the region II. Formally, on the left-hand side of both equations one has the free-particle Dirac Hamiltonian, so that one may look for solutions of the type $\psi(\vec{x}) = u(k)\exp(i\vec{k}\vec{x})$. It makes sense to choose the momentum variable as $\vec{k} = (0, 0, k)$; with such a choice, solving equations (11.3), (11.4) then turns out to be effectively a one-dimensional problem, since the action of $\partial/\partial x$ and $\partial/\partial y$ on the wave function becomes trivial. Thus, one is left with

$$\left(-i\alpha^3 \frac{\partial}{\partial z} + \beta m\right) \psi_{\mathrm{I}}(z) = E \psi_{\mathrm{I}}(z) \tag{11.5}$$

and

$$\left(-i\alpha^3 \frac{\partial}{\partial z} + \beta m\right) \psi_{\mathrm{II}}(z) = (E - V) \psi_{\mathrm{II}}(z) . \tag{11.6}$$

Now we are in a position to specify the physical boundary conditions to be imposed on the wave function in question. The setting we have in mind corresponds to one-dimensional scattering, where one has to consider incident and reflected waves in the region I, along with the transmitted wave in the region II. The incident wave can be written as

$$\psi_{\mathrm{inc.}}(z) = u(p)\, e^{ipz} = \begin{pmatrix} \varphi \\ \chi \end{pmatrix} e^{ipz} , \tag{11.7}$$

with $p > 0$ (it corresponds to the particle motion toward the potential step). The four-component amplitude $u(p)$ is conveniently described in terms of its two-component upper and lower parts $\varphi$ and $\chi$; the $\varphi$ can be chosen e.g. as

$$\varphi = \begin{pmatrix} 1 \\ 0 \end{pmatrix}, \tag{11.8}$$

and $\chi$ is then obtained by solving Eq. (11.5). Working in the standard representation for Dirac matrices, one has

$$\alpha^3 = \begin{pmatrix} 0 & \sigma_3 \\ \sigma_3 & 0 \end{pmatrix}, \tag{11.9}$$

and Eq. (11.5) becomes

$$p\sigma_3 \chi + m\varphi = E\varphi , \tag{11.10}$$
$$p\sigma_3 \varphi - m\chi = E\chi . \tag{11.11}$$

Thus, Eq. (11.11) yields

$$\chi = \frac{p}{E + m} \sigma_3 \varphi , \tag{11.12}$$

and substituting this into Eq. (11.10), one of course gets the expected relation $p^2 = E^2 - m^2$. So, using (11.8) and (11.12), one may write finally

$$\psi_{\mathrm{inc.}}(z) = \begin{pmatrix} 1 \\ 0 \\ \dfrac{p}{E + m} \\ 0 \end{pmatrix} e^{ipz} . \tag{11.13}$$

Note that our choice (11.8) thus corresponds to an incident particle with positive helicity.

Next, a solution corresponding to the reflected particle should carry the momentum $-p$, and one cannot exclude, *a priori*, the helicity flip; thus, a general form of the reflected wave can be written as

$$\psi_{\mathrm{refl.}} = a\, e^{-ipz} \begin{pmatrix} \varphi_1 \\ \chi_1 \end{pmatrix} + b\, e^{-ipz} \begin{pmatrix} \varphi_2 \\ \chi_2 \end{pmatrix}, \tag{11.14}$$

73

where

$$\varphi_1 = \begin{pmatrix} 1 \\ 0 \end{pmatrix}, \qquad \varphi_2 = \begin{pmatrix} 0 \\ 1 \end{pmatrix}, \tag{11.15}$$

and $\chi_1$, $\chi_2$ are obtained, as before, by solving Eq. (11.5). In analogy with (11.12) one gets here

$$\chi_{1,2} = \frac{-p}{E+m}\sigma_3\varphi_{1,2}, \tag{11.16}$$

so that (11.14) becomes

$$\psi_{\text{refl.}}(z) = a\,e^{-ipz}\begin{pmatrix} 1 \\ 0 \\ \dfrac{-p}{E+m} \\ 0 \end{pmatrix} + b\,e^{-ipz}\begin{pmatrix} 0 \\ 1 \\ 0 \\ \dfrac{p}{E+m} \end{pmatrix}. \tag{11.17}$$

Finally, let us consider the transmitted wave that represents the particle penetrating the potential step for $z \geq 0$. Obviously, one may employ an Ansatz

$$\psi_{\text{trans.}} = c\,e^{iqz}\begin{pmatrix} \varphi_1 \\ \chi_1 \end{pmatrix} + d\,e^{iqz}\begin{pmatrix} \varphi_2 \\ \chi_2 \end{pmatrix}, \tag{11.18}$$

which is supposed to satisfy Eq. (11.6). The relevant equations for any pair $\varphi$, $\chi$ are now obtained from (11.10), (11.11) by replacing $E$ with $E - V$. So, we have

$$q\sigma_3\chi + m\varphi = (E - V)\varphi, \tag{11.19}$$
$$q\sigma_3\varphi - m\chi = (E - V)\chi. \tag{11.20}$$

Eq. (11.20) yields immediately

$$\chi = \frac{q}{E - V + m}\sigma_3\varphi, \tag{11.21}$$

and substituting this back into Eq. (11.19) one gets, after a trivial manipulation,

$$q^2 = (E - V)^2 - m^2 \tag{11.22}$$

(as expected, it corresponds to the substitution $E \to E - V$ in the previous relation for $p^2$). Eq. (11.18) now reads

$$\psi_{\text{trans.}}(z) = c\,e^{iqz}\begin{pmatrix} 1 \\ 0 \\ \dfrac{q}{E-V+m} \\ 0 \end{pmatrix} + d\,e^{iqz}\begin{pmatrix} 0 \\ 1 \\ 0 \\ \dfrac{-q}{E-V+m} \end{pmatrix}, \tag{11.23}$$

with $q$ given by (11.22). With the expressions (11.13), (11.17) and (11.23) at hand, one may impose the natural condition of continuity of the wave function at $z = 0$, the edge of the potential step. Explicitly, the condition reads

$$\psi_{\text{I}}(0) = \psi_{\text{II}}(0), \tag{11.24}$$

where

$$\psi_{\text{I}}(z) = \psi_{\text{inc.}}(z) + \psi_{\text{refl.}}(z) \tag{11.25}$$

and
$$\psi_{\text{II}}(z) = \psi_{\text{trans.}}(z) . \tag{11.26}$$

The "matching" condition (11.24) then becomes, after a trivial manipulation,

$$1 + a = c , \tag{11.27}$$

$$b = d , \tag{11.28}$$

$$\frac{p}{E + m} + a \frac{-p}{E + m} = c \frac{q}{E - V + m} , \tag{11.29}$$

$$\frac{p}{E + m} b = \frac{-q}{E - V + m} d . \tag{11.30}$$

It is easy to see that the equations (11.28) and (11.30) imply $b = d = 0$, i.e. the particle helicity is not flipped in a reflection or transmission event. The equations (11.27) and (11.29) yield

$$a = \frac{1 - r}{1 + r} , \qquad c = \frac{2}{1 + r} , \tag{11.31}$$

where

$$r = \frac{E + m}{E - V + m} \frac{q}{p} . \tag{11.32}$$

Let us now examine the character of the solution in the region II (i.e. the transmitted wave), in dependence on the particle energy $E$ and the potential step height $V$. For this purpose, the key relation is Eq. (11.22). First, if $|E - V| < m$, $q^2$ is negative, and $q$ is thus purely imaginary. It is reasonable to chose $q = i|q|$; the transmitted wave then decreases exponentially inside the step (the choice $q = -i|q|$ would lead to exponential growth of the wave function, which is an unacceptable behaviour). Note that the condition $|E - V| < m$ means

$$V - m < E < V + m . \tag{11.33}$$

Thus, $E - m < V$, i.e. the kinetic energy is less than $V$; the exponential decay of the wave function inside the step is in agreement with the familiar result known from the analysis of the non-relativistic Schrödinger equation in an analogous situation.

The second possibility is $|E - V| > m$, which means either

$$E - V > m , \qquad \text{i.e.} \qquad E - m > V \tag{11.34}$$

or

$$E - V < -m , \qquad \text{i.e.} \qquad E < V - m . \tag{11.35}$$

Equivalently, (11.35) yields a condition for $V$:

$$V > E + m . \tag{11.36}$$

In this case, (11.22) tells us that $q^2 > 0$, so that the transmitted wave has oscillatory character: the solution in question behaves as $\exp(\pm i|q|z)$. If (11.34) holds, such a result is quite transparent and corresponds to the non-relativistic description (the kinetic energy $E - m$ is greater than the potential step height and the free motion inside is classically feasible). On the other hand, the situation corresponding to the condition (11.35) is rather different. Here the kinetic energy is certainly less than $V$, but the transmitted wave still has an oscillatory character! Such a result is astonishing, if one relies on an earlier experience based on classical physics and non-relativistic quantum mechanics. It simply means that an incident particle with kinetic energy less than potential energy for $z \geq 0$ can get arbitrarily far inside the step, i.e. for any $z \to +\infty$ (because of an undamped wave function). In this context, the relation (11.36) is also remarkable; according to this condition, the height of the potential barrier is quite big, certainly at least $2m$, twice the rest energy of the incident particle. The counterintuitive behaviour of the wave function described above is the core of the effect called commonly the **Klein paradox**, since Oskar Klein was the first who came across such a conundrum in 1929.

Preceding considerations can be supplemented with the evaluation of the corresponding reflection and transmission coefficients characterizing the considered dynamical process. As we have noted at the beginning of this chapter, we are in fact studying a one-dimensional scattering, and in such a case the above-mentioned coefficients describe the event probabilities, instead of a scattering cross section used in three dimensions. The coefficients in question can be defined by means of the relevant probability density currents (recall the basic formula $\vec{j} = \psi^\dagger \vec{\alpha} \psi$ derived in Chapter 2). The probability currents are expressed in terms of the incident, reflected and transmitted waves shown above; using the result (11.34) one gets

$$T = \frac{j_{\text{trans.}}}{j_{\text{inc.}}} = \frac{4r}{(1+r)^2} \tag{11.37}$$

for the transition coefficient, and the reflection coefficient becomes

$$R = \frac{j_{\text{refl.}}}{j_{\text{inc.}}} = \frac{(1-r)^2}{(1+r)^2}, \tag{11.38}$$

where $r$ is given by (11.32) (the reader is encouraged to reproduce these results independently). Now, the question is what is in fact the value of $q$ appearing in the expression for the quantity $r$. Using the relation (11.22), and taking mechanically the value of $q$ as the positive square root $q = \left[ (E - V)^2 - m^2 \right]^{1/2}$, the kinematical factor $r$ becomes negative, when $E$ and $V$ satisfy the condition (11.35). Then $T$ is negative, while $R$ becomes greater than 1, but we still have $R + T = 1$. This strange fact is often taken as an additional attribute of the Klein paradox (see e.g. [1, 6, 11]). However, a closer look reveals that it is not quite so. Important explanatory remarks can be found e.g. in the book [5] or in the nice review paper [36]. The point is that one should be more careful when defining the boundary condition for the transmitted wave. For that purpose, one may consider the (group) velocity $v = dE/dq$. This is computed easily; differentiating the relation (11.22) with respect to $q$, one has

$$2q = 2(E - V)\frac{dE}{dq},$$

so that

$$\frac{dE}{dq} = \frac{q}{E - V}. \tag{11.39}$$

Since $E < V$ in the considered case (11.35), for a motion of the particle from $z = 0$ to $+\infty$ (from left to right), i.e. with positive $v$, one should take $q$ to be the negative square root

$$q = -\sqrt{(E - V)^2 - m^2}. \tag{11.40}$$

Then $r$ is positive,

$$r = \sqrt{\frac{(V - E + m)(E + m)}{(V - E - m)(E - m)}}, \tag{11.41}$$

so that for $R$ and $T$ given by (11.37) and (11.38) one has $0 < T < 1$, $0 < R < 1$ and, of course, $R + T = 1$. As noted in ref. [36], a hint concerning the velocity was given to O. Klein by W. Pauli (thus, the seeming paradox referring to $T < 0$ and $R > 1$ was not an issue in the original Klein's paper).

Anyway, the physical essence of the Klein paradox persists. Once again: its content is that the solution of the Dirac equation for transmitted wave is not exponentially damped inside the potential step, in a situation that is classically forbidden.[6] Having found that the Klein paradox is real, one may wonder how to live with it. A natural reaction is that such a puzzling observation reveals a limit of applicability of the Dirac equation as a one-particle equation of relativistic quantum mechanics. In particular, the Klein paradox shows that one may run into controversy in the presence of a very strong field; let us recall that in the considered case we have $V > 2m$. The "critical" value $V = 2m$ suggests a seemingly crazy idea that such a huge potential energy might give rise to two more particles with the mass $m$ and thus the whole picture of the scattering process would be substantially altered. In fact, such an idea is not quite crazy (after all, there is an equivalence of the energy and mass in relativistic physics!). Quantum field theory does enable one to describe the creation of particle-antiparticle pairs in a sufficiently strong external field (the so-called Schwinger process); the condition $V > 2m$ corresponds precisely to such a situation. So, the resolution of the conundrum, called traditionally Klein paradox, should rely on abandoning the one-particle interpretation of the Dirac equation in favour of quantum field theory, which can describe naturally the processes with changing number and even the type of particles involved. We will not pursue this topic further now; referring to the Klein paradox has served us here just as another hint at possible difficulties of the one-particle relativistic quantum mechanics. An interested reader may find a detailed discussion of this issue e.g. in [36] and some references therein, or in a more recent paper [37].

---

[6]Moreover, from (11.41) it is clear that $r \to 1$ for $V \to \infty$ and $E$ fixed. Thus, according to (11.37), $T \to 1$ in such a limit. It means that an arbitrarily high barrier is penetrable for the Dirac particle; this is certainly a paradoxical behaviour.

# Chapter 12

# Relativistic equation for spin-1 particle

Up to now we have considered relativistic equations for particles with spin 0 (Klein–Gordon) and spin 1/2 (Dirac or Weyl). Next in the hierarchy are particles with spin 1 (which also play a crucial role in the present-day standard model of particle physics). So, it certainly makes sense to devote one chapter to a brief discussion of the relevant relativistic equation describing a massive spin-1 particle (the case of massless photon is deferred, for convenience, to one of the later chapters on the field theory).

In fact, finding such an equation is a non-trivial task. The point is that a massive spin-1 particle has three independent states, which correspond to spin projections ±1 and 0 onto a given space direction in its rest system; thus, one would like to describe it in terms of a wave function with three independent components. However, if one wants to maintain the relativistic covariance, writing a pertinent equation for such an object is not straightforward. The simplest and rather natural way to achieve this goal is to employ a wave function that is Lorentz four-vector, and impose one covariant constraint on its components. Since we wish to maintain the standard relativistic relation for energy, momentum and mass of a free particle, it is clear that the components of the wave function should satisfy the Klein–Gordon equation. So, denoting the wave function in question as $A^\mu(x)$, $\mu = 0, 1, 2, 3$, one has

$$(\Box + m^2)A^\mu(x) = 0. \tag{12.1}$$

A simple choice of the envisaged Lorentz covariant constraint, linear in $A^\mu$, reads

$$\partial_\mu A^\mu(x) = 0. \tag{12.2}$$

In this way, one arrives at covariant equations for a wave function with three independent components; these can be chosen as e.g. $A^j(x)$, $j = 1, 2, 3$, and $A^0(x)$ is supposed to be evaluated using the constraint (12.2).

It turns out that there is another elegant way leading to the equations (12.1), (12.2). The idea is to use, as a starting point, Maxwell equations in the covariant form and add an appropriate mass term (sure, one might object that Maxwell equations describe a classical field, but it does not matter — the Lorentz covariance is the main point here). Thus, one may write, tentatively,

$$\partial_\mu F^{\mu\nu} + m^2 A^\nu = 0, \tag{12.3}$$

where $F^{\mu\nu} = \partial^\mu A^\nu - \partial^\nu A^\mu$. In order to check whether we are on the right track, let us work out the left-hand side of Eq. (12.3). One gets

$$\partial_\mu F^{\mu\nu} + m^2 A^\nu = \partial_\mu(\partial^\mu A^\nu - \partial^\nu A^\mu) + m^2 A^\nu = \Box A^\nu - \partial^\nu(\partial_\mu A^\mu) + m^2 A^\nu.$$

Thus, Eq. (12.3) can be recast as

$$(\Box + m^2)A^\nu - \partial^\nu(\partial_\mu A^\mu) = 0. \tag{12.4}$$

Differentiating this equation by means of $\partial_\nu$ and denoting $\partial_\mu A^\mu$ as $\partial \cdot A$, one obtains

$$\Box(\partial \cdot A) + m^2 \partial \cdot A - \Box(\partial \cdot A) = 0. \tag{12.5}$$

It means that we are left with $m^2 \partial \cdot A = 0$, so that $\partial \cdot A = 0$, which is precisely the constraint (12.2). As a result, the equation (12.4) is reduced to (12.1). Thus we see that the equation (12.3) is equivalent to the pair (12.1), (12.2): Eq. (12.3) implies the validity of (12.1) along with (12.2) and from (12.4) it is obvious that the converse is true as well. A remark is in order here. The equation (12.2) is formally identical with the Lorenz condition in the Maxwell theory of electromagnetism. However, in the considered case, this identity emerges as a consequence of the basic equation (12.3), while in the Maxwell theory it must be imposed by hand. Of course, it is the mass $m \neq 0$ that makes the difference.

The construction described above is due to A. Proca,[7] who came up with it in 1936 within the framework of field theory (trying to extend a theory of nuclear forces); Eq. (12.3) is therefore usually called the **Proca equation**. Of course, such an equation can be utilized equally well within relativistic quantum mechanics; as noted before, this is precisely what we are doing here — it is just the Lorentz covariance that matters.

Let us now examine some basic properties of solutions of the equations (12.1), (12.2). Of course, our primary goal is to verify that these equations provide us indeed with a proper description of a spin-1 particle. To this end, we are going to consider the solutions of the plane-wave form, i.e. those corresponding to the states with a given energy and momentum. In analogy with our previous treatment of relativistic equations, an appropriate Ansatz for such a solution reads

$$A_\mu(x) = \varepsilon_\mu(k) e^{-ikx}. \tag{12.6}$$

Since Eqs. (12.1), (12.2) are invariant under complex conjugation, another independent solution then would be

$$A_\mu^*(x) = \varepsilon_\mu^*(k) e^{+ikx}. \tag{12.7}$$

Note that the form (12.6) is an analogue of the Dirac plane wave, with $\varepsilon_\mu(k)$ being a counterpart of $u(k)$; the $\varepsilon_\mu^*(k)$ is then analogous to $v(k)$ (cf. Chapter 6). Substituting (12.6) into Eq. (12.1), one gets immediately $k^2 = m^2$, i.e. $k_0^2 = \vec{k}^2 + m^2$. Choosing $k_0 > 0$, the solution (12.6) carries positive energy $k_0 = (\vec{k}^2 + m^2)^{1/2}$ and momentum $\vec{k}$. Similarly, (12.7) then corresponds to negative energy $-k_0$ and momentum $-\vec{k}$. Further, Eq. (12.2) amounts to the requirement

$$k^\mu \varepsilon_\mu(k) = 0. \tag{12.8}$$

A terminological remark is perhaps in order here. Since the form of the solutions (12.6), (12.7) including the condition (12.8) is so similar to the description of plane waves in Maxwell theory, the amplitude $\varepsilon_\mu(k)$ is called generally the **polarization vector**.

So, now the question is what are the independent polarization vectors satisfying the constraint (12.8). In order to clarify this point, one may start in the rest system, i.e. with $k = k^{(0)} = (m, \vec{0})$. Writing $\varepsilon^\mu = (\varepsilon^0, \vec{\varepsilon})$, from (12.8) one then gets readily $\varepsilon^0 = 0$, so that

$$\varepsilon^\mu(k^{(0)}) = (0, \vec{\varepsilon}). \tag{12.9}$$

---

[7] Alexandru Proca (1897-1955) was an eminent Romanian theoretical physicist who spent most of his professional career in France. His work belongs to the "golden age" of the quantum theory in the first half of the 20th century.

Thus, $\varepsilon^\mu$ is space-like. Obviously, there are three independent 3-vectors $\vec{\varepsilon}$; for convenience, one may write them as columns, and as a particular example one can choose

$$\vec{\varepsilon}^{(1)} = \begin{pmatrix} 1 \\ 0 \\ 0 \end{pmatrix}, \qquad \vec{\varepsilon}^{(2)} = \begin{pmatrix} 0 \\ 1 \\ 0 \end{pmatrix}, \qquad \vec{\varepsilon}^{(3)} = \begin{pmatrix} 0 \\ 0 \\ 1 \end{pmatrix}. \tag{12.10}$$

Having in mind our earlier remark concerning the independent components of the wave function (i.e. that these are just $A^j$, $j = 1, 2, 3$), one may observe that the set (12.10) defines three possible solutions for the particle at rest, which are the eigenstates of the spin matrix

$$S_3 = \begin{pmatrix} 1 & 0 & 0 \\ 0 & 0 & 0 \\ 0 & 0 & -1 \end{pmatrix}, \tag{12.11}$$

with eigenvalues $+1, 0$ and $-1$. This, of course, is an anticipated result for a spin-1 particle.

Let us now extend our analysis to the case of a particle in motion, i.e. $\vec{k} \neq 0$ in the relation (12.8). There is a natural way of constructing a triplet of space-like polarization vectors $\varepsilon^\mu(k, \lambda)$, $\lambda = 1, 2, 3$, satisfying (12.8). First, one can use

$$\varepsilon^\mu(k, 1) = \left(0, \vec{\varepsilon}(\vec{k}, 1)\right),$$
$$\varepsilon^\mu(k, 2) = \left(0, \vec{\varepsilon}(\vec{k}, 2)\right), \tag{12.12}$$

where $\vec{\varepsilon}(\vec{k}, 1)$ and $\vec{\varepsilon}(\vec{k}, 2)$ are two linearly independent vectors lying in the plane perpendicular to $\vec{k}$, i.e. satisfying

$$\vec{k} \cdot \vec{\varepsilon}(\vec{k}, \lambda) = 0, \qquad \lambda = 1, 2. \tag{12.13}$$

Following (12.10), one may normalize conventionally $\vec{\varepsilon}(\vec{k}, \lambda)$ to be mutually orthogonal unit vectors, so that then

$$\varepsilon^\mu(k, \lambda)\varepsilon_\mu(k, \lambda') = -\delta_{\lambda\lambda'}, \qquad \lambda, \lambda' = 1, 2. \tag{12.14}$$

For the third member of the sought triplet one may try an Ansatz

$$\varepsilon^\mu(k, 3) = \left(\varepsilon^0, a\frac{\vec{k}}{|\vec{k}|}\right), \tag{12.15}$$

where the numbers $\varepsilon^0$ and $a$ are to be determined by means of (12.8) and the normalization condition indicated in (12.14) (needless to say, $\varepsilon(k, 3)$ is orthogonal to $\varepsilon(k, 1)$ and $\varepsilon(k, 2)$). From (12.8) one gets readily

$$\varepsilon_0 = a\frac{|\vec{k}|}{k_0}, \tag{12.16}$$

and the normalization condition $\varepsilon^\mu(k, 3)\varepsilon_\mu(k, 3) = -1$ then yields

$$a^2 = \frac{-1}{\dfrac{\vec{k}^2}{k_0^2} - 1} = \frac{k_0^2}{m^2}. \tag{12.17}$$

Conventionally, we will choose $a > 0$, i.e. $a = k_0/m$. Thus, we have finally

$$\varepsilon^\mu(k, 3) = \left(\frac{|\vec{k}|}{m}, \frac{k_0}{m}\frac{\vec{k}}{|\vec{k}|}\right). \tag{12.17}$$

80

Since the spatial part of $\varepsilon^\mu(k, 3)$ is directed along $\vec{k}$, it is called **longitudinal polarization**; correspondingly, $\varepsilon^\mu(k, \lambda)$ with $\lambda = 1, 2$ are **transverse polarizations**. There is a remarkable technical point that should be mentioned here. It is seen that the four-vector $\varepsilon(k, 3)$ given by (12.17) has, mathematically, the same form as the spin four-vector corresponding to the helicity of a spin-$\frac{1}{2}$ Dirac particle (cf. (8.7)). Of course, such an accidental coincidence is due to the fact that the defining conditions for $\varepsilon(k, 3)$ and $s_R(k)$ are the same, but one should keep in mind that the physical roles of these two quantities are quite different.

So, we have arrived at a rather detailed form of the plane-wave solutions of the Proca equation and now one may try to identify the spin states in terms of the polarization vectors $\varepsilon(k, \lambda)$, $\lambda = 1, 2, 3$. Hopefully, the concept of helicity could be relevant here, similarly to the case of the Dirac equation. To this end, let us first specify the relevant spin matrices $S^j$, $j = 1, 2, 3$. It turns out that a triplet $\vec{S}$ best suited for our purpose is

$$S^1 = \begin{pmatrix} 0 & 0 & 0 \\ 0 & 0 & -i \\ 0 & i & 0 \end{pmatrix}, \qquad S^2 = \begin{pmatrix} 0 & 0 & i \\ 0 & 0 & 0 \\ -i & 0 & 0 \end{pmatrix}, \qquad S^3 = \begin{pmatrix} 0 & -i & 0 \\ i & 0 & 0 \\ 0 & 0 & 0 \end{pmatrix}. \qquad (12.18)$$

It is easy to check that the matrices (12.18) satisfy indeed the commutation relations

$$[S^j, S^k] = i\varepsilon^{jkl} S^l. \qquad (12.19)$$

The reader may recognize in (12.18) the familiar antisymmetric generators of rotations in three-dimensional space. Notice also that this is a representation different from that mentioned before (cf. (12.11)); please don't worry about it, it is just a matter of choice of the algebraic basis in the space of $3 \times 3$ matrices (more precisely, the basis of the Lie algebra of the rotation group). One defines the operator (matrix) of helicity as before, i.e. as the spin projection onto the direction of particle momentum. So, one has

$$\widehat{h}(\vec{k}) = \vec{n} \cdot \vec{S}, \qquad \vec{n} = \vec{k}/|\vec{k}|. \qquad (12.20)$$

Substituting the matrices (12.18) into the definition (12.20), one gets

$$\widehat{h}(\vec{k}) = i \begin{pmatrix} 0 & -n_3 & n_2 \\ n_3 & 0 & -n_1 \\ -n_2 & n_1 & 0 \end{pmatrix}. \qquad (12.21)$$

Writing the spatial part of a polarization vector $\varepsilon(k, \lambda)$ as a three-component column, one gets from (12.21)

$$\widehat{h}(\vec{k})\vec{\varepsilon} = i(\vec{n} \times \vec{\varepsilon}). \qquad (12.22)$$

Now, the two vectors $\vec{\varepsilon}(k, 1)$, $\vec{\varepsilon}(k, 2)$ perpendicular to $\vec{k}$ (the transverse polarizations (12.12)) can be chosen conventionally, so that

$$\vec{\varepsilon}(k, 2) = \vec{n} \times \vec{\varepsilon}(k, 1). \qquad (12.23)$$

Then one also has

$$\vec{n} \times \vec{\varepsilon}(k, 2) = \vec{n} \times (\vec{n} \times \vec{\varepsilon}(k, 1)) = \vec{n}(\vec{n} \cdot \vec{\varepsilon}(k, 1)) - \vec{\varepsilon}(k, 1)(\vec{n} \cdot \vec{n}) = -\vec{\varepsilon}(k, 1). \qquad (12.24)$$

Thus, using (12.22), (12.23) and (12.24) one gets

$$\begin{aligned} \widehat{h}(\vec{k})\vec{\varepsilon}(k, 1) &= i\vec{\varepsilon}(k, 2), \\ \widehat{h}(\vec{k})\vec{\varepsilon}(k, 2) &= -i\vec{\varepsilon}(k, 1), \\ \widehat{h}(\vec{k})\vec{\varepsilon}(k, 3) &= 0. \end{aligned} \qquad (12.25)$$

81

Note that the last identity in (12.25) is obtained readily from (12.22), taking into account that $\vec{\varepsilon}(k,3)$ is parallel to $\vec{k}$ (see (12.17)). With (12.25) at hand, a natural next step is to introduce complex combinations

$$
\begin{aligned}
\vec{\varepsilon}(k,+) &= \frac{1}{\sqrt{2}} \left[ \vec{\varepsilon}(k,1) + i\vec{\varepsilon}(k,2) \right], \\
\vec{\varepsilon}(k,-) &= \frac{1}{\sqrt{2}} \left[ \vec{\varepsilon}(k,1) - i\vec{\varepsilon}(k,2) \right]
\end{aligned}
\tag{12.26}
$$

(one may notice that the transformation (12.26) represents the passage from linear to circular polarizations). From (12.25) and (12.26) one then gets immediately

$$
\begin{aligned}
\widehat{h}(\vec{k})\vec{\varepsilon}(k,+) &= \vec{\varepsilon}(k,+), \\
\widehat{h}(\vec{k})\vec{\varepsilon}(k,-) &= -\vec{\varepsilon}(k,-).
\end{aligned}
\tag{12.27}
$$

In this way we have achieved the desired goal: the results (12.27) along with the last identity (12.25) tell us that the transverse polarization vectors $\varepsilon(k,+)$ and $\varepsilon(k,-)$ correspond to helicities $+1$ and $-1$, respectively, and the longitudinal polarization $\varepsilon(k,3)$ represents the state with zero helicity. A minor technical remark is in order here: It is easy to realize that the polarization vectors, which now are in general complex, satisfy the orthonormality relation

$$
\varepsilon^{\mu}(k,\lambda)\varepsilon_{\mu}^{*}(k,\lambda') = -\delta_{\lambda\lambda'}, \tag{12.28}
$$

where $\lambda$, $\lambda'$ take on values $+$, $-$ and $3$ (it follows readily from the relations mentioned above for the real $\varepsilon$'s and from (12.26)).

As we have stressed before, the polarization vectors $\varepsilon(k,\lambda)$ are in fact counterparts of the plane-wave amplitudes $u(k,s)$ or $v(k,s)$ known in the Dirac theory of spin-$\frac{1}{2}$ particles. We know that the spinorial amplitudes $u$ and $v$ satisfy the completeness relations (6.24), (6.30), and one may thus wonder how to obtain an analogous relation for the polarization vectors we are working with. For such a purpose, one may utilize the following trick: Obviously, the triplet of space-like vectors $\vec{\varepsilon}^{\mu}(k,\lambda)$, $\lambda = 1, 2, 3$, along with the time-like vector $k^{\mu}/m$ form an orthonormal basis in the Minkowski space and thus it is not difficult to guess that the corresponding completeness relation can be written as

$$
-\sum_{\lambda=1}^{3} \varepsilon_{\mu}(k,\lambda)\varepsilon_{\nu}^{*}(k,\lambda) + \frac{k_{\mu}}{m}\frac{k_{\nu}}{m} = g_{\mu\nu}. \tag{12.29}
$$

The reader is encouraged to verify this identity; for those who would be reluctant to take up this task, a hint may be helpful: Multiply both sides of Eq. (12.29) consecutively by the basis vectors and employ the orthonormality relation (12.28). So, Eq. (12.29) leads to the desired polarization sum

$$
\sum_{\lambda=1}^{3} \varepsilon_{\mu}(k,\lambda)\varepsilon_{\nu}^{*}(k,\lambda) = -g_{\mu\nu} + \frac{k_{\mu}k_{\nu}}{m^2}. \tag{12.30}
$$

We will stop our discussion of the Proca equation here; what we have done up to now will be sufficient for later applications. In particular, the properties of the plane waves and polarization vectors that we have found will be utilized in our future perturbative calculations of decay and scattering processes. Of course, it is the same story as our previous detailed treatment of the plane waves for Dirac particles (remember how we tried to find excuses for pursuing such a rather boring topic). One should also add that the Proca equation, as an equation of relativistic quantum mechanics, is not so interesting as the Dirac equation, as regards phenomenological applications. The reason is that there is no stable massive elementary particle with spin 1, so that e.g. the problem of finding bound states of a spin-1 particle in external Coulomb field is not so important as the same problem for the Dirac's electron. Strictly speaking, the Proca equation is more important for the quantum field theory; as we have noted before, this was the framework, in which A. Proca formulated originally his idea.

# Chapter 13

# Splendors and miseries
# of relativistic quantum mechanics

The present chapter is a sort of epilogue, in which we would like to review the successes and failures of the relativistic quantum mechanics; its specific character is reflected in a literary style of the title (the connoisseurs of the classical French literature would perhaps appreciate the original expression "splendeurs et misères").

In a recapitulation of successes (splendors), the prominent position certainly belongs to the Dirac equation, which brought a real breakthrough in the early years of quantum theory in 1920s. As we have seen, it assigns very naturally the half-integer intrinsic angular momentum (spin $\frac{1}{2}$) to the electron. Furthermore, it leads to a simple and elegant explanation of the value of the spin magnetic moment. In its time, this was actually a prediction, in a limited sense; we will return to this point shortly. Another highlight of the Dirac theory of the electron is the famous Darwin–Gordon formula for the energy levels of the hydrogen atom; though we have not discussed its derivation before, let us reproduce the result here for reader's convenience. It is written as an expression for the full electron energy, which is the sum of the rest energy and the (negative) binding energy. It reads

$$E = m \left[ 1 + \frac{\alpha^2}{(n_r + \gamma)^2} \right]^{-1/2}, \tag{13.1}$$

where $\alpha$ is the fine-structure constant ($\alpha \doteq 1/137$), $n_r$ is the radial quantum number, $n_r = 0, 1, 2, \ldots$, and $\gamma = [(j + \frac{1}{2})^2 - \alpha^2]^{1/2}$, with $j$ denoting full angular momentum (including both the orbital momentum and spin, so that $j$ takes on half-integer values). The formula (13.1) becomes more transparent when expanded in powers of $\alpha^2$; one thus gets

$$E = m \left[ 1 - \frac{\alpha^2}{2n^2} - \frac{\alpha^4}{2n^3} \left( \frac{1}{j + \frac{1}{2}} - \frac{3}{4n} \right) + \mathcal{O}(\alpha^6) \right], \tag{13.2}$$

where $n$ now denotes the principal quantum number, $n = n_r + j + \frac{1}{2}$. It is easy to see that the second term in square brackets reproduces the familiar Balmer formula for hydrogen levels obtained in the non-relativistic theory (recall that $\alpha = e^2/\hbar c$ in ordinary units, so that $m\alpha^2/2n^2$ then becomes $me^4/2n^2\hbar^2$). The term proportional to $\alpha^4$ represents the relativistic corrections that bring about the famous "fine structure" of the energy levels. So, the natural explanation of such subtle effects in the hydrogen atom spectrum was certainly a stunning success of the Dirac equation at the end of 1920s. However, twenty years later, more accurate measurements (performed by W. Lamb and R. Retherford) revealed that there is a tiny difference between energies of two states carrying identical values of $n$ and $j$ (recall that according to (13.1), such a difference should be exactly zero). In the common spectroscopic notation, the difference in question is $E(2S_{1/2}) - E(2P_{1/2})$ and it is called, traditionally, the **Lamb shift**. Numerically, it is of the order of $10^{-6}$ eV and the Dirac equation alone is not capable of explaining such an effect. However, soon after this experimental discovery it was clarified within the framework of quantum electrodynamics (QED) and this computational success had become one of the powerful arguments in favour of QED as a physically relevant model of quantum field theory.[8]

Let us now come back to the electron magnetic moment. As we have already noted in Chapter 2, the successful prediction of the value $\mu_e = \mu_B = e/2m$ depends on the assumption that the electromagnetic interaction has the "minimal" form, obtained by replacing $\partial_\mu$ in the Dirac equation with the covariant derivative $\partial_\mu + ieA_\mu$ (where $A_\mu$ is the relevant electromagnetic four-potential). In fact, the interaction may have a more general form that is easily incorporated in the Dirac equation. Let us now explain briefly how it can be done. The above-mentioned minimal scheme relies on the Dirac equation written as

$$\left[i\gamma^\mu(\partial_\mu + ieA_\mu) - m\right]\psi = 0,\tag{13.3}$$

and it is the covariant derivative that guarantees here the desired gauge invariance. Now, the point is that apart from the four-potential $A_\mu$, there is another electromagnetic quantity that may be utilized as well, namely the field strength tensor $F_{\mu\nu} = \partial_\mu A_\nu - \partial_\nu A_\mu$. It is gauge invariant and antisymmetric, so one may couple it to the familiar antisymmetric combination of the $\gamma$-matrices, $\sigma_{\mu\nu} = \frac{i}{2}[\gamma_\mu, \gamma_\nu]$. A simple extension of Eq. (13.3) can then be written as

$$\left(i\slashed{\partial} - e\slashed{A} - m - \kappa\frac{e}{4m}\sigma_{\mu\nu}F^{\mu\nu}\right)\psi = 0,\tag{13.4}$$

where we have used the slash notation for brevity, and the form of the prefactor in the last term has been chosen for convenience. The constant $\kappa$ is an arbitrary dimensionless coefficient, and the whole expression involving $F_{\mu\nu}$ is traditionally called the **Pauli term**. Choosing the potential $A_\mu$ so that it represents a magnetic field and employing the non-relativistic approximation, one may proceed in analogy with what we have done in Chapter 2. We will not go into the details of the calculation (it might be a non-trivial challenge for a diligent reader), and rather show the final result. It turns out that the particle described by Eq. (13.4) carries the spin magnetic moment with the magnitude

$$\mu = (1 + \kappa)\frac{e}{2m}.\tag{13.5}$$

In this way, one may arrive correctly at any value of the spin magnetic moment; in this sense, the result obtained previously for the electron is not a true prediction (rather, it is a "conditional prediction" relying on the assumption that $\kappa = 0$). On the other hand, Eq. (13.4) enables one to describe also the other spin-$\frac{1}{2}$ particles like proton or neutron; as we know, these are not pointlike objects and their magnetic moments differ substantially from the simple Dirac's values.

In fact, the problem of the Pauli term (to be or not to be) has one more aspect. As we will see later, the presence of such a term would change dramatically the properties of QED; in particular, it would spoil the conventional perturbative renormalizability, which is generally recognized as a user-friendly feature of any model of quantum field theory (QFT).

So, for the electron, the value $\kappa = 0$ seems to be the right option to start with, but this does not mean that it is the end of the story. Similarly as in the case of the Lamb shift, accurate

---

[8]The experimental result of Lamb and Retherford was published in 1947 and Willis Lamb (1913 – 2008) received for it the Nobel Prize in 1955. The first successful theoretical calculation of the Lamb shift was performed also in 1947 by Hans Bethe (1906 – 2005), who received the Nobel Prize in 1967 for something else (the so-called CNO cycle in astrophysics).

measurements performed in the late 1940s uncovered a tiny deviation from the value $\mu_B$ assigned to the electron within the Dirac theory.[9] Such a difference, which is of the relative order of $10^{-3}$, was explained successfully by means of QED (the first theorist who did the calculation was J. Schwinger). This achievement had become another milestone in establishing the QED as a realistic physical theory at the fundamental level.

Let us now turn to the difficulties ("miseries") of relativistic quantum mechanics. As we have seen in the preceding chapters, a generic feature of all considered cases is the appearance of negative energy solutions for the free particle. This leads to specific interpretation problems that we are not going to summarize here, but perhaps the most serious flaw, from the physical point of view, is the potential instability of matter that would be caused by spontaneous transitions of particles to the available (unoccupied) states with negative energy. Worried by such an obvious drawback of his theory, Dirac came up with an original and rather surprising solution. He proposed that what we perceive as the "vacuum" (the no-particle state) is in fact a state, in which all negative energy levels are occupied, so that no spontaneous transitions are possible. Such an infinite continuum of negative energy particles is supposed to be a sort of unobservable background and is called, traditionally, the **Dirac sea**. Under the influence of an external force, a particle (an electron, for definiteness) can be excited from the negative energy level to become an ordinary electron with positive energy. Removing a negative energy particle from the sea amounts to creating a "hole" and, effectively, such a state carries a positive energy (since subtracting a negative value is tantamount to adding a positive one); moreover, its charge is opposite to that of the electron. After a short period of confusion, the picture of a possible hole in the Dirac sea was recognized as the prediction of an **antiparticle** of the electron. The origin of such a term is obvious; furthermore, since the electron charge is conventionally taken to be negative, the (by now familiar) name **positron** for such an antiparticle has been established soon after its theoretical prediction. The proposal made by Dirac has turned into a triumph, when the positron was observed experimentally in cosmic rays (Carl Anderson in 1932).

The prediction of antiparticles (or, more generally, antimatter) was certainly one of the most spectacular successes of the Dirac theory, but it must be taken with a grain of salt. The concept of the filled sea of the negative energy states is in fact highly controversial. Why is it so? We know that electrons are fermions, which means, in particular, that they satisfy the Pauli exclusion principle — there can be just one particle in a given state. So, in such a case one may, in principle, imagine a filled sea that would not absorb any other particle falling into it as a result of a spontaneous quantum transition. However, we also know that the spinless particles described by the Klein–Gordon equation, or spin-1 particles obeying the Proca equation are bosons, which means that their number in a given state is unlimited. Thus, a would-be sea of negative energy bosonic states could never be filled, so that this concept thus fails completely (for an early critique see e.g. [39]). On the other hand, we know that the existence of antiparticles is an universal phenomenon — they do exist both for bosons and fermions. So, strictly speaking, the amazing theoretical construction of the positron was a right prediction based on a wrong (or at least dubious) argument. The reader following the above exposition may have noticed how the "splendors" and "miseries" are sometimes remarkably intertwined in the achievements of relativistic quantum mechanics. Fortunately, as we know now, there is a truly universal and consistent treatment of antiparticles within the QFT framework, in which the idea of a filled "sea" is totally irrelevant. One may thus conclude that the original concept of the Dirac sea is obsolete and in fact may be forgotten (or, more politely, deferred to a history chapter), since it has been superseded by the QFT methods. Having in mind our ultimate goal, it is in order to

---

[9]The corresponding experimental result was published in 1947 by P. Kusch and H. Foley. Polykarp Kusch (1911 – 1993) received for it the Nobel Prize in 1955 together with Willis Lamb.

offer the reader some references to sources describing the QFT genesis and history: apart from the papers [39] quoted above, instructive discussion of the subject can be found e.g. in [9, 13] and [54]. In any case, the existence of negative energy solutions is a fundamental difficulty of relativistic quantum mechanics in general, so it is reassuring that this problem is eradicated naturally within QFT.

As a last item on the list of "miseries" one should mention the puzzling behaviour of a particle in a strong external field, as exemplified by the so-called Klein paradox that we have discussed in Chapter 11. This also points to limitations of the relativistic quantum mechanics as a theory of essentially one-particle systems. As we have already noted before, in contrast to this, QFT is able to describe very naturally physical processes, in which the number and type of participating particles is variable; this in fact is the most important QFT asset as regards the description of processes involving subatomic particles.

Of course, one could say more about the issues mentioned above, but the main message is clear. Equations of relativistic quantum mechanics certainly yielded some wonderful or even astonishing predictions, though sometimes the employed arguments are not quite waterproof. In fact, these equations represent a precursor of a deeper approach, which is the quantum field theory; as we have seen, there is certainly more than one reason to proceed in this way. A bonus drawn from the preceding chapters is that many mathematical formulae and relations derived there will be utilized without change in our forthcoming discussion of the QFT models.

# Chapter 14

# Interlude:
# Lagrangian formalism for classical fields

Before proceeding to the study of quantum fields, it is necessary to discuss first an appropriate formalism for their classical counterparts. Similarly as in the ordinary classical mechanics of point particles, the key concept is a Lagrange function, or briefly Lagrangian.

At the end of the preceding chapter we have mentioned a certain bonus descending from our previous effort. A substantial part of this bonus is that we in fact already know all relevant equations of motion for the classical fields we would like to deal with. Indeed, such relativistically covariant wave equations coincide with the equations of relativistic quantum mechanics considered previously; so, we will discuss a Klein–Gordon field, Dirac field, etc. Although the physical content is different (earlier we have worked with quantum mechanical wave functions, while now we have in mind classical fields), the equations are the same, since their form is determined uniquely by the requirement of Lorentz covariance.

Since the Lagrangian formalism for fields is developed in close analogy with classical mechanics, we would like to recall first the basic principles of the latter. Let us consider a mechanical system with $f$ degrees of freedom, described in terms of the generalized coordinates $q_i = q_i(t)$, $i = 1, \ldots, f$. The Lagrange function depends on the coordinates and on velocities $\dot{q}_i(t)$,

$$L = L\big(q_i(t), \dot{q}_i(t)\big) \tag{14.1}$$

(from the mathematical point of view, $L$ is a functional depending on the time variable $t$). Equations of motion (Lagrange equations) are obtained from the variational principle of stationary action

$$\delta S = 0, \tag{14.2}$$

where the functional of action is given by

$$S = \int_{t_1}^{t_2} L\big(q_i(t), \dot{q}_i(t)\big) \, dt. \tag{14.3}$$

The resulting Lagrange equations then read

$$\frac{d}{dt} \frac{\partial L}{\partial \dot{q}_i} - \frac{\partial L}{\partial q_i} = 0, \qquad i = 1, \ldots, f. \tag{14.4}$$

Now, how about fields? A classical field is described, in general, by a multiplet of functions $\varphi_r(x) = \varphi_r(t, \vec{x})$, $r = 1, \ldots, n$. A natural counterpart of the index $i$ in (14.1) or (14.4) is the continuous "index" $\vec{x}$. Since we have in mind relativistically covariant models (where all spacetime coordinates are treated on an equal footing), it is natural to introduce the density of Lagrange function (briefly, Lagrangian density), depending on the fields and their first derivatives, i.e.

$$\mathscr{L} = \mathscr{L}\big(\varphi_r(x), \partial_\mu \varphi_r(x)\big), \tag{14.5}$$

so that the Lagrange function is the integral over the whole three-dimensional space

$$L = \int \mathscr{L}\, \mathrm{d}^3 x. \tag{14.6}$$

The action is then conveniently defined as

$$S = \int_{-\infty}^{+\infty} L\, \mathrm{d}x_0 = \int \mathscr{L}\, \mathrm{d}^4 x, \tag{14.7}$$

and the variational principle of stationary action then yields

$$
\begin{aligned}
0 = \delta S = \int \mathrm{d}^4 x\, \delta \mathscr{L} &= \int \mathrm{d}^4 x \left[ \frac{\partial \mathscr{L}}{\partial \varphi_r} \delta \varphi_r + \frac{\partial \mathscr{L}}{\partial(\partial_\mu \varphi_r)} \delta(\partial_\mu \varphi_r) \right] \\
&= \int \mathrm{d}^4 x \left[ \frac{\partial \mathscr{L}}{\partial \varphi_r} \delta \varphi_r + \frac{\partial \mathscr{L}}{\partial(\partial_\mu \varphi_r)} \partial_\mu(\delta \varphi_r) \right],
\end{aligned}
\tag{14.8}
$$

where in the last step we have used the obvious rule that the variation of a derivative of a function is the derivative of the variation. Now, in the last expression we may carry out partial integration; in order to get rid of the surface terms, let us assume that the variation $\delta \varphi_r$ vanishes at the spacetime infinity (this is a usual constraint on the trial functions in the considered variational problem). The expression (14.8) is then recast as

$$0 = \delta S = \int \mathrm{d}^4 x \left[ \frac{\partial \mathscr{L}}{\partial \varphi_r} - \partial_\mu \left( \frac{\partial \mathscr{L}}{\partial(\partial_\mu \varphi_r)} \right) \right] \delta \varphi_r, \tag{14.9}$$

and one thus arrives at the Lagrange equations (or, if you want, Euler–Lagrange equations)

$$\partial_\mu \frac{\partial \mathscr{L}}{\partial(\partial_\mu \varphi_r)} - \frac{\partial \mathscr{L}}{\partial \varphi_r} = 0, \qquad r = 1, \dots, n. \tag{14.10}$$

Notice that (14.10) is a "covariantized" form of the familiar equations (14.4) of classical mechanics, so it is easy to remember.

Now we are in a position to find some examples of Lagrangian densities corresponding to the relativistic wave equations we already know (note that for the sake of brevity, we will usually use the term "Lagrangian" instead of "Lagrangian density"). Let us start with a real Klein–Gordon field $\varphi(x)$, which, as we know, satisfies the equation

$$(\Box + m^2)\varphi(x) = 0. \tag{14.11}$$

It is easy to verify that an appropriate Lagrangian can be written as

$$\mathscr{L} = \frac{1}{2} \partial_\mu \varphi\, \partial^\mu \varphi - \frac{1}{2} m^2 \varphi^2. \tag{14.12}$$

Indeed, from (14.12) one gets readily

$$\frac{\partial \mathscr{L}}{\partial(\partial_\mu \varphi)} = \partial^\mu \varphi, \qquad \frac{\partial \mathscr{L}}{\partial \varphi} = -m^2 \varphi, \tag{14.13}$$

89

and substituting this into (14.10), the Klein–Gordon equation (14.11) is recovered (needless to say, (14.13) amounts just to elementary differentiation of quadratic functions). Note that the factors $1/2$ standing in (14.12) are conventional; it is clear that any Lagrangian proportional to $\partial_\mu \varphi \partial^\mu \varphi - m^2 \varphi^2$ would yield Eq. (14.11) as well. The option (14.12) leads to a convenient normalization of the quantities to be discussed later on (e.g. the field energy and momentum).

Next, let us consider a complex Klein–Gordon field. Of course, the equation of motion is (14.11), but now the field has in fact two components; one may describe it in terms of its real and imaginary part, or, equivalently, $\varphi$ and $\varphi^*$ are considered as the two independent field variables. It is easy to see that an appropriate Lagrangian reads

$$\mathscr{L} = \partial_\mu \varphi \, \partial^\mu \varphi^* - m^2 \varphi \varphi^* . \tag{14.14}$$

The relevant derivatives are

$$\frac{\partial \mathscr{L}}{\partial(\partial_\mu \varphi)} = \partial^\mu \varphi^* , \qquad \frac{\partial \mathscr{L}}{\partial \varphi} = -m^2 \varphi^* , \tag{14.15}$$

and similarly for the complex conjugate quantities. From (14.10) one thus gets readily the Klein–Gordon equations for $\varphi^*$ and $\varphi$, respectively. Concerning the conventional overall factor embodied in (14.14), the same remark holds as in the preceding case.

As another example, let us consider a real Proca field $A_\mu(x)$, $\mu = 0, 1, 2, 3$. This is defined by means of the equations of motion

$$\partial_\mu F^{\mu\nu} + m^2 A^\nu = 0 , \tag{14.16}$$

with $F^{\mu\nu} = \partial^\mu A^\nu - \partial^\nu A^\mu$. As we know, Eq. (14.16) is equivalent to the pair

$$(\Box + m^2) A_\mu = 0 , \qquad \partial^\mu A_\mu = 0 . \tag{14.17}$$

It turns out that the corresponding (properly normalized) Lagrangian reads

$$\mathscr{L} = -\frac{1}{4} F_{\mu\nu} F^{\mu\nu} + \frac{1}{2} m^2 A_\mu A^\mu . \tag{14.18}$$

Let us verify that it is indeed the case. The derivative of $\mathscr{L}$ with respect to $A_\mu$ is obtained immediately; obviously, it is

$$\frac{\partial \mathscr{L}}{\partial A_\mu} = m^2 A^\mu . \tag{14.19}$$

The differentiation with respect to the field derivatives is slightly more complicated, so let us now carry out the calculation in detail. One has

$$\frac{\partial \mathscr{L}}{\partial(\partial_\rho A_\sigma)} = \frac{\partial}{\partial(\partial_\rho A_\sigma)} \left( -\frac{1}{4} F_{\mu\nu} F^{\mu\nu} \right) , \tag{14.20}$$

and the expression in parentheses is worked out as

$$-\frac{1}{4} F_{\mu\nu} F^{\mu\nu} = -\frac{1}{4} (\partial_\mu A_\nu - \partial_\nu A_\mu)(\partial^\mu A^\nu - \partial^\nu A^\mu)$$

$$= -\frac{1}{2} (\partial_\mu A_\nu)(\partial^\mu A^\nu) + \frac{1}{2} (\partial_\mu A_\nu)(\partial^\nu A^\mu) . \tag{14.21}$$

So, the basic ingredients are the derivatives

$$\frac{\partial(\partial_\mu A_\nu)}{\partial(\partial_\rho A_\sigma)} = g_\mu^\rho g_\nu^\sigma \tag{14.22}$$

and

$$\frac{\partial(\partial^\mu A^\nu)}{\partial(\partial_\rho A_\sigma)} = g^{\mu\rho} g^{\nu\sigma} . \tag{14.23}$$

The relations (14.22), (14.23) are easily understood: basically, they represent a straightforward generalization of the trivial identity $\partial x^\mu / \partial x^\nu = g^\mu_\nu$ (please remember that $\delta^\mu_\nu = g^\mu_\nu$). In this way, one gets

$$\begin{aligned}
\frac{\partial\mathscr{L}}{\partial(\partial_\rho A_\sigma)} &= -\frac{1}{2}(g^\rho_\mu g^\sigma_\nu \partial^\mu A^\nu + \partial_\mu A_\nu g^{\rho\mu} g^{\sigma\nu}) \\
&\quad + \frac{1}{2}(g^\rho_\mu g^\sigma_\nu \partial^\nu A^\mu + \partial_\mu A_\nu g^{\rho\nu} g^{\sigma\mu}) \\
&= -(\partial^\rho A^\sigma - \partial^\sigma A^\rho) .
\end{aligned}$$

So, the result is

$$\frac{\partial\mathscr{L}}{\partial(\partial_\rho A_\sigma)} = -F^{\rho\sigma} . \tag{14.24}$$

Using Eq. (14.10) and the identities (14.19), (14.24) one obtains

$$\partial_\rho F^{\rho\sigma} + m^2 A^\sigma = 0 , \tag{14.25}$$

i.e. the Proca equation is thereby recovered (just to be sure, the label $r$ in the general equations (14.10) coincides here with $\sigma$).

As a last example, let us consider the Dirac field satisfying the familiar equation

$$i\gamma^\mu \partial_\mu \psi - m\psi = 0 . \tag{14.26}$$

The function $\psi$ is in general complex, so the set of dynamical field variables in Eq. (14.10) should also include an appropriate conjugation of $\psi$, e.g. $\psi^\dagger$ or $\bar\psi$. The Dirac conjugation $\bar\psi = \psi^\dagger \gamma_0$ is the right option, since, as we know, it is instrumental in constructing bilinear covariants like $\bar\psi\psi$, etc. One possible form of a Lagrangian yielding Eq. (14.26) is then guessed easily. It reads

$$\mathscr{L} = \bar\psi(i\gamma^\mu \partial_\mu - m)\psi . \tag{14.27}$$

Indeed, in (14.27) there is no derivative of $\bar\psi$, and the left-hand side of Eq. (14.26) appears there manifestly as a (column) factor. Thus, following the general recipe (14.10), one Lagrange equation is simply

$$\frac{\partial\mathscr{L}}{\partial\bar\psi} = 0 . \tag{14.28}$$

This, using (14.27), gives immediately Eq. (14.26). A technical remark is in order here. So as to be absolutely correct, one should work with bispinor components of $\psi$ and $\bar\psi$, i.e. use $\psi_j$, $j = 1, 2, 3, 4$ and similarly for $\bar\psi$, when working out Eq. (14.10). Hopefully, the compact matrix notation employed here will not cause any confusion.

Further, differentiating (14.27) with respect to $\psi$ and $\partial_\mu\psi$, one has

$$\frac{\partial\mathscr{L}}{\partial(\partial_\mu\psi)} = i\bar\psi\gamma^\mu , \qquad \frac{\partial\mathscr{L}}{\partial\psi} = -m\bar\psi . \tag{14.29}$$

So, apart from (14.28), one gets from (14.10) and (14.29) also

$$i\partial_\mu\bar\psi\gamma^\mu + m\bar\psi = 0 . \tag{14.30}$$

It is easy to see that Eq. (14.30) is, as expected, just the Dirac conjugation of Eq. (14.26).

The form (14.27) is simple and user-friendly in further applications (as we will see later on). However, an aesthetically minded reader might worry about the asymmetric way, in which $\psi$ and $\bar{\psi}$ appear there. It is a legitimate observation, so one may try to find a more sophisticated form of the Lagrangian density. Such a more symmetric solution does exist, namely

$$\overline{\mathscr{L}} = \frac{i}{2}\bar{\psi}\gamma^{\mu}\overset{\leftrightarrow}{\partial}_{\mu}\psi - m\bar{\psi}\psi \qquad (14.31)$$

(cf. e.g. the book [6]), where the symbol $\leftrightarrow$ means the both-sided derivative, defined here conventionally as

$$f\overset{\leftrightarrow}{\partial}_{\mu}g = f\,\partial_{\mu}g - \partial_{\mu}f\,g\,. \qquad (14.32)$$

So, the expression (14.31) may be recast as

$$\overline{\mathscr{L}} = \frac{i}{2}\bar{\psi}\gamma^{\mu}\partial_{\mu}\psi - \frac{i}{2}\partial_{\mu}\bar{\psi}\gamma^{\mu}\psi - m\bar{\psi}\psi\,. \qquad (14.33)$$

The derivatives to be utilized in the general Eq. (14.10) are then e.g.

$$\frac{\partial\overline{\mathscr{L}}}{\partial(\partial_{\mu}\psi)} = \frac{i}{2}\bar{\psi}\gamma^{\mu}\,, \qquad \frac{\partial\overline{\mathscr{L}}}{\partial\psi} = -\frac{i}{2}\partial_{\mu}\bar{\psi}\gamma^{\mu} - m\bar{\psi}\,. \qquad (14.34)$$

From (14.34) and (14.10) one gets immediately Eq. (14.30). Similarly, using the derivatives involving $\bar{\psi}$, one obtains directly Eq. (14.26).

Thus, one may say that the Lagrangians (14.27) and (14.31) are equivalent, in the sense that they lead to the same equation of motion for the field $\psi$. In fact, it is a nice demonstration of the ambiguity of the Lagrangian density, which is not determined uniquely by the equations of motion. We have already noticed an ambiguity that is basically trivial, corresponding to the rescaling of $\mathscr{L}$ by a constant factor. The relation between (14.27) and (14.31) is more intriguing. As the reader may verify easily, it holds

$$\mathscr{L} = \overline{\mathscr{L}} + \frac{i}{2}\partial_{\mu}\left(\bar{\psi}\gamma^{\mu}\psi\right)\,. \qquad (14.35)$$

The essential point is that the difference $\mathscr{L} - \overline{\mathscr{L}}$ is equal to a four-divergence (total derivative) of a particular expression made of $\psi$ and $\bar{\psi}$. This in turn means that the action is not changed when passing from $\mathscr{L}$ to $\overline{\mathscr{L}}$ (adding a total derivative leads, upon the spacetime integration, to an extra surface term that vanishes because of boundary conditions). Thus, the equations of motion following from the principle of stationary action (cf. (14.8)) should be left unchanged as well.

The next chapter will be devoted to the important issue of the conservation laws for classical fields and their relation to symmetries. A conservation law is, basically, a consequence of the equations of motion (an integral of motion). So, as a preliminary step, it may be useful to carry out here a straightforward derivation of the conservation law for the energy and momentum, in the simplest case of the real Klein–Gordon field.

We start with Eq. (14.11), i.e.

$$\partial_{\mu}\partial^{\mu}\varphi + m^2\varphi = 0\,.$$

Multiplying it by $\partial^{\nu}\varphi$, one has

$$\partial_{\mu}\partial^{\mu}\varphi\partial^{\nu}\varphi + m^2\varphi\partial^{\nu}\varphi = 0\,,$$

and this can be recast as

$$0 = \partial_\mu(\partial^\mu\varphi\partial^\nu\varphi) - \partial^\mu\varphi\partial_\mu\partial^\nu\varphi + \frac{1}{2}m^2\partial^\nu(\varphi^2)$$

$$= \partial_\mu(\partial^\mu\varphi\partial^\nu\varphi) - \frac{1}{2}\partial^\nu(\partial_\mu\varphi\partial^\mu\varphi) + \frac{1}{2}m^2\partial^\nu(\varphi^2) \, . \tag{14.36}$$

Taking into account (14.12), the last identity may be rewritten as

$$\partial_\mu(\partial^\mu\varphi\partial^\nu\varphi) - g^{\mu\nu}\partial_\mu\mathcal{L} = 0 \, . \tag{14.37}$$

Thus, we have arrived at the four-divergence identity ("continuity equation")

$$\partial_\mu\mathcal{T}^{\mu\nu} = 0 \, , \tag{14.38}$$

where

$$\mathcal{T}^{\mu\nu} = \partial^\mu\varphi\partial^\nu\varphi - g^{\mu\nu}\mathcal{L} \, . \tag{14.39}$$

By the way, an observant reader may have noticed that our derivation has been quite similar to what one does in classical mechanics when deriving the energy conservation for linear harmonic oscillator (LHO). Indeed, starting with the LHO equation of motion $\ddot{x} + \omega^2 x = 0$ and multiplying it by $\dot{x}$, one gets immediately

$$\frac{\mathrm{d}}{\mathrm{d}t}\left(\frac{1}{2}\dot{x}^2 + \frac{1}{2}\omega^2 x^2\right) = 0 \, ,$$

and that's it.

Now, with such an analogy in mind, one may wonder what is in fact the meaning of Eq. (14.38). It is not difficult to realize that the tensor $\mathcal{T}^{\mu\nu}$ is closely related, *per analogiam*, to an energy. Indeed, let us consider the component $\mathcal{T}^{00}$. According to (14.39), this is

$$\mathcal{T}^{00} = \partial_0\varphi\partial_0\varphi - \mathcal{L} \, . \tag{14.40}$$

However, using (14.13) one has

$$\partial_0\varphi = \frac{\partial\mathcal{L}}{\partial(\partial_0\varphi)} \, , \tag{14.41}$$

so that the relation (14.40) is analogous to the Legendre transformation

$$H = \dot{q}\frac{\partial L}{\partial\dot{q}} - L \tag{14.42}$$

that in classical mechanics leads from the Lagrange function to the Hamiltonian function (i.e. the energy). Thus, it is quite natural to interpret $\mathcal{T}^{00}$ as the energy density (for a given field configuration $\varphi(x)$). It remains to be shown that the energy, i.e. the integrated energy density, is indeed conserved, i.e. constant in time. To this end, let us set $\nu = 0$ in Eq. (14.38). One then has

$$\partial_0\mathcal{T}^{00} = -\partial_j\mathcal{T}^{j0} \, . \tag{14.43}$$

When Eq. (14.43) is integrated over the three-dimensional space and one employs the Gauss theorem to work out the right-hand side (i.e. transform it to a surface integral), this is seen to vanish (because of the boundary conditions at infinity); thus, we are left with

$$\partial_0 \int \mathcal{T}^{00}\,\mathrm{d}^3x = 0 \, , \tag{14.44}$$

and this is the desired result. Please notice also that using (14.40) along with (14.12) one gets

$$\mathscr{T}^{00} = \frac{1}{2}\partial_0\varphi\partial_0\varphi + \frac{1}{2}\partial^j\varphi\partial^j\varphi + \frac{1}{2}m^2\varphi^2 . \tag{14.45}$$

So, since $\varphi(x)$ is taken to be real, $\mathscr{T}^{00} \geq 0$.

Having identified the energy density, we may also call the components $\mathscr{T}^{0j}$, $j = 1, 2, 3$, the momentum density, since, repeating the preceding procedure, one gets first

$$\partial_0\mathscr{T}^{0j} = -\partial_k\mathscr{T}^{kj} , \tag{14.46}$$

and subsequently

$$\partial_0 \int \mathscr{T}^{0j}\,\mathrm{d}^3x = 0 . \tag{14.47}$$

Thus, we conclude that the four-component quantity

$$P^\mu = \int \mathscr{T}^{0\mu}\,\mathrm{d}^3x , \qquad \mu = 0, 1, 2, 3 , \tag{14.48}$$

may be consistently interpreted as the conserved four-momentum of the Klein–Gordon field. The tensor $\mathscr{T}^{\mu\nu}$ is called the **energy–momentum tensor**; as we have seen, its components represent the relevant densities.

# Chapter 15

# Conservation laws from symmetries

A direct derivation of the energy–momentum tensor from the equations of motion is not always so easy and straightforward as in the case of the Klein–Gordon field described in the preceding chapter. Fortunately, there is an elegant general method of obtaining conservation laws, which relies on the symmetry properties of the Lagrangian (or the action) of the considered field theory model. Such a method originates in the work of Emmy Noether published more than 100 years ago; so, what we are going to present in this chapter is a particular implementation of the famous Noether's theorem.

In general, the analysis of conservation laws à la Noether starts usually by considering the invariance of the action that defines the dynamics of a given physical system. Since our playground is the classical field theory, one may try to employ directly the Lagrangian density as the basic quantity, instead of the action. It turns out that such a simplified approach does work, if the notion of the Lagrangian symmetry is defined properly. So, let us consider a general Lagrangian density of the form (14.5). There are basically two possible types of the symmetries: either a transformation of spacetime coordinates is involved (which induces an appropriate field transformation), or the coordinates are left unchanged, and only the functional form of the field is transformed (this is the case of the so-called **internal symmetry**). Let us start with the former case, corresponding to symmetries of geometric origin. We consider the transformation of coordinates and fields

$$
\begin{aligned}
x_\mu &\longrightarrow x'_\mu \,, \\
\varphi_r(x) &\longrightarrow \varphi'_r(x') \,,
\end{aligned}
\tag{15.1}
$$

and assume that the form of the Lagrangian does not change under (15.1), i.e. that it holds

$$
\mathcal{L}\left(\varphi'_r(x'), \frac{\partial \varphi'_r(x')}{\partial x'^\mu}\right) = \mathcal{L}\left(\varphi_r(x), \frac{\partial \varphi_r(x)}{\partial x^\mu}\right).
\tag{15.2}
$$

For obvious reasons, we will call such a relation the **form-invariance** of the Lagrangian.[10] It is not difficult to realize that the Lagrangians discussed in the preceding chapter satisfy the condition (15.2) when the coordinate transformation in (15.1) is a spacetime translation or Lorentz transformation; more about this later. For the purpose of further discussion, let us

---

[10]Note that such a term is not quite common in current literature, but occasionally it is used explicitly, see e.g. the book [12].

introduce the following notation:

$$\mathcal{L}'(x) = \mathcal{L}\left(\varphi'_r(x), \frac{\partial \varphi'_r(x)}{\partial x^\mu}\right),$$

$$\delta \mathcal{L}(x) = \mathcal{L}'(x) - \mathcal{L}(x), \tag{15.3}$$

$$\delta \varphi_r(x) = \varphi'_r(x) - \varphi_r(x),$$

$$\delta x_\mu = x'_\mu - x_\mu.$$

Note that the term "form-invariance" has been introduced so as to distinguish the relation (15.2) from a plain invariance that would mean, in the shorthand notation (15.3),

$$\mathcal{L}'(x) = \mathcal{L}(x), \tag{15.4}$$

i.e.

$$\delta \mathcal{L}(x) = 0. \tag{15.5}$$

Obviously, such a relation would be relevant in the case of an internal symmetry (to be discussed later on).

To proceed, we will consider continuous transformations of the type (15.1) (i.e. those depending on some continuous parameters), and these may be restricted to the infinitesimal form. Utilizing the notation (15.3), the form-invariance condition (15.2) may be written as

$$\mathcal{L}'(x') = \mathcal{L}(x), \tag{15.6}$$

i.e.

$$\mathcal{L}'(x') = \mathcal{L}(x' - \delta x). \tag{15.7}$$

Neglecting terms of higher order in $\delta x$, one then gets

$$\mathcal{L}'(x) = \mathcal{L}(x - \delta x) \tag{15.8}$$

(the reader is encouraged to check independently the last statement). The relation (15.8) may be further recast in the form

$$\mathcal{L}'(x) - \mathcal{L}(x) = \mathcal{L}(x - \delta x) - \mathcal{L}(x) \tag{15.9}$$

that is worth noting: while its left-hand side represents the variation of $\mathcal{L}(x)$ due to the variation $\varphi_r(x) \to \varphi'_r(x)$, the right-hand side is the change of the value of $\mathcal{L}(x)$ due to the (infinitesimal) coordinate transformation. Thus, the identity (15.9) may be expressed as

$$\frac{\partial \mathcal{L}}{\partial \varphi_r} \delta \varphi_r + \frac{\partial \mathcal{L}}{\partial (\partial_\mu \varphi_r)} \delta(\partial_\mu \varphi_r) = -\delta x^\mu \frac{\partial \mathcal{L}}{\partial x^\mu} \tag{15.10}$$

(where one is summing over the index $r$ on the left-hand side).

As we know, one has to invoke the equations of motion in order to obtain a conservation law. So, one employs the Euler–Lagrange equations (14.10) to express the derivative $\partial \mathcal{L}/\partial \varphi_r$, and one may also use the obvious identity $\delta(\partial_\mu \varphi_r) = \partial_\mu(\delta \varphi_r)$. One thus gets

$$\partial_\mu \left[\frac{\partial \mathcal{L}}{\partial (\partial_\mu \varphi_r)}\right] \delta \varphi_r + \frac{\partial \mathcal{L}}{\partial (\partial_\mu \varphi_r)} \partial_\mu(\delta \varphi_r) = -\delta x^\mu \frac{\partial \mathcal{L}}{\partial x^\mu},$$

and this becomes, finally,

$$\frac{\partial}{\partial x^\mu}\left(\frac{\partial \mathcal{L}}{\partial (\partial_\mu \varphi_r)} \delta \varphi_r\right) + \delta x^\mu \frac{\partial \mathcal{L}}{\partial x^\mu} = 0. \tag{15.11}$$

The general relation (15.11) may be called the **Noether's identity** and one expects that it will yield some relevant conservation laws if the appropriate transformations (15.1) are considered.

Let us start with spacetime translations. The transformation of coordinates is simply

$$x'^\mu = x^\mu + a^\mu , \tag{15.12}$$

where $a^\mu$, $\mu = 0, 1, 2, 3$, are some constants. It is easy to see that the Lagrangians we have discussed in the preceding chapter are form-invariant if the transformed fields are defined by

$$\varphi'_r(x') = \varphi_r(x) . \tag{15.13}$$

For infinitesimal translations, we write

$$x'^\mu = x^\mu + \varepsilon^\mu ,$$

i.e. $\delta x^\mu = \varepsilon^\mu$, and

$$\varphi'_r(x) = \varphi_r(x - \varepsilon) = \varphi_r(x) - \varepsilon^\mu \frac{\partial \varphi_r}{\partial x^\mu} .$$

Thus, one has

$$\delta \varphi_r(x) = -\varepsilon^\mu \frac{\partial \varphi_r}{\partial x^\mu} . \tag{15.14}$$

Substituting these expressions into the relation (15.11), one gets first

$$\frac{\partial}{\partial x^\mu} \left( -\frac{\partial \mathscr{L}}{\partial(\partial_\mu \varphi_r)} \frac{\partial \varphi_r}{\partial x_\nu} \varepsilon_\nu \right) + \varepsilon_\nu \frac{\partial \mathscr{L}}{\partial x_\nu} = 0 ,$$

and this can be recast as

$$\partial_\mu \mathscr{T}^{\mu\nu} \varepsilon_\nu = 0 , \tag{15.15}$$

where

$$\mathscr{T}^{\mu\nu} = \frac{\partial \mathscr{L}}{\partial(\partial_\mu \varphi_r)} \partial^\nu \varphi_r - g^{\mu\nu} \mathscr{L} . \tag{15.16}$$

The identity (15.15) holds for an arbitrary infinitesimal $\varepsilon_\nu$, so one may conclude that

$$\partial_\mu \mathscr{T}^{\mu\nu} = 0 , \qquad \nu = 0, 1, 2, 3 . \tag{15.17}$$

The quantity $\mathscr{T}^{\mu\nu}$ is called the **canonical energy–momentum tensor**. It is reassuring that using the general formula (15.16) for the real Klein–Gordon field one recovers the expression (14.39) we have found before by means of a direct manipulation with the equation of motion. One may notice that (15.16) looks like a covariantized form of the Legendre transformation $L \to H$ in classical mechanics, which we have recalled in the preceding chapter (cf. (14.42)). This observation may serve as a helpful mnemonics for remembering the important result (15.16).

We already know that the four-divergence equations (15.17) for $\nu = 0, 1, 2, 3$ lead to four constants of motion, which can be identified with the energy and momentum. Let us emphasize the nice feature of the above derivation, namely that we have thereby confirmed an expected connection between the four-momentum conservation and the invariance under spacetime translations (in a more philosophical parlance, the energy and momentum conservation is a consequence of the homogeneity of the flat four-dimensional spacetime).

The canonical form (15.16) is certainly suitable for computing the energy and momentum (defined as the corresponding space integrals of densities), but the tensor $\mathscr{T}^{\mu\nu}$ itself does not always possess the properties we would like to have. In particular, for some applications it is important to have a symmetric tensor (recall that it stands e.g. on the right-hand side of the Einstein's gravitational equations), but, as it turns out, the formula (15.6) does not always guarantee such a property. A prominent example is the Maxwell field, whose Lagrangian is given by (14.18) with $m = 0$. In general, the way out of such a problem is to realize that $\mathscr{T}^{\mu\nu}$ is not determined uniquely by Eq. (15.17); one may always add to the canonical expression (15.16) an appropriate term $X^{\mu\nu}$ such that $\partial_\mu X^{\mu\nu} = 0$.

Let us now discuss this point for the aforementioned case of the Maxwell field. One has

$$\mathscr{L} = -\frac{1}{4} F_{\mu\nu} F^{\mu\nu} , \qquad (15.18)$$

with $F_{\mu\nu} = \partial_\mu A_\nu - \partial_\nu A_\mu$. The canonical energy–momentum tensor is thus given by

$$\mathscr{T}^{\mu\nu}_{\text{can.}} = \frac{\partial \mathscr{L}}{\partial(\partial_\mu A_\rho)} \partial^\nu A_\rho - g^{\mu\nu} \mathscr{L} \qquad (15.19)$$

(so that $\rho$ now plays the role of the index $r$ in (15.16)). According to (14.24) it holds

$$\frac{\partial \mathscr{L}}{\partial(\partial_\mu A_\rho)} = -F^{\mu\rho} . \qquad (15.20)$$

Substituting this into (15.19) and using (15.18), one gets

$$\mathscr{T}^{\mu\nu}_{\text{can.}} = -F^{\mu\rho} \partial^\nu A_\rho + \frac{1}{4} g^{\mu\nu} F_{\alpha\beta} F^{\alpha\beta} . \qquad (15.21)$$

It is clear that the expression (15.21) has two flaws: it is not gauge invariant (since its first term depends explicitly on the potential $A_\rho$) and it is not symmetric under $\mu \leftrightarrow \nu$ (also because of the first term). To get some hint how to proceed further, let us examine the variation of (15.21) under the gauge transformation

$$A_\rho \longrightarrow A_\rho + \partial_\rho f , \qquad (15.22)$$

where $f$ is an arbitrary differentiable function. Carrying out the change (15.22) in (15.21) one gets

$$\begin{aligned} \mathscr{T}^{\mu\nu\prime}_{\text{can.}} &= -F^{\mu\rho} (\partial^\nu A_\rho + \partial^\nu \partial_\rho f) + \frac{1}{4} g^{\mu\nu} F_{\alpha\beta} F^{\alpha\beta} \\ &= \mathscr{T}^{\mu\nu}_{\text{can.}} - F^{\mu\rho} \partial^\nu \partial_\rho f \\ &= \mathscr{T}^{\mu\nu}_{\text{can.}} - \left[ \partial_\rho (F^{\mu\rho} \partial^\nu f) - \partial_\rho F^{\mu\rho} \partial^\nu f \right] . \end{aligned} \qquad (15.23)$$

In the last term one can use the equation of motion, i.e. $\partial_\rho F^{\mu\rho} = 0$, and we are thus left with

$$\mathscr{T}^{\mu\nu\prime}_{\text{can.}} = \mathscr{T}^{\mu\nu}_{\text{can.}} - \partial_\rho (F^{\mu\rho} \partial^\nu f) . \qquad (15.24)$$

The desired hint is now clear: one may add to the expression (15.21) the term

$$\Delta^{\mu\nu} = \partial_\rho (F^{\mu\rho} A^\nu) \qquad (15.25)$$

that is designed to compensate the gauge dependence of the original $\mathscr{T}^{\mu\nu}_{\text{can.}}$. At the same time, it holds

$$\partial_\mu \Delta^{\mu\nu} = 0$$

(because of the antisymmetry of $F^{\mu\rho}$), so that the divergence equation (15.17) remains valid for the modified energy–momentum tensor. So, instead of (15.21) one may consider

$$
\begin{aligned}
\mathscr{T}^{\mu\nu} &= \mathscr{T}^{\mu\nu}_{\text{can.}} + \Delta^{\mu\nu} \\
&= -F^{\mu\rho}\partial^\nu A_\rho + \frac{1}{4}g^{\mu\nu}F_{\alpha\beta}F^{\alpha\beta} + \partial_\rho(F^{\mu\rho}A^\nu) \\
&= -F^{\mu\rho}\partial^\nu A_\rho + F^{\mu\rho}\partial_\rho A^\nu + \frac{1}{4}g^{\mu\nu}F_{\alpha\beta}F^{\alpha\beta} \\
&= F^\mu{}_\rho F^{\rho\nu} + \frac{1}{4}g^{\mu\nu}F_{\alpha\beta}F^{\alpha\beta} ,
\end{aligned}
$$

(15.26)

where we have used the equation of motion $\partial_\rho F^{\mu\rho} = 0$. Now the gauge invariance of the new tensor is manifest and there is even an additional bonus: the expression (15.26) is also symmetric under the interchange $\mu \leftrightarrow \nu$! Indeed, one has

$$
F^\mu{}_\rho F^{\rho\nu} \xrightarrow{\mu\leftrightarrow\nu} F^\nu{}_\rho F^{\rho\mu} = -F^\nu{}_\rho F^{\mu\rho} = -F^{\nu\rho}F^\mu{}_\rho = F^{\rho\nu}F^\mu{}_\rho
$$

(of course, the second term in (15.26) is manifestly symmetric). Thus, one may conclude that our achievement consists in identifying a symmetric, gauge invariant energy–momentum tensor for the Maxwell field, which reads

$$
\mathscr{T}^{\mu\nu}_{\text{sym.}} = F^\mu{}_\rho F^{\rho\nu} + \frac{1}{4}g^{\mu\nu}F_{\alpha\beta}F^{\alpha\beta} .
$$

(15.27)

It is also possible to check that the energy density corresponding to (15.27) has the familiar value

$$
\mathscr{T}^{00}_{\text{sym.}} = \frac{1}{2}(\vec{E}^2 + \vec{B}^2) ,
$$

(15.28)

where $\vec{E}$ and $\vec{B}$ denote the electric and magnetic field strength, respectively. To this end, one can use the relations $F^{j0} = E^j$ and $F^{jk} = -\varepsilon^{jkl}B^l$.

As a last example, let us consider the case of the Dirac field. According to (15.16) one has, in general

$$
\mathscr{T}^{\mu\nu} = \frac{\partial \mathscr{L}}{\partial(\partial_\mu \psi)}\partial^\nu \psi + \partial^\nu \bar{\psi}\frac{\partial \mathscr{L}}{\partial(\partial_\mu \bar{\psi})} - g^{\mu\nu}\mathscr{L} .
$$

(15.29)

For the computation of the energy and momentum it is convenient to employ the simple form (14.27) for the Lagrangian, since it vanishes for the solutions of the Dirac equation, and, moreover, it does not contain derivatives of $\bar{\psi}$. Thus, the expression (15.29) is then reduced to

$$
\mathscr{T}^{\mu\nu} = \frac{\partial \mathscr{L}}{\partial(\partial_\mu \psi)}\partial^\nu \psi = i\bar{\psi}\gamma^\mu \partial^\nu \psi .
$$

(15.30)

We will utilize this simple formula later on, when discussing the quantization of the Dirac field.

Next, we are going to examine conservation laws associated with the Lorentz symmetry. It is not difficult to realize that the Lagrangians we have considered up to now possess the form-invariance under proper Lorentz transformations $x' = \Lambda x$, where $\Lambda$ is, in general, a six-parametric matrix encompassing spatial rotations and boosts. Indeed, for the Klein–Gordon field (which is a Lorentz scalar) one may use the trivial transformation $\varphi'(x') = \varphi(x)$. For the Dirac field (which is a bispinor), an appropriate choice is $\psi'(x') = S(\Lambda)\psi(x)$ with $S(\Lambda)$ being the $4 \times 4$ matrix that we have uncovered in Chapter 4 (cf. Eq. (4.9)). Similarly, for the Proca and Maxwell field one may assume that $A_\mu(x)$ is transformed as a four-vector (i.e. through the

99

matrix $\Lambda$ itself). These remarks justify the usual statement that the Lorentz form-invariance of the Lagrangians in question is manifest.

So, let us consider an infinitesimal Lorentz transformation. As we know, it can be written as

$$x'_\mu = x_\mu + \delta\omega_{\mu\nu}x^\nu, \tag{15.31}$$

where $\delta\omega_{\mu\nu} = -\delta\omega_{\nu\mu}$ are the six independent parameters for rotations and boosts. The field transformation is, in general, given by

$$\varphi'_r(x') = \left(\mathbb{1} - \frac{i}{4}\delta\omega_{\alpha\beta}\Sigma^{\alpha\beta}\right)_{rs}\varphi_s(x), \tag{15.32}$$

where the generators $\Sigma^{\alpha\beta}$ stand for the corresponding representation of the Lorentz algebra. For instance,

$$\Sigma^{\alpha\beta} = \sigma^{\alpha\beta} = \frac{i}{2}[\gamma^\alpha, \gamma^\beta] \tag{15.33}$$

for the Dirac bispinor field. Referring to the notation introduced in (15.3), the relation (15.31) means that

$$\delta x_\mu = \delta\omega_{\mu\nu}x^\nu, \tag{15.34}$$

and for the field variation $\delta\varphi_r$ one gets

$$\begin{aligned}
\delta\varphi_r(x) &= \varphi'_r(x) - \varphi_r(x) = \varphi'_r(x') - \varphi_r(x) + \varphi'_r(x) - \varphi'_r(x') \\
&= -\frac{i}{4}\delta\omega_{\alpha\beta}\left(\Sigma^{\alpha\beta}\right)_{rs}\varphi_s(x) - \frac{\partial\varphi'_r}{\partial x_\alpha}\delta x_\alpha \\
&= -\frac{i}{4}\delta\omega_{\alpha\beta}\left(\Sigma^{\alpha\beta}\right)_{rs}\varphi_s(x) - \frac{\partial}{\partial x_\alpha}(\varphi_r + \delta\varphi_r)\delta x_\alpha.
\end{aligned}$$

Thus, neglecting the higher-order term involving $\delta\varphi_r\delta x_\alpha$, one is left with

$$\delta\varphi_r = -\frac{i}{4}\delta\omega_{\alpha\beta}\left(\Sigma^{\alpha\beta}\right)_{rs}\varphi_s(x) - \frac{\partial\varphi_r}{\partial x_\alpha}\delta\omega_{\alpha\beta}x^\beta. \tag{15.35}$$

For later convenience, the last relation can be rewritten as

$$\delta\varphi_r(x) = -\frac{i}{2}\delta\omega_{\alpha\beta}\left[\frac{1}{2}\left(\Sigma^{\alpha\beta}\right)_{rs}\varphi_s(x) + i(x^\alpha\partial^\beta - x^\beta\partial^\alpha)\varphi_r(x)\right], \tag{15.36}$$

where we have utilized the antisymmetry of the parameters $\delta\omega_{\alpha\beta}$.

Now, the Noether's relation (15.11) reads

$$\frac{\partial}{\partial x^\mu}\left(\frac{\partial\mathcal{L}}{\partial(\partial_\mu\varphi_r)}\delta\varphi_r\right) + \delta\omega_{\alpha\beta}x^\beta\frac{\partial\mathcal{L}}{\partial x_\alpha} = 0. \tag{15.37}$$

In order to turn this into a "continuity equation" analogous to (15.17), one has to work out the second term:

$$\delta\omega_{\alpha\beta}x^\beta\frac{\partial\mathcal{L}}{\partial x_\alpha} = \delta\omega_{\alpha\beta}\left[\frac{\partial}{\partial x_\alpha}(x^\beta\mathcal{L}) - g^{\alpha\beta}\mathcal{L}\right] = \delta\omega_{\alpha\beta}\frac{\partial}{\partial x_\alpha}(x^\beta\mathcal{L}). \tag{15.38}$$

Note that for arriving at (15.38) we have taken into account that $\delta\omega_{\alpha\beta}g^{\alpha\beta} = 0$ because of the antisymmetry of $\delta\omega_{\alpha\beta}$. After some simple manipulations, the expression (15.38) can be finally recast as

$$\delta\omega_{\alpha\beta}x^\beta\frac{\partial\mathcal{L}}{\partial x_\alpha} = \frac{1}{2}\delta\omega_{\alpha\beta}\frac{\partial}{\partial x^\mu}\left[(g^{\alpha\mu}x^\beta - g^{\beta\mu}x^\alpha)\mathcal{L}\right]. \tag{15.39}$$

So, using (15.36) and (15.39), the relation (15.37) yields

$$\frac{\partial}{\partial x^\mu} \cdot \mathcal{M}^{\mu\alpha\beta} = 0 \,,$$ (15.40)

where

$$\mathcal{M}^{\mu\alpha\beta} = \frac{\partial \mathcal{L}}{\partial(\partial_\mu \varphi_r)} \left[ -\frac{i}{2} \Sigma_{rs}^{\alpha\beta} \varphi_s + (x^\alpha \partial^\beta - x^\beta \partial^\alpha) \varphi_r \right] + (g^{\alpha\mu} x^\beta - g^{\beta\mu} x^\alpha) \mathcal{L} \,.$$ (15.41)

Eq. (15.40) is the desired "continuity equation", or, if you want, "current conservation" (in fact, a set of six such equations) corresponding to the Lorentz symmetry. It is interesting that the formula (15.41) can also be recast in terms of the canonical energy–momentum tensor (15.16), namely

$$\mathcal{M}^{\mu\alpha\beta} = x^\alpha \mathcal{T}_{\text{can.}}^{\mu\beta} - x^\beta \mathcal{T}_{\text{can.}}^{\mu\alpha} - \frac{i}{2} \frac{\partial \mathcal{L}}{\partial(\partial_\mu \varphi_r)} \Sigma_{rs}^{\alpha\beta} \varphi_s$$ (15.42)

(checking (15.42) is an easy task).

Let us now consider some particular examples. In the simplest case of the scalar Klein–Gordon field one has $\Sigma^{\alpha\beta} = 0$, so that

$$\mathcal{M}_{\text{scalar}}^{\mu\alpha\beta} = x^\alpha \mathcal{T}_{\text{can.}}^{\mu\beta} - x^\beta \mathcal{T}_{\text{can.}}^{\mu\alpha} \,.$$ (15.43)

The identity (15.40) then yields

$$\begin{aligned}
0 &= \partial_\mu \left( x^\alpha \mathcal{T}_{\text{can.}}^{\mu\beta} \right) - \partial_\mu \left( x^\beta \mathcal{T}_{\text{can.}}^{\mu\alpha} \right) \\
&= g^\alpha{}_\mu \mathcal{T}_{\text{can.}}^{\mu\beta} + x^\alpha \partial_\mu \mathcal{T}_{\text{can.}}^{\mu\beta} - g^\beta{}_\mu \mathcal{T}_{\text{can.}}^{\mu\alpha} - x^\beta \partial_\mu \mathcal{T}_{\text{can.}}^{\mu\alpha} \\
&= \mathcal{T}_{\text{can.}}^{\alpha\beta} - \mathcal{T}_{\text{can.}}^{\beta\alpha} \,.
\end{aligned}$$ (15.44)

Thus we see that there is a simple general consequence of the equations (15.40) and (15.42): for a scalar field the energy–momentum tensor is always symmetric. An elementary illustration of this fact is provided by our formula (14.39) for the free field, but the result (15.44) is quite general (i.e. it is valid also for interacting scalar fields described by Lagrangians involving terms beyond the quadratic form (14.12)).

As an example of a situation involving non-trivial $\Sigma^{\alpha\beta}$ in Eq. (15.42), it is quite instructive to consider the Dirac field. Before working out this particular example, a general remark is in order. One might guess that the most interesting components of the quantity $\mathcal{M}^{\mu\alpha\beta}$ are those with $(\alpha\beta) = (jk)$, $j, k = 1, 2, 3$. Why? The reason is that such a combination of indices $\alpha$, $\beta$ corresponds to spatial rotations, and one knows from other theories (classical and quantum mechanics) that the rotational invariance is intimately connected with the conservation of angular momentum. So, one should be ready to identify the corresponding quantities

$$M^{jk} = \int \mathrm{d}^3 x \, \mathcal{M}^{0jk} \,,$$ (15.45)

with components of the field angular momentum (note that owing to the antisymmetry in the indices $j$, $k$ there are just three independent components $M^{jk}$). Now, let us turn to the case of the Dirac field. According to (15.42) and taking into account (15.33), one has

$$\mathcal{M}^{0jk} = x^j \mathcal{T}^{0k} - x^k \mathcal{T}^{0j} - \frac{i}{2} \cdot i \bar{\psi} \gamma_0 \frac{i}{2} [\gamma^j, \gamma^k] \psi \,.$$ (15.46)

101

Components of the energy–momentum tensor are given by (15.30), so that e.g.

$$\mathcal{T}^{0j} = i\bar{\psi}\gamma_0\partial^j\psi = -i\psi^\dagger\frac{\partial}{\partial x^j}\psi .$$ (15.47)

One thus gets, finally,

$$\mathcal{M}^{0jk} = \psi^\dagger(x^j p^k - x^k p^j + \frac{1}{2}\sigma^{jk})\psi ,$$ (15.48)

where we have used a suggestive notation

$$p^n = -i\frac{\partial}{\partial x^n} , \qquad n = j, k .$$ (15.49)

Furthermore, it holds $\sigma^{jk} = \varepsilon^{jkl}\Sigma^l$, where $\Sigma^l$ (or, more precisely, $\frac{1}{2}\Sigma^l$), $l = 1, 2, 3$, are the spin matrices that we know from the chapters on the relativistic quantum mechanics. So one has, for instance,

$$\mathcal{M}^{012} = \psi^\dagger(x^1 p^2 - x^2 p^1 + \frac{1}{2}\Sigma^3)\psi ,$$ (15.50)

and this offers another suggestive notation, namely $x^1 p^2 - x^2 p^1 = L^3$. In general, one may introduce the vector $\vec{L}$ defined by

$$x^j p^k - x^k p^j = \varepsilon^{jkl}L^l ,$$ (15.51)

and in this way one is led to the vector dual of the antisymmetric tensor (15.48) (tensor under spatial rotations), i.e. $\mathcal{M}^{0jk} \to \vec{\mathcal{M}}$, where

$$\vec{\mathcal{M}} = \psi^\dagger\left(\vec{L} + \frac{1}{2}\vec{\Sigma}\right)\psi .$$ (15.52)

The conserved angular momentum for the Dirac field is then

$$\vec{M} = \int \vec{\mathcal{M}}\, d^3x .$$ (15.53)

We should also recall that the conserved field momentum is, according to (15.47),

$$\vec{P} = \int d^3x\, \psi^\dagger\left(-i\vec{\nabla}\right)\psi .$$ (15.54)

　　　An astute reader might now say that our results (15.52) through (15.54) are a sort of "déjà vu", in the sense that they involve quantum mechanical operators of the momentum and angular momentum. In fact, it is not surprising; within the relativistic quantum mechanics, the expressions (15.53), (15.54) would represent the relevant expectation values and these are certainly constant in time. As we have stressed repeatedly, the classical field considered here is a completely different system, but it satisfies the same Dirac equation, and this is what really matters.

　　　As a last item of our program we are going to consider the case of conservation laws associated with internal symmetries. This, in a sense, is the simplest situation, since the coordinate transformations are not involved here. As we have noted earlier, one may then say that the Lagrangian is truly invariant under the field transformation in question (cf. Eq. (15.4)). So, suppose that the Lagrangian is invariant under a continuous transformation $\varphi_r(x) \to \varphi'_r(x)$;

its infinitesimal form is given by a variation $\delta\varphi_r(x)$ and, at the same time, one sets $\delta x = 0$. The Noether's identity (15.11) thus reads

$$\partial_\mu \left[ \frac{\partial \mathscr{L}}{\partial(\partial_\mu \varphi_r)} \delta\varphi_r \right] = 0 \,. \tag{15.55}$$

It is clear that such an identity yields immediately the desired conserved currents (or "continuity equations"); it is sufficient to disentangle the dependence of the variations $\delta\varphi_r$ on the transformation parameters. Let us show, on some instructive examples, how it works.

Our first example will refer to the complex Klein–Gordon field. Its Lagrangian

$$\mathscr{L} = (\partial_\mu \varphi)(\partial^\mu \varphi^*) - m^2 \varphi\varphi^*$$

is obviously invariant under phase transformations

$$\begin{aligned} \varphi'(x) &= e^{i\alpha} \varphi(x) \,, \\ \varphi^{*\prime}(x) &= e^{-i\alpha} \varphi^*(x) \,, \end{aligned} \tag{15.56}$$

where $\alpha$ is a real constant parameter. Infinitesimal variations $\delta\varphi$, $\delta\varphi^*$ are then

$$\delta\varphi(x) = i\delta\alpha\varphi(x) \,, \qquad \delta\varphi^*(x) = -i\delta\alpha\varphi^*(x) \,. \tag{15.57}$$

Substituting this into the identity (15.55) one gets

$$0 = \partial_\mu \left[ \frac{\partial \mathscr{L}}{\partial(\partial_\mu \varphi)} \delta\varphi + \frac{\partial \mathscr{L}}{\partial(\partial_\mu \varphi^*)} \delta\varphi^* \right] = \partial_\mu \left[ (\partial^\mu \varphi^*) i\delta\alpha\varphi - (\partial^\mu \varphi) i\delta\alpha\varphi^* \right] \,.$$

It means that the conserved current satisfying $\partial_\mu J^\mu = 0$ has the form

$$J^\mu = i \left[ \varphi(\partial^\mu \varphi^*) - (\partial^\mu \varphi)\varphi^* \right] \tag{15.58}$$

(an attentive reader may recall that we have in fact arrived at such a result in Chapter 1 by means of direct manipulations with the Klein–Gordon equation).

Next, one may repeat an analogous exercise with the Dirac field. In this case, the infinitesimal phase transformations are written as

$$\begin{aligned} \delta\psi(x) &= i\delta\alpha\psi(x) \,, \\ \delta\bar\psi(x) &= -i\delta\alpha\bar\psi(x) \,. \end{aligned} \tag{15.59}$$

The Noether's identity (15.11) then leads immediately to the current conservation equation $\partial_\mu J^\mu = 0$ with

$$J^\mu(x) = \bar\psi(x)\gamma^\mu\psi(x) \tag{15.60}$$

(again, we have thus recovered the expression found earlier in Chapter 4).

There is another interesting point concerning the Dirac field that is worth mentioning here. It turns out that for $m = 0$ its Lagrangian possesses an extra internal symmetry, in addition to the phase invariance discussed above. The corresponding transformations read

$$\begin{aligned} \psi'(x) &= e^{i\alpha\gamma_5}\psi(x) \,, \\ \bar\psi'(x) &= \bar\psi(x) e^{i\alpha\gamma_5} \end{aligned} \tag{15.61}$$

(where $\alpha$ is a real constant parameter), and usually they are called "chiral", because the matrix $\gamma_5$ is involved here. The reader is urged to check that the transformations for $\psi$ and $\bar\psi$ are mutually compatible (it is easy, one has just to utilize the anticommutativity property of $\gamma_5$). The infinitesimal form of (15.61) is immediately clear, and from the Noether's identity (15.11) one thus gets readily the current conservation

$$\partial^\mu J_{\mu 5} = 0 , \qquad (15.62)$$

with

$$J_{\mu 5} = \bar{\psi} \gamma_\mu \gamma_5 \psi . \qquad (15.63)$$

Recall that whereas the current (15.60) is a true Lorentz four-vector, the expression (15.63) is an axial vector (cf. (5.15)). One might wonder what happens with such an axial vector current for $m \neq 0$. The answer is straightforward: employing the Dirac equation, one gets

$$\partial^\mu J_{\mu 5} = 2im\bar{\psi}\gamma_5\psi . \qquad (15.64)$$

So, in the limit $m \to 0$ Eq. (15.62) is recovered. Discovering the extra conservation law for the massless Dirac field, embodied in (15.62), has been an elementary exercise at this level. In fact, the concept of the chiral symmetry had been immensely important in the development of particle physics in 20th century (but a more detailed treatment of this topic would go far beyond the scope of these lecture notes).

Let us now summarize briefly the main results we have achieved here. First of all, we have found ten conservation laws, embodied in the identities (15.17) and (15.40). Four of the "conserved currents", which coincide with the canonical energy–momentum tensor, correspond to spacetime translations, and the remaining six are related to the Lorentz symmetry. All together they reflect the notorious **Poincaré invariance**, which is supposed to be a universal property of the laws of relativistic physics. Apart from this, for some models there are also "accidental" internal symmetries that lead to additional conservation laws. In principle, the conserved quantities may be derived directly from the field equations of motion; however, the procedure à la Emmy Noether is certainly much more efficient. So, the general identity (15.11) is indeed an invaluable tool in this context.

There is another point that should be stressed. Most of the classical fields we have considered here are rather abstract entities that are not realized as physical objects in the nature; an obvious exception is the Maxwell field (and, of course, the gravitational field, which, however, is beyond our scope in the present context). In particular, one can certainly never encounter a classical Dirac field in our Universe. Nevertheless, this is by no means a drawback of our approach. The real meaning of our effort is that we have developed a basic formalism for the description of relativistic field, and these may now be quantized. Historically, the first attempt in such a direction concerned the electromagnetic Maxwell field, and it resulted in a successful description of the light quanta — photons. The procedure of the field quantization was subsequently extended to encompass other fields and the corresponding particles, and it marked the birth of the methods of the quantum field theory that we are using today. Precisely this, i.e. the theory of quantized fields and their relation to relativistic particles, constitutes the contents of the forthcoming chapters.

# Chapter 16

# Canonical quantization of real scalar field

We are going to start our tour of the field quantization schemes with the simplest possible case, which is, of course, a real Klein–Gordon field. By the way, maybe it is not entirely impossible that you come across such a classical field when wandering around the universe, since it does appear in some cosmological models (though, admittedly, it is still a highly speculative matter). Anyway, it is not our concern here. Our goal is to find out what is the physical picture that emerges upon quantizing such a field.

So, let us take up this task. The real scalar field we have in mind is described, classically, by means of a single function $\varphi(x) = \varphi(\vec{x}, t)$ that may be understood as a "generalized coordinate" within the framework of the Lagrangian formalism developed in chapters 14 and 15. Using the familiar Lagrangian density

$$\mathcal{L} = \frac{1}{2}\partial_\mu\varphi\partial^\mu\varphi - \frac{1}{2}m^2\varphi^2 \,, \tag{16.1}$$

one may define, in analogy with the classical particle mechanics, the "conjugate momentum"

$$\pi(x) = \frac{\partial\mathcal{L}}{\partial(\partial_0\varphi)} = \partial_0\varphi(x) \,. \tag{16.2}$$

Now, with the "canonical variables" (coordinate and momentum) at hand, one may consider $\varphi(x)$ and $\pi(x)$ as operators (in an as yet unspecified Hilbert space), and try to imitate the commutation relations of the ordinary quantum mechanics. Before doing this, one should realize that $\varphi(x) = \varphi(\vec{x}, t)$ and $\pi(x) = \pi(\vec{x}, t)$ are to be understood as operators in the Heisenberg picture, due to their time dependence. Canonical quantum mechanical commutation relations read (in the Schrödinger picture)

$$[x_j, p_k] = i\delta_{jk} \,,$$
$$[x_j, x_k] = 0 \,, \tag{16.3}$$
$$[p_j, p_k] = 0 \,.$$

It is easy to see that in the Heisenberg picture, where

$$x_j(t) = e^{iHt} x_j e^{-iHt} \,, \qquad p_k(t) = e^{iHt} p_k e^{-iHt} \,,$$

the commutators (16.3) remain the same. So, the equations (16.3) are equivalent to the corresponding set of **equal-time commutation relations** (ETCR). Inspired by these observations, one may postulate the following "canonical" commutation relations for the field variables $\varphi(x)$ and $\pi(x)$:

$$[\varphi(\vec{x}, t), \pi(\vec{y}, t)] = i\delta^{(3)}(\vec{x} - \vec{y}) \,,$$
$$[\varphi(\vec{x}, t), \varphi(\vec{y}, t)] = 0 \,, \tag{16.4}$$
$$[\pi(\vec{x}, t), \pi(\vec{y}, t)] = 0 \,.$$

Note that we have used here $\delta^{(3)}(\vec{x} - \vec{y})$ as a counterpart of the Kronecker delta appearing in (16.3). This seems to be quite natural replacement, since, as we have noted in Chapter 14, in field theory the spatial coordinates $\vec{x}$ play the role of the discrete indices used for labelling dynamical variables in the particle mechanics. One can also check that mass dimensions come out right in the commutator of $\varphi$ and $\pi$. It may not be obvious at first sight, so let us explain it in detail. Within our system of units with $\hbar = c = 1$ the dimension of the Lagrangian density $\mathscr{L}$ is $M^4$ (where $M$ is an arbitrary mass), simply because the action (14.7) is dimensionless (due to $\hbar = 1$) and $\mathrm{d}^4x$ has the dimension (length)$^4$, which is $M^{-4}$. Thus, the dimension of the field $\varphi$ in (16.1) is $M$, and, subsequently, $\pi = \partial_0\varphi$ must have dimension $M^2$ (why?). It means that the commutator of $\varphi$ and $\pi$ has dimension $M^3$. On the other hand, the delta function $\delta^{(3)}(\vec{x} - \vec{y})$ has dimension (length)$^{-3}$ (why?), which is $M^3$, and this completes the argument.

A general solution of the Klein–Gordon equation for the classical field $\varphi(x)$ can be written as

$$\varphi(x) = \int \mathrm{d}^3k\, N_k \left[a(k)\, e^{-ikx} + a^*(k)\, e^{ikx}\right], \tag{16.5}$$

where $k = (k_0, \vec{k})$ with $k_0 = (\vec{k}^2 + m^2)^{1/2}$, $N_k = (2k_0)^{-1/2}(2\pi)^{-3/2}$, and $a(k)$ (in fact $a(\vec{k})$) are some arbitrary expansion coefficients. The normalization factor $N_k$ is conventional and has been chosen so for later convenience. Note also that in (16.5) we have taken into account that $\varphi(x)$ is supposed to be real. Upon quantization, a classical real $\varphi(x)$ becomes a Hermitian operator written as

$$\varphi(x) = \int \mathrm{d}^3k\, N_k \left[a(k)\, e^{-ikx} + a^\dagger(k)\, e^{ikx}\right], \tag{16.6}$$

where the expansion coefficients $a(k)$, $a^\dagger(k)$ are operators (of course, $a^\dagger(k)$ denotes the Hermitian conjugation of $a(k)$). Having postulated the canonical commutation relations (16.4), we may now find out what are the corresponding formulae for $a(k)$, $a^\dagger(k)$. These operator coefficients can be extracted from $\varphi(x)$ by utilizing appropriate orthogonality relations for the normalized exponential functions

$$f_k(x) = N_k\, e^{-ikx}, \qquad f_k^*(x) = N_k\, e^{ikx} \tag{16.7}$$

appearing in (16.6). It holds

$$\int f_k^*(\vec{x}, t) i \overleftrightarrow{\partial_0} f_{k'}(\vec{x}, t)\, \mathrm{d}^3x = \delta^{(3)}(\vec{k} - \vec{k}'),$$

$$\int f_k(\vec{x}, t) i \overleftrightarrow{\partial_0} f_{k'}(\vec{x}, t)\, \mathrm{d}^3x = 0, \tag{16.8}$$

where we have used the symbol for the both-sided derivative defined earlier (cf. (14.32)). The proof of the identities (16.8) is not difficult and can be left to the reader as an instructive exercise. Employing the formulae (16.8), one gets from (16.6)

$$a(k) = i \int \mathrm{d}^3x\, f_k^*(\vec{x}, t) \overleftrightarrow{\partial_0} \varphi(\vec{x}, t),$$

$$a^\dagger(k) = -i \int \mathrm{d}^3x\, f_k(\vec{x}, t) \overleftrightarrow{\partial_0} \varphi(\vec{x}, t). \tag{16.9}$$

Now we are in a position to work out the commutators of the operators $a(k)$, $a^\dagger(k)$. One has

$$[a(k), a^\dagger(k')] = \int \mathrm{d}^3x\, \mathrm{d}^3y \left[f_k^*(\vec{x}, t) \overleftrightarrow{\partial_0} \varphi(\vec{x}, t), f_{k'}(\vec{y}, t) \overleftrightarrow{\partial_0} \varphi(\vec{y}, t)\right]$$

$$= \int \mathrm{d}^3x\, \mathrm{d}^3y \left[f_k^*(\vec{x}, t)\dot\varphi(\vec{x}, t) - \dot f_k^*(\vec{x}, t)\varphi(\vec{x}, t), f_{k'}(\vec{y}, t)\dot\varphi(\vec{y}, t) - \dot f_{k'}(\vec{y}, t)\varphi(\vec{y}, t)\right].$$
$$\tag{16.10}$$

Now, using (16.2), the last expression may be recast in terms of the canonical commutators (16.4). Discarding immediately the vanishing commutators of the type $[\varphi, \varphi]$ and $[\pi, \pi]$, one is left with

$$
\begin{aligned}
[a(k), a^\dagger(k')] &= \int d^3x \, d^3y \left( -f_k^*(\vec{x}, t) \dot{f}_{k'}(\vec{y}, t) [\pi(\vec{x}, t), \varphi(\vec{y}, t)] - \dot{f}_k^*(\vec{x}, t) f_{k'}(\vec{y}, t) [\varphi(\vec{x}, t), \pi(\vec{y}, t)] \right) \\
&= \int d^3x \, d^3y \left( i\delta^{(3)}(\vec{x} - \vec{y}) f_k^*(\vec{x}, t) \dot{f}_{k'}(\vec{y}, t) - i\delta^{(3)}(\vec{x} - \vec{y}) \dot{f}_k^*(\vec{x}, t) f_{k'}(\vec{y}, t) \right) \\
&= i \int d^3x \, f_k^*(\vec{x}, t) \overset{\leftrightarrow}{\partial_0} f_{k'}(\vec{x}, t) \\
&= \delta^{(3)}(\vec{k} - \vec{k}'),
\end{aligned}
\tag{16.11}
$$

where we have utilized (16.8) in the final step. In a similar way, it can be shown that $[a(k), a(k')] = 0$ (which immediately implies also $[a^\dagger(k), a^\dagger(k')] = 0$). Thus, our results can be summarized as

$$
\begin{aligned}
[a(k), a^\dagger(k')] &= \delta^{(3)}(\vec{k} - \vec{k}'), \\
[a(k), a(k')] &= 0, \\
[a^\dagger(k), a^\dagger(k')] &= 0.
\end{aligned}
\tag{16.12}
$$

As a next step towards the physical interpretation of the quantized scalar field, let us calculate the energy and momentum according to the formulae derived in the preceding chapters. Needless to say, these quantities now become operators (hopefully, time-independent) expressed eventually in terms of $a(k)$ and $a^\dagger(k)$. We are going to start with (the operator of) the energy. Anticipating its role as the Hamiltonian of the quantized field, we denote it as $H$. One has

$$
H = \int \mathscr{T}^{00} \, d^3x = \int \left( \frac{1}{2} \partial_0\varphi \, \partial_0\varphi + \frac{1}{2} \vec{\nabla}\varphi \, \vec{\nabla}\varphi + \frac{1}{2} m^2 \varphi^2 \right) d^3x,
\tag{16.13}
$$

where we have employed the formula (14.45). The calculation is straightforward, but rather lengthy, so please be patient (at subsequent stages of our quantization tour we will be able to proceed faster, utilizing the skills gained here).

Substituting the expression (16.6) into (16.13), we will evaluate consecutively the individual terms appearing there. First,

$$
\begin{aligned}
H_1 &= \frac{1}{2} \int d^3x \, \partial_0\varphi \partial_0\varphi \\
&= \frac{1}{2} \iint d^3k \, d^3l \, N_k N_l \Big[ (-ik_0)(-il_0) \, e^{-i(k_0+l_0)x_0} (2\pi)^3 \delta^{(3)}(\vec{k} + \vec{l}) a(k)a(l) \\
&\quad + (-ik_0)(il_0) \, e^{-i(k_0-l_0)x_0} (2\pi)^3 \delta^{(3)}(\vec{k} - \vec{l}) a(k)a^\dagger(l) \\
&\quad + (ik_0)(-il_0) \, e^{i(k_0-l_0)x_0} (2\pi)^3 \delta^{(3)}(\vec{k} - \vec{l}) a^\dagger(k)a(l) \\
&\quad + (ik_0)(il_0) \, e^{i(k_0+l_0)x_0} (2\pi)^3 \delta^{(3)}(\vec{k} + \vec{l}) a^\dagger(k)a^\dagger(l) \Big].
\end{aligned}
\tag{16.14}
$$

Note that in arriving at (16.14) we have performed the integration with $d^3x$ producing the delta functions $\delta^{(3)}(\vec{k} \pm \vec{l})$. Taking into account how $k_0$, $l_0$ depend on $\vec{k}$, $\vec{l}$, the expression (16.14) is readily simplified to

$$
\begin{aligned}
H_1 = \frac{1}{2} \int d^3k \, N_k^2 (2\pi)^3 \Big[ &-k_0^2 \, e^{-2ik_0x_0} a(k)a(-k) + k_0^2 a(k)a^\dagger(k) \\
&+ k_0^2 a^\dagger(k)a(k) - k_0^2 \, e^{2ik_0x_0} a^\dagger(k)a^\dagger(-k) \Big].
\end{aligned}
\tag{16.15}
$$

Just to be sure, let us stress that $a(-k)$, $a^\dagger(-k)$ here in fact mean $a(-\vec{k})$, $a^\dagger(-\vec{k})$.

In a similar way, for the second term in (16.13) one obtains

$$
\begin{aligned}
H_2 &= \frac{1}{2} \int d^3 x\, \vec{\nabla}\varphi \vec{\nabla}\varphi \\
&= \frac{1}{2} \iint d^3 k\, d^3 l\, N_k N_l \Big[ -\vec{k}\cdot\vec{l}\, e^{-i(k_0+l_0)x_0} (2\pi)^3 \delta^{(3)}(\vec{k}+\vec{l}) a(k) a(l) \\
&\qquad + \vec{k}\cdot\vec{l}\, e^{-i(k_0-l_0)x_0} (2\pi)^3 \delta^{(3)}(\vec{k}-\vec{l}) a(k) a^\dagger(l) \\
&\qquad + \vec{k}\cdot\vec{l}\, e^{i(k_0-l_0)x_0} (2\pi)^3 \delta^{(3)}(\vec{k}-\vec{l}) a^\dagger(k) a(l) \\
&\qquad - \vec{k}\cdot\vec{l}\, e^{i(k_0+l_0)x_0} (2\pi)^3 \delta^{(3)}(\vec{k}+\vec{l}) a^\dagger(k) a^\dagger(l) \Big].
\end{aligned}
\tag{16.16}
$$

The integration over the variable $\vec{l}$ then obviously yields

$$
\begin{aligned}
H_2 &= \frac{1}{2} \int d^3 k\, N_k^2 (2\pi)^3 \Big[ \vec{k}^2\, e^{-2ik_0 x_0} a(k) a(-k) + \vec{k}^2 a(k) a^\dagger(k) \\
&\qquad + \vec{k}^2 a^\dagger(k) a(k) + \vec{k}^2\, e^{2ik_0 x_0} a^\dagger(k) a^\dagger(-k) \Big].
\end{aligned}
\tag{16.17}
$$

Finally,

$$
\begin{aligned}
H_3 &= \frac{1}{2} \int d^3 x\, m^2 \varphi^2 \\
&= \frac{1}{2} \iint d^3 k\, d^3 l\, N_k N_l \Big[ e^{-i(k_0+l_0)x_0} (2\pi)^3 \delta^{(3)}(\vec{k}+\vec{l}) a(k) a(l) \\
&\qquad + e^{-i(k_0-l_0)x_0} (2\pi)^3 \delta^{(3)}(\vec{k}-\vec{l}) a(k) a^\dagger(l) \\
&\qquad + e^{i(k_0-l_0)x_0} (2\pi)^3 \delta^{(3)}(\vec{k}-\vec{l}) a^\dagger(k) a(l) \\
&\qquad + e^{i(k_0+l_0)x_0} (2\pi)^3 \delta^{(3)}(\vec{k}+\vec{l}) a^\dagger(k) a^\dagger(l) \Big],
\end{aligned}
\tag{16.18}
$$

and this becomes

$$
\begin{aligned}
H_3 &= \frac{1}{2} \int d^3 k\, N_k^2 (2\pi)^3 m^2 \Big[ e^{-2ik_0 x_0} a(k) a(-k) + a(k) a^\dagger(k) \\
&\qquad + a^\dagger(k) a(k) + e^{2ik_0 x_0} a^\dagger(k) a^\dagger(-k) \Big].
\end{aligned}
\tag{16.19}
$$

Putting all this together, it is seen that the sum of the time-dependent terms in (16.15), (16.17) and (16.19) includes the overall coefficient $-k_0^2 + \vec{k}^2 + m^2$, which is zero. It is gratifying, since we have anticipated a time-independent energy as a result of correct calculation. On the other hand, the sum of coefficients in the time-independent terms amount to $k_0^2 + \vec{k}^2 + m^2 = 2k_0^2$. Taking also into account that $N_k^2 = (2\pi)^{-3}(2k_0)^{-1}$ one then gets, after a simple manipulation,

$$
H = H_1 + H_2 + H_3 = \frac{1}{2} \int d^3 k\, k_0 \left[ a^\dagger(k) a(k) + a(k) a^\dagger(k) \right].
\tag{16.20}
$$

An observant reader may have noticed already that the integrand in the last expression bears a striking resemblance to the Hamiltonian of the linear harmonic oscillator (written in terms of the popular "ladder" operators). It is indeed so and we will elaborate on this point in the next chapter.

Further, let us consider the (operator of) momentum. As we know, it is given by the space integral of the components $\mathcal{T}^{0j}$ of the energy–momentum tensor; according to (14.39) this is

$$P^j = \int d^3x\, \partial_0\varphi\, \partial^j\varphi\,. \tag{16.21}$$

For convenience, we will switch to the notation involving the gradient symbol $\vec{\nabla}$, similarly to (15.54). Recall that $\vec{\nabla}$ represents conventionally the derivatives $\partial/\partial x^j\ (\equiv \partial_j)$. Thus, (16.21) can be recast as

$$\vec{P} = -\int d^3x\, \partial_0\varphi\, \vec{\nabla}\varphi\,. \tag{16.22}$$

Substituting the expression (16.6) into (16.22) and proceeding similarly as before, one gets first

$$
\begin{aligned}
\vec{P} = \iint d^3k\, d^3l\, N_k N_l \Big[ &-k_0\vec{l}\, e^{-i(k_0+l_0)x_0}(2\pi)^3\delta^{(3)}(\vec{k}+\vec{l})a(k)a(l) \\
&+ k_0\vec{l}(2\pi)^3\delta^{(3)}(\vec{k}-\vec{l})a(k)a^\dagger(l) \\
&+ k_0\vec{l}(2\pi)^3\delta^{(3)}(\vec{k}-\vec{l})a^\dagger(k)a(l) \\
&- k_0\vec{l}\, e^{i(k_0+l_0)x_0}(2\pi)^3\delta^{(3)}(\vec{k}+\vec{l})a^\dagger(k)a^\dagger(l) \Big]\,,
\end{aligned} \tag{16.23}
$$

and this is readily turned into the form

$$
\begin{aligned}
\vec{P} = \int d^3k\, N_k^2 (2\pi)^3 \Big[ &k_0\vec{k}\, e^{-2ik_0x_0}a(k)a(-k) + k_0\vec{k}\, e^{2ik_0x_0}a^\dagger(k)a^\dagger(-k) \\
&+ k_0\vec{k}a(k)a^\dagger(k) + k_0\vec{k}a^\dagger(k)a(k) \Big]\,.
\end{aligned} \tag{16.24}
$$

Now, one would like to get rid of the time dependent terms in (16.24). This is done quite easily. Indeed, these contributions to the integrand are odd functions under $\vec{k} \to -\vec{k}$, since all factors appearing there are even, except $\vec{k}$; in particular, the products $a(k)a(-k)$ and $a^\dagger(k)a^\dagger(-k)$ are even, due to the commutation relations (16.12). Thus, the first two terms in (16.24) vanish upon the integration. The final result can be then written, after a simple manipulation, as

$$\vec{P} = \frac{1}{2}\int d^3k\, \vec{k}\, [a^\dagger(k)a(k) + a(k)a^\dagger(k)]\,. \tag{16.25}$$

We will study the properties of the operators $H$ and $\vec{P}$ in detail in the next chapter. However, there is at least one simple observation that should be mentioned already here, to convince ourselves that we are on the right track. One should rightly expect that the energy operator (16.20) is a correct quantum Hamiltonian, in the sense that it controls the time evolution of the field operator $\varphi(\vec{x}, t)$. To see this, let us evaluate the relevant commutator $[H, \varphi(x)]$. First, one can show that

$$
\begin{aligned}
[H, a(k)] &= -k_0 a(k)\,, \\
[H, a^\dagger(k)] &= k_0 a^\dagger(k)\,.
\end{aligned} \tag{16.26}
$$

Proving (16.26) is not difficult. Indeed, using the basic commutation relations (16.12), it is easy to see that

$$
\begin{aligned}
[a^\dagger(l)a(l), a(k)] &= [a(l)a^\dagger(l), a(k)] = -\delta^{(3)}(\vec{k}-\vec{l})a(l)\,, \\
[a^\dagger(l)a(l), a^\dagger(k)] &= [a(l)a^\dagger(l), a^\dagger(k)] = \delta^{(3)}(\vec{k}-\vec{l})a^\dagger(l)\,.
\end{aligned} \tag{16.27}
$$

Then, employing the formula (16.20), one has

$$[H, a(k)] = \int d^3l \left[ \frac{1}{2} l_0 (a^\dagger(l) a(l) + a(l) a^\dagger(l)), a(k) \right]$$

$$= \int d^3l \, \frac{1}{2} l_0 \left( -\delta^{(3)}(\vec{k} - \vec{l}) a(l) - \delta^{(3)}(\vec{k} - \vec{l}) a(l) \right)$$

$$= -k_0 a(k) \,.$$

Similarly, one gets the second identity (16.26). With the identities (16.26) at hand, the calculation of the commutator $[H, \varphi(x)]$ is straightforward. One has

$$i[H, \varphi(x)] = i \int d^3k \, N_k \left( [H, a(k)] e^{-ikx} + [H, a^\dagger(k)] e^{ikx} \right)$$

$$= \int d^3k \, N_k \left( -ik_0 a(k) e^{-ikx} + ik_0 a^\dagger(k) e^{ikx} \right). \tag{16.28}$$

However, the expression (16.28) is precisely what one obtains for the time derivative $\partial_0 \varphi$ by using (16.6). Thus, we have arrived at the equation of motion (in the Heisenberg picture)

$$\partial_0 \varphi(x) = i[H, \varphi(x)] \,, \tag{16.29}$$

which is certainly reassuring.

In a similar way, using (16.25), one gets easily

$$\partial^j \varphi(x) = i[P^j, \varphi(x)] \,, \qquad j = 1, 2, 3 \,. \tag{16.30}$$

Thus, (16.29) and (16.30) may be written summarily as a covariant relation

$$\partial^\mu \varphi(x) = i[P^\mu, \varphi(x)] \,, \qquad \mu = 0, 1, 2, 3 \,, \tag{16.31}$$

for the commutator of quantized field with the four-momentum operator. Obviously, the identity (16.31) indicates that $P^\mu$ is the generator of spacetime shifts (of course, this is in agreement with our expectations). Now, employing the well-known operator identity

$$e^A B e^{-A} = B + \frac{1}{1!} [A, B] + \frac{1}{2!} [A, [A, B]] + \dots \tag{16.32}$$

along with (16.31), it is not difficult to obtain an elegant formula for the shift of the spacetime coordinates in $\varphi(x)$, namely

$$\varphi(x + a) = e^{iP \cdot a} \varphi(x) e^{-iP \cdot a} \,. \tag{16.33}$$

Proving (16.33) is left to an interested reader as an instructive exercise.

In concluding this chapter, let us add one more remark. It is clear that along with the above treatment of the quantized field $\varphi(x)$, one gets immediately the results for the energy and momentum of the classical Klein–Gordon field configuration (16.5); it suffices to replace the operators $a(k)$, $a^\dagger(k)$ in the formulae (16.20) and (16.25) with the numerical coefficients $a(k)$, $a^*(k)$ from (16.5). One thus gets readily

$$E_{\text{class.}} = \int d^3k \, k_0 |a(k)|^2 \,,$$

$$\vec{P}_{\text{class.}} = \int d^3k \, \vec{k} |a(k)|^2 \,.$$

It is seen, as we have noted earlier, that $E_{\text{class.}} \geq 0$ (this is obvious already at the level of the energy density appearing in (16.13)). Thus, the problem of negative energies, which is detrimental to the relativistic quantum mechanics, is absent in the theory of classical Klein–Gordon field. Of course, the point is that the energy here and in the one-particle quantum mechanics has a different meaning; a comparison of the two cases is left to the reader as a useful revision of the old stuff.

# Chapter 17

# Particle interpretation of quantized field

Now we are on the way to establish an adequate description of the physical states of the quantized scalar field; in particular, we have in mind the eigenstates of the energy and momentum. The formulae (16.20) and (16.25) could be employed for such a purpose, but it is more convenient to develop a slightly different technical language, or rather a "dialect", which turns out to be more user-friendly (and, moreover, it becomes highly useful also in other situations).

The idea of the technique we are going to introduce is that one may consider the field in question (classical or quantum) to be confined within a finite spatial box, which is e.g. a cube with the side length $L$, i.e. the volume $V = L^3$. The auxiliary parameter $L$ is essentially arbitrary, but the reader can rest assured that no relevant physical result will depend on it. To make such a scheme workable, one should also choose appropriate boundary conditions; one possible option is to require that the field has the same value for $x^j = 0$ and $x^j = L$, $j = 1, 2, 3$ (briefly, one thus requires the periodicity with respect to the box side). Such a choice turns out to be convenient with regard to the ubiquitous integrals of the exponential functions, which thus get simplified in a desirable way. Let us emphasize that the restrictions we impose on the field in question are just technical — there is no deeper principle that would enforce a "correct" formulation of the whole set-up.

The field $\varphi(\vec{x}, t)$ placed in the box should of course satisfy the Klein–Gordon equation, so that a general solution can again by represented as a superposition of the exponentials $\exp(\pm ikx)$ with $k_0 = (\vec{k}^2 + m^2)^{1/2}$, but these must now satisfy the above-mentioned periodic boundary conditions in the 3-dimensional space. It is easy to see that the variables $k_j$, $j = 1, 2, 3$, in $\exp(\pm i\vec{k}\vec{x})$ then can take on just the discrete values

$$k_j = \frac{2\pi n_j}{L}, \qquad n_j = 0, \pm 1, \pm 2, \ldots \tag{17.1}$$

So, the quantized Klein–Gordon field $\varphi(x) = \varphi(\vec{x}, t)$ can be written as an infinite sum

$$\varphi(x) = \sum_{\vec{k}} N_k \left( a_k \, e^{-ikx} + a_k^\dagger \, e^{ikx} \right), \tag{17.2}$$

with $\vec{k}$ given by (17.1). It turns out that an appropriate value of the normalization factor $N_k$ is now

$$N_k = \frac{1}{(2k_0 V)^{1/2}}. \tag{17.3}$$

Let us see why the formula (17.3) is a good choice. First, because of our boundary conditions it holds, taking into account (17.1),

$$\int d^3x \, e^{-i\vec{k}\cdot\vec{x}} e^{i\vec{l}\cdot\vec{x}} = V\delta_{\vec{k},\vec{l}}, \tag{17.4}$$

where the integration domain is just the box. Then, defining the functions $f_k(x)$, $f_k^*(x)$ in the same manner as in (16.7), and utilizing Eq. (17.4), one gets the orthogonality relations

$$\int d^3x\, f_k^*(\vec{x},t) i \overset{\leftrightarrow}{\partial_0} f_{k'}(\vec{x},t) = \delta_{\vec{k},\vec{k}'},$$

$$\int d^3x\, f_k(\vec{x},t) i \overset{\leftrightarrow}{\partial_0} f_{k'}(\vec{x},t) = 0. \tag{17.5}$$

The option (17.3) is thereby justified: the identities (17.5) are directly analogous to (16.8) and one may thus expect that the commutation relations for the operators $a_k$, $a_k^\dagger$ will have a simple structure similar to (16.12). Indeed, owing to the identities (17.5), $a_k$ and $a_k^\dagger$ can be expressed in terms of $\varphi(x)$ in the same way as before (cf. (16.9)); employing then the commutators (16.4), one gets readily

$$[a_k, a_l^\dagger] = \delta_{kl},$$
$$[a_k, a_l] = 0, \tag{17.6}$$
$$[a_k^\dagger, a_l^\dagger] = 0$$

(again, the commutators (17.6) are eventually determined by the orthogonality relations (17.5)). Recall that we use the notation $k$, $l$ instead of $\vec{k}$, $\vec{l}$ whenever it cannot cause a confusion.

The calculation of the energy and momentum proceeds along the same lines as in Chapter 16; the only difference here is that for the integrals involving the exponentials like $\exp\left[i(\vec{k} - \vec{l})\vec{x}\right]$ one gets $V\delta_{\vec{k},\vec{l}}$ instead of $(2\pi)^3 \delta^{(3)}(\vec{k} - \vec{l})$. But this matches precisely the expression (17.3) for the normalization factors and one thus immediately obtains

$$H = \frac{1}{2} \sum_{\vec{k}} k_0 \left(a_k^\dagger a_k + a_k a_k^\dagger\right),$$

$$\vec{P} = \frac{1}{2} \sum_{\vec{k}} \vec{k} \left(a_k^\dagger a_k + a_k a_k^\dagger\right). \tag{17.7}$$

We have already noticed in the preceding chapter that the structure of the field Hamiltonian $H$ reminds one of the linear harmonic oscillator (LHO). Such a picture becomes even clearer now, when we focus on the formula (17.7). Indeed, from (17.6) one gets, in particular,

$$[a_k, a_k^\dagger] = 1 \tag{17.8}$$

for any $k$ satisfying (17.1). By the way, the validity of the simple relation (17.8) is the reason why the present "discrete" formalism can be considered as more user-friendly than the "continuous" one used in Chapter 16: it is seen that from (16.12) one would get

$$[a(k), a^\dagger(k)] = \delta^{(3)}(0), \tag{17.9}$$

and the (notoriously ill-defined) expression on the right-hand side of (17.9) is surely a nuisance. Utilizing (17.8), the expression for $H$ in (17.7) may be recast as

$$H = \sum_{\vec{k}} k_0 \left(a_k^\dagger a_k + \frac{1}{2}\right), \tag{17.10}$$

and the "oscillator picture" of the quantized field should now be obvious. The formula (17.10) represents an infinite sum of LHO Hamiltonians written in terms of the "ladder operators" (or, raising and lowering operators), where the coefficient $k_0$ corresponds to the oscillator frequency (recall that in ordinary units the coefficient of $a^\dagger a + \frac{1}{2}$ is $\hbar\omega$). Thus, the field Hamiltonian (17.10) incorporates an infinite number of oscillator modes with different frequencies; this is why the term "decomposition into oscillators" is traditionally used in connection with the field quantization.

Obviously, the constant

$$\frac{1}{2} \sum_{\vec{k}} k_0 = \frac{1}{2} \sum_{\vec{k}} (\vec{k}^2 + m^2)^{1/2} \tag{17.11}$$

is infinite, in view of the formula (17.1). However, such a constant does not influence the equation of motion (16.29), since it commutes with anything. So, let us discard the infinite constant (17.11) from (17.10); such a redefinition of $H$ will be called "normal ordering" (of the field operators) from now on (one might also say that it is the first example of a "renormalization" in quantum field theory). The formula (17.7) for the momentum $\vec{P}$ may be recast in the same way; thus, in what follows we are going to work with the normal ordered expressions

$$H = \sum_{\vec{k}} k_0 a_k^\dagger a_k \,,$$
$$\vec{P} = \sum_{\vec{k}} \vec{k} a_k^\dagger a_k \,. \tag{17.12}$$

Now we are in a position to examine the eigenvalues and eigenstates of the operators $H$ and $\vec{P}$. First of all, let us note that $[H, \vec{P}] = 0$ (as expected *a priori*). Indeed, to see this, it suffices to prove that

$$[a_k^\dagger a_k, a_l^\dagger a_l] = 0 \tag{17.13}$$

for any $k$, $l$. The proof of (17.13) is easy: for $k = l$ it holds trivially, and for $k \neq l$ one should take into account that the commutator in question involves only mutually commuting operators (cf. (17.6)).

To proceed, we will invoke the formulae (16.26). Obviously, such identities are valid within our "discrete formalism" as well, so let us reproduce them here for reader's convenience:

$$[H, a_k] = -k_0 a_k \,,$$
$$[H, a_k^\dagger] = k_0 a_k^\dagger \,. \tag{17.14}$$

Now, suppose that $|\Psi\rangle$ is a common eigenvector of $H$ and $\vec{P}$, so that

$$H|\Psi\rangle = E|\Psi\rangle \,, \qquad \vec{P}|\Psi\rangle = \vec{p}|\Psi\rangle \,. \tag{17.15}$$

Let us find out what happens if the operators $a_k$ and $a_k^\dagger$ act on $|\Psi\rangle$. One gets, using (17.14),

$$H a_k^\dagger |\Psi\rangle = (a_k^\dagger H + k_0 a_k^\dagger)|\Psi\rangle$$
$$= (E + k_0) a_k^\dagger |\Psi\rangle \tag{17.16}$$

and

$$H a_k |\Psi\rangle = (a_k H - k_0 a_k)|\Psi\rangle$$
$$= (E - k_0) a_k |\Psi\rangle \,. \tag{17.17}$$

113

Similarly, the results for the momentum operator become

$$\vec{P}a_k^\dagger|\Psi\rangle = (\vec{p} + \vec{k})a_k^\dagger|\Psi\rangle \tag{17.18}$$

and

$$\vec{P}a_k|\Psi\rangle = (\vec{p} - \vec{k})a_k|\Psi\rangle. \tag{17.19}$$

The relations (17.16) through (17.19) are quite remarkable. They show that the action of $a_k^\dagger$ amounts to adding the momentum $\vec{k}$ and the energy $k_0 = (\vec{k}^2+m^2)^{1/2}$, while $a_k$ removes the same momentum and energy when acting on a four-momentum eigenstate. Thus, such a behaviour may be described as adding a particle with the four-momentum $k$, $k^2 = m^2$ to the original state $|\Psi\rangle$, or removing such a particle from it. This leads naturally to the usual terminology: $a_k^\dagger$ is called the **creation operator** (of a particle) and $a_k$ is the **annihilation operator**. Recall that in the theory of LHO, $a^\dagger$ and $a$ is called the raising and lowering operator, respectively (referring to the corresponding shift of the oscillator energy).

Now we are very close to our final goal. In view of the structure of the Hamiltonian (17.12), one may construct its eigenstates as the tensor products of the eigenvectors of the operators $a_k^\dagger a_k$ for different values of $k$ (these commute, according to (17.13)!). Owing to the commutation relation (17.8) it is clear that the action of $a_k$ decreases the eigenvalue of $a_k^\dagger a_k$ by 1. In full analogy with the LHO theory one may then observe that for a given annihilation operator $a_k$ there must be an eigenvector $|\Omega_k\rangle$ of $a_k^\dagger a_k$ such that $a_k|\Omega_k\rangle = 0$. Indeed, if there were no such $|\Omega_k\rangle$, the repeated action of $a_k$ would produce a negative eigenvalue of $a_k^\dagger a_k$, but this is in contradiction with the fact that $a_k^\dagger a_k$ is, obviously, a positive operator. So, let us define the vector $|0\rangle$ such that

$$a_k|0\rangle = 0 \quad \text{for any } k \tag{17.20}$$

($|0\rangle$ might be represented formally as a tensor product of the vectors $|\Omega_k\rangle$ for all possible values of $k$). It is clear that one then has

$$H|0\rangle = 0, \qquad \vec{P}|0\rangle = 0. \tag{17.21}$$

Thus, $|0\rangle$ may be considered as a no-particle state and called the **vacuum**. It is the ground state of the Hamiltonian (17.12).

Taking into account what we already know about the properties of the creation operators (see (17.16), (17.18)), one may construct one-particle, two-particle, etc. states by means of appropriate actions of $a_k^\dagger$ on the vacuum. In general, an $n$-particle state can be built as

$$\left(a_{k_1}^\dagger\right)^{n_1} \dots \left(a_{k_r}^\dagger\right)^{n_r} |0\rangle, \tag{17.22}$$

with $n_1 + \dots + n_r = n$. Of course, the corresponding eigenvalue of $H$ is the sum of particle energies, i.e. $n_1 E(k_1)+\dots+n_r E(k_r)$. The numbers $n_1, \dots, n_r$ are naturally called the **occupation numbers**, since they tell us how many particles occupy a given state. Consequently, the description of the field states shown in (17.22) is called the **representation of occupation numbers** (a knowledgeable reader may recall that such a scheme is employed also in the theory of many-body systems, e.g. in the condensed matter physics or nuclear physics). The linear span of all possible state vectors of the type (17.22) can be rightly called the **$n$-particle subspace** of the whole space of states of the quantized Klein–Gordon field, and denoted e.g. as $\mathcal{H}^{(n)}$. The full space is then the direct sum of such subspaces, i.e.

$$\mathcal{H} = \mathcal{H}^{(0)} \oplus \mathcal{H}^{(1)} \oplus \mathcal{H}^{(2)} \oplus \dots, \tag{17.23}$$

where $\mathcal{H}^{(0)}$ is just the vacuum state. $\mathcal{H}$ is called the **Fock space**, in honour of the Russian theorist V. A. Fock (as we have already noted in Chapter 1, the transcription "Fok" would perhaps be more appropriate).

It is important to notice that the number of particles sharing the same state is, in principle, unlimited, i.e. the occupation numbers $n_1, \ldots, n_r$ in (17.22) are arbitrary; the point is that the commutation relations for the creation and annihilation operators do not imply any constraint here. This means that the Klein–Gordon scalar particles are **bosons**, i.e. they obey the Bose–Einstein statistics. Referring to our earlier experience with the Klein–Gordon equation, one may guess that these particles have zero spin. Thus, we have an explicit example of the famous connection between the **spin and statistics** (namely that particles with an integer spin are bosons and those with half-integer spin are fermions), which is a general result of relativistic quantum field theory (many details of the history of this problem and several proofs of this remarkable theorem can be found in the book [20]; for an instructive comprehensive treatment of the subject see also the book [27]).

A last remark: One might wonder if a scalar particle described here really exists in nature. Yes, it does; it is the celebrated Higgs boson (see e.g. [22]).

Let us summarize our results. We have arrived at the explicit construction of a space of states corresponding to the quantized real scalar field. Such a space is spanned by the common eigenvectors of the field energy and momentum, and the most remarkable finding is that such states may be described in terms of particles satisfying the standard relativistic relation $k^2 = m^2$ for the four-momentum (the condition of mass shell). This is why one says, in the common physical folklore, that "particles are the field quanta". Obviously, the "oscillator decomposition" of the quantized field is instrumental in the whole treatment based on the notion of creation and annihilation operators (this seems to support the popular wisdom that most of the quantum theory relies on the linear harmonic oscillator). The expansion of the quantized field in terms of the creation and annihilation operators may be implemented either in the infinite 3-dimensional space ("continuous formalism") or in a finite spatial box ("discrete formalism"). The origin of such provisional names should be clear when comparing (16.5) and (17.2) along with (17.1). Obviously, the discrete scheme is more convenient for establishing the correspondence with the oscillator picture, but in the subsequent chapters we will use freely both formalisms. In fact, these two "dialects" of the same language could also be compared, in musical terms, to playing the same tune on two different instruments, e.g. violin or piano, according to one's taste.

In concluding this chapter, a historical remark is in order. The method of canonical quantization that we have demonstrated here and in Chapter 16 on the example of the real Klein–Gordon field is due to Werner Heisenberg and Wolfgang Pauli, who published their original work in 1929. Another important contribution appeared in the paper by Wolfgang Pauli and Victor Weisskopf published in 1934. The latter work treated the problem of antiparticles and it was, partly, an opposition to the Dirac's idea of a filled sea of negative energy states. For a brief review of the QFT history, see e.g. [13]. As we know, it is indeed the field theory approach, which provides a consistent description of particles and antiparticles. This fundamental QFT theme will be discussed in the following chapter.

# Chapter 18

# Complex scalar field. Antiparticles

---

Let us now consider a complex Klein–Gordon field. As we have seen in Chapter 14, its Lagrangian can be written as

$$\mathcal{L} = \partial_\mu \varphi \partial^\mu \varphi^* - m^2 \varphi \varphi^* , \tag{18.1}$$

and the independent dynamical variables ("generalized coordinates") are $\varphi(x)$ and $\varphi^*(x)$. The corresponding conjugate momenta are then

$$\varphi(x) \quad \longrightarrow \quad \pi(x) = \frac{\partial \mathcal{L}}{\partial(\partial_0 \varphi)} = \partial_0 \varphi^*(x) ,$$

$$\varphi^*(x) \quad \longrightarrow \quad \pi^*(x) = \frac{\partial \mathcal{L}}{\partial(\partial_0 \varphi^*)} = \partial_0 \varphi(x) . \tag{18.2}$$

In the quantum case, $\varphi(x)$ and $\varphi^*(x)$ become operators $\varphi(x)$ and $\varphi^\dagger(x)$ related by means of Hermitian conjugation. Thus, there are two pairs of canonical variables, namely

$$\left( \varphi(x), \dot{\varphi}^\dagger(x) \right) , \qquad \left( \varphi^\dagger(x), \dot{\varphi}(x) \right) , \tag{18.3}$$

and it is easy to realize that the relevant equal-time commutation relations to be postulated are

$$[\varphi(\vec{x}, t), \dot{\varphi}^\dagger(\vec{y}, t)] = i\delta^{(3)}(\vec{x} - \vec{y}) ,$$
$$[\varphi(\vec{x}, t), \dot{\varphi}(\vec{y}, t)] = 0 ,$$
$$[\varphi(\vec{x}, t), \varphi(\vec{y}, t)] = 0 ,$$
$$[\varphi(\vec{x}, t), \varphi^\dagger(\vec{y}, t)] = 0 ,$$
$$[\dot{\varphi}(\vec{x}, t), \dot{\varphi}(\vec{y}, t)] = 0 ,$$
$$[\dot{\varphi}(\vec{x}, t), \dot{\varphi}^\dagger(\vec{y}, t)] = 0 . \tag{18.4}$$

Other possible canonical commutators are reduced to the Hermitian conjugation of those shown above. Needless to say, to arrive at the relations (18.4), we have followed consistently the paradigm of quantum mechanics, formulated succinctly in Chapter 16 (see (16.3)). In brief, one could also say that

$$[\varphi(\vec{x}, t), \dot{\varphi}^\dagger(\vec{y}, t)] = i\delta^{(3)}(\vec{x} - \vec{y}) , \qquad [\varphi^\dagger(\vec{x}, t), \dot{\varphi}(\vec{y}, t)] = i\delta^{(3)}(\vec{x} - \vec{y}) , \tag{18.5}$$

and all remaining equal-time commutators vanish.

As a solution of the Klein–Gordon equation, the classical field $\varphi(x)$ can be written as

$$\varphi(x) = \int d^3k \, N_k \left[ b(k) \, e^{-ikx} + d^*(k) \, e^{ikx} \right] , \tag{18.6}$$

116

where $k = (k_0, \vec{k})$, $k_0 = (\vec{k}^2 + m^2)^{1/2}$ and $N_k$ has the familiar form $(2\pi)^{-3/2}(2k_0)^{-1/2}$. We have denoted the coefficient at $\exp(ikx)$ deliberately as $d^*(k)$, which may seem rather artificial. However, the quantum counterpart of (18.6) is then naturally written as

$$\varphi(x) = \int d^3k \, N_k \left[ b(k) \, e^{-ikx} + d^\dagger(k) \, e^{ikx} \right], \tag{18.7}$$

and, consequently,

$$\varphi^\dagger(x) = \int d^3k \, N_k \left[ b^\dagger(k) \, e^{ikx} + d(k) \, e^{-ikx} \right]. \tag{18.8}$$

The conventional notation $d^\dagger(k)$ in (18.7) gets a deeper meaning: it turns out that it will be possible to interpret it as a specific creation operator.

So, let us find out what are the commutation relations for the operator coefficients in (18.6), (18.7). To this end, one may express them in analogy with the formulae (16.9); one gets

$$b(k) = i \int f_k^*(\vec{x}, t) \overleftrightarrow{\partial_0} \varphi(\vec{x}, t) \, d^3x,$$

$$b^\dagger(k) = -i \int f_k(\vec{x}, t) \overleftrightarrow{\partial_0} \varphi^\dagger(\vec{x}, t) \, d^3x,$$

$$d(k) = i \int f_k^*(\vec{x}, t) \overleftrightarrow{\partial_0} \varphi^\dagger(\vec{x}, t) \, d^3x, \tag{18.9}$$

$$d^\dagger(k) = -i \int f_k(\vec{x}, t) \overleftrightarrow{\partial_0} \varphi(\vec{x}, t) \, d^3x.$$

The computation of the commutators in question relies on (18.9) and the canonical relations (18.4), and it is essentially the same procedure as before, in Chapter 16. The results can be summarized as follows:

$$[b(k), b^\dagger(l)] = \delta^{(3)}(\vec{k} - \vec{l}),$$
$$[d(k), d^\dagger(l)] = \delta^{(3)}(\vec{k} - \vec{l}),$$
$$[b(k), b(l)] = 0,$$
$$[b(k), d(l)] = 0, \tag{18.10}$$
$$[d(k), d(l)] = 0,$$
$$[b(k), d^\dagger(l)] = 0.$$

Again, the succinct statement could be that just $[b, b^\dagger]$ and $[d, d^\dagger]$ are non-zero, having the standard value as in (16.12), and the rest is zero. According to the findings presented in Chapter 17 it is clear that $b$, $d$ and $b^\dagger$, $d^\dagger$ are candidates for the annihilation and creation operators, respectively. The derivation of the above results is straightforward, but, just to be sure, let us verify the second relation in (18.10), which we have envisaged before, in connection with the notation introduced in (18.6), (18.7). One has, using (18.9),

$$[d(k), d^\dagger(l)] = \iint d^3x \, d^3y \left[ f_k^*(\vec{x}, t)\dot{\varphi}^\dagger(\vec{x}, t) - \dot{f}_k^*(\vec{x}, t)\varphi^\dagger(\vec{x}, t), \right.$$

$$\left. f_l(\vec{y}, t)\dot{\varphi}(\vec{y}, t) - \dot{f}_l(\vec{y}, t)\varphi(\vec{y}, t) \right]. \tag{18.11}$$

Discarding the terms involving vanishing equal-time commutators according to (18.4), one is left with

$$[d(k), d^\dagger(l)] = \iint d^3x \, d^3y \left( -f_k^*(\vec{x}, t)\dot{f}_l(\vec{y}, t)[\dot{\varphi}^\dagger(\vec{x}, t), \varphi(\vec{y}, t)] \right.$$

$$\left. - \dot{f}_k^*(\vec{x}, t)f_l(\vec{y}, t)[\varphi^\dagger(\vec{x}, t), \dot{\varphi}(\vec{y}, t)] \right). \tag{18.12}$$

117

Then, employing again (18.4), one gets finally

$$[d(k), d^\dagger(l)] = \int d^3x \, f_k^*(\vec{x}, t) i \overset{\leftrightarrow}{\partial_0} f_l(\vec{x}, t) = \delta^{(3)}(\vec{k} - \vec{l}) \,, \tag{18.13}$$

and that's it.

The next step is the evaluation of the energy and momentum. Using the general formula (15.16) for the energy–momentum tensor, one may write

$$H = \int d^3x \left( \partial_0 \varphi^\dagger \partial_0 \varphi + \vec{\nabla} \varphi^\dagger \vec{\nabla} \varphi + m^2 \varphi^\dagger \varphi \right),$$
$$\vec{P} = -\int d^3x \left( \partial_0 \varphi^\dagger \vec{\nabla} \varphi + \partial_0 \varphi \vec{\nabla} \varphi^\dagger \right). \tag{18.14}$$

Substituting there the expressions (18.7), (18.8), one may follow basically the same boring procedure that in Chapter 16 led from (16.13) and (16.22) to (16.20) and (16.25). So, let us skip the calculational details and announce immediately the relevant results. One gets

$$H = \int d^3k \, k_0 \left[ b^*(k) b(k) + d(k) d^\dagger(k) \right],$$
$$\vec{P} = \int d^3k \, \vec{k} \left[ b^*(k) b(k) + d(k) d^\dagger(k) \right]. \tag{18.15}$$

From the preceding chapter we know that it is quite convenient to play the same tune on a different instrument; so, shrinking the 3-dimensional space to a finite box, the formulae (18.15) are recast in the form

$$H = \sum_{\vec{k}} k_0 \left( b_k^\dagger b_k + d_k d_k^\dagger \right),$$
$$\vec{P} = \sum_{\vec{k}} \vec{k} \left( b_k^\dagger b_k + d_k d_k^\dagger \right) \tag{18.16}$$

in analogy with (17.7). The commutation relations (18.10) now become

$$[b_k, b_l^\dagger] = \delta_{kl} \,, \qquad [d_k, d_k^\dagger] = \delta_{kl} \,, \tag{18.17}$$

and one may thus repeat the analysis presented in the preceding chapter. As a result, (18.16) can be rewritten in normal ordered form

$$H = \sum_{\vec{k}} k_0 \left( b_k^\dagger b_k + d_k^\dagger d_k \right),$$
$$\vec{P} = \sum_{\vec{k}} \vec{k} \left( b_k^\dagger b_k + d_k^\dagger d_k \right), \tag{18.18}$$

and one may then construct the common eigenstates of the operators (18.18) in the same manner like in the preceding chapter, recovering the architecture of the Fock space. The resulting picture is that one can generate multiparticle states by means of the creation operators of two different types, $b_k^\dagger$ and $d_k^\dagger$, but the corresponding particles carry the same mass; therefore, one would like to figure out a way of discerning the "$d$-particles" from "$b$-particles".

Fortunately, the solution is at hand, and is rather profound from the physical point of view. As we have seen in Chapter 15, for complex Klein–Gordon field there is an extra integral of motion, corresponding to the conserved current $J^\mu$, which can be written, classically, as

$$J^\mu = i \left[ (\partial^\mu \varphi) \varphi^* - \varphi (\partial^\mu \varphi^*) \right] \tag{18.19}$$

(cf. (15.58); the choice of sign here is purely conventional). From the identity $\partial_\mu J^\mu = 0$ one gets the time-independent quantity

$$Q = \int d^3x \, J_0(\vec{x}, t) \tag{18.20}$$

that may be called, conventionally, the **charge**. The quantum counterpart of the expression (18.20) is the operator

$$Q = i \int d^3x \left( \partial_0 \varphi \, \varphi^\dagger - \varphi \, \partial_0 \varphi^\dagger \right). \tag{18.21}$$

Substituting (18.7) and (18.8) into (18.21), one gets the result

$$Q = \int d^3k \left[ b(k)b^\dagger(k) - d^\dagger(k)d(k) \right]. \tag{18.22}$$

Note that to arrive at (18.22), one employs basically the same manipulations that we have described in detail in Chapter 16 when deriving the formulae (16.20) and (16.25). So, the verification of (18.22) is left to the reader as an instructive exercise. In the "discrete dialect", corresponding to the field quantized in a finite box, the formula (18.22) becomes, as usual,

$$Q = \sum_{\vec{k}} (b_k b_k^\dagger - d_k^\dagger d_k). \tag{18.23}$$

Using the commutation relation $[b_k, b_k^\dagger] = 1$ and discarding an infinite constant, the expression for $Q$ is recast in the normal ordered form

$$Q = \sum_{\vec{k}} (b_k^\dagger b_k - d_k^\dagger d_k). \tag{18.24}$$

Now, it is easy to see that the commutation relations (18.17) imply

$$[Q, b_k^\dagger] = b_k^\dagger, $$
$$[Q, d_k^\dagger] = -d_k^\dagger. \tag{18.25}$$

Of course, the vacuum state $|0\rangle$ is defined by

$$b_k|0\rangle = 0, \qquad d_k|0\rangle = 0 \tag{18.26}$$

for any $k$. Then also $Q|0\rangle = 0$ and from (18.25) one gets readily

$$Qb_k^\dagger|0\rangle = b_k^\dagger|0\rangle, $$
$$Qd_k^\dagger|0\rangle = -d_k^\dagger|0\rangle. \tag{18.27}$$

Thus, the one-particle state $b_k^\dagger|0\rangle$ can be distinguished from $d_k^\dagger|0\rangle$ by means of the eigenvalue of the charge operator $Q$: for the "$b$-particle" one has $Q = +1$, and the "$d$-particle" carries the value $Q = -1$. In this way one arrives naturally at the concept of **antiparticles**: the one-particle state created by means of $d_k^\dagger$ can be conventionally called the antiparticle with respect to the particle state produced by the action of $b_k^\dagger$.

Let us summarize our main results. The lesson to be learnt from the above simple analysis is that the theory of quantized complex scalar field leads naturally to the appearance of particles and antiparticles; the ground state of the whole system is the vacuum, in which particles and antiparticles are totally absent; at the same time, there is no question of negative energies. Such a picture should be appreciated as a real achievement of the quantum field theory. It differs radically from the older approach formulated within the framework of the relativistic quantum mechanics, which relied on the idea of the vacuum as a filled sea of negative energy states; as we have noted earlier (cf. the end of Chapter 13), such an idea is in a sense absurd and fails completely just in the case of bosonic particles. Thus, it is reassuring that the theory of quantized complex Klein–Gordon field accommodates bosons and leads to a natural prediction of antiparticles and their consistent description.

# Chapter 19

# Quantization of Dirac field. Anticommutators

The next stop on our quantization tour is the Dirac field. We will see that this case exhibits some fundamentally new features that go far beyond our previous experience (the most important point has already been indicated in the title of this chapter).

Let us start with the conventional Lagrangian (14.27) that we reproduce here for the reader's convenience:

$$\mathcal{L} = i\bar{\psi}\gamma^\mu \partial_\mu \psi - m\bar{\psi}\psi . \tag{19.1}$$

As we know, the independent dynamical variables are $\psi(x)$ and $\bar{\psi}(x)$. One may try to develop the canonical formalism in accordance with the paradigm established in previous chapters, but it is clear that one immediately runs into a difficulty. Indeed, while the conjugate momentum for $\psi$ is

$$\frac{\partial \mathcal{L}}{\partial(\partial_0 \psi)} = i\bar{\psi}\gamma_0 = i\psi^\dagger , \tag{19.2}$$

for $\bar{\psi}$ one obviously gets zero. Well, an astute reader might object that one could invoke the alternative form (14.31), which is more symmetric in $\psi$ and $\bar{\psi}$. However, in fact, the real problem lies deeper. It turns out that when sticking to commutators of field operators, one would reach an impasse!

To elucidate this crucial point, we are going to evaluate the energy and momentum of the Dirac field. For such a purpose, the Lagrangian (19.1) can be used safely. Let us recall that according to our earlier results (15.30), (15.54) we have

$$H = \int d^3x \, \psi^\dagger i \, \partial_0 \psi , \tag{19.3}$$

$$\vec{P} = \int d^3x \, \psi^\dagger \left(-i\vec{\nabla}\right)\psi . \tag{19.4}$$

Now, one may employ the formula (10.19) for a general solution of the Dirac equation (we have considered it originally in the context of relativistic quantum mechanics, but, as we have stressed repeatedly, it is equally valid for the classical field). So, one has

$$\psi(x) = \int d^3p \sum_s N_p \left[ b(p,s)u(p,s)\,e^{-ipx} + d^*(p,s)v(p,s)\,e^{ipx} \right], \tag{19.5}$$

where all used symbols have the standard meaning defined in preceding chapters. Concerning the conventional notation $d^*(p,s)$, it is the same story as in the case of the complex Klein–Gordon field: the quantum version of the formula (19.5) reads

$$\psi(x) = \int d^3p \sum_s N_p \left[ b(p,s)u(p,s) e^{-ipx} + d^\dagger(p,s)v(p,s) e^{ipx} \right], \qquad (19.6)$$

where $b(p,s)$, $d(p,s)$ are operators, and we anticipate the role of $d^\dagger(p,s)$ as a creation operator. Of course, the Hermitian conjugation of the formula (19.6) then reads

$$\psi^\dagger(x) = \int d^3p \sum_s N_p \left[ b^\dagger(p,s)u^\dagger(p,s) e^{ipx} + d(p,s)v^\dagger(p,s) e^{-ipx} \right]. \qquad (19.7)$$

Substituting the expressions (19.6), (19.7) into (19.3), one integrates first the exponentials and then the resulting delta functions are used up. When these routine manipulations are carried out, we are left with

$$H = \int d^3p \sum_{s,s'} N_p^2 (2\pi)^3 p_0 \Big[ b^\dagger(p,s)b(p,s')u^\dagger(p,s)u(p,s')$$

$$- d(p,s)d^\dagger(p,s')v^\dagger(p,s)v(p,s')$$

$$- b^\dagger(p,s)d^\dagger(\tilde{p},s')u^\dagger(p,s)v(\tilde{p},s') e^{2ip_0x_0}$$

$$+ d(p,s)b(\tilde{p},s')v^\dagger(p,s)u(\tilde{p},s') e^{-2ip_0x_0} \Big], \qquad (19.8)$$

where $\tilde{p} = (p_0, -\vec{p})$ if $p = (p_0, \vec{p})$. Now one may utilize the results mentioned in Chapter 10 (cf. the text following Eq. (10.20)), namely

$$u^\dagger(p,s)u(p,s') = 2p_0\delta_{ss'}, \qquad v^\dagger(p,s)v(p,s') = 2p_0\delta_{ss'} \qquad (19.9)$$

and

$$u^\dagger(p,s)v(\tilde{p},s') = 0, \qquad v^\dagger(p,s)u(\tilde{p},s') = 0 \qquad (19.10)$$

(let us recall that these relations are derived by using the Gordon identity (10.5) and its siblings summarized in Appendix C). Of course, it is reassuring that the relations (19.10) hold, since the time-dependent terms in (19.8) are thereby eliminated. Thus, the final result for the Hamiltonian reads

$$H = \int d^3p \sum_s p_0 \left[ b^\dagger(p,s)b(p,s) - d(p,s)d^\dagger(p,s) \right]. \qquad (19.11)$$

In a similar way, for the momentum (19.4) one gets

$$\vec{P} = \int d^3p \sum_s \vec{p} \left[ b^\dagger(p,s)b(p,s) - d(p,s)d^\dagger(p,s) \right]. \qquad (19.12)$$

Note that the formulae (19.11) and (19.12) can be modified readily so as to get the energy and momentum of the classical Dirac field configuration defined by (19.5). In particular, for the energy one gets, *mutatis mutandis*,

$$E = \int d^3p \sum_s p_0 \left( |b(p,s)|^2 - |d(p,s)|^2 \right). \qquad (19.13)$$

So, in contrast to the case of the scalar field, the energy of Dirac field is, in general, not positive.

Now we are in a position to take up our main task, i.e. to formulate a meaningful theory of quantized Dirac field, including its particle interpretation. For convenience, let us first switch to the familiar alternative dialect, describing the field confined in a finite box. Of course, it means that the formulae (19.11), (19.12) are recast, as usual, in the form of the sums over the admissible discrete values of $\vec{p}$:

$$H = \sum_{\vec{p}} \sum_s p_0 \left( b_{p,s}^\dagger b_{p,s} - d_{p,s} d_{p,s}^\dagger \right),$$

$$\vec{P} = \sum_{\vec{p}} \sum_s \vec{p} \left( b_{p,s}^\dagger b_{p,s} - d_{p,s} d_{p,s}^\dagger \right). \tag{19.14}$$

It is seen that the expressions (19.14) have the right "oscillator-like" structure, so that $b$, $b^\dagger$, etc. are good candidates for possible annihilation and creation operators. According to our previous experience, for achieving a natural particle interpretation of the eigenstates of $H$ and $\vec{P}$, appropriate commutation relations for creation and annihilation operators are instrumental. However, in the present case the commutation relations are useless, since the problem of negative energy would persist; the point is that $H$ is a difference of two positive operators and the minus sign inside (19.14) is not removed upon commuting $d$ and $d^\dagger$. The radical way out, historically due to Pascual Jordan and Eugene Wigner, is to introduce **anticommutation relations** for $d$ and $d^\dagger$ (and, subsequently, for $b$, $b^\dagger$ as well). So, let us proceed by postulating the set of anticommutation relations

$$\{b_{p,s}, b_{q,s'}^\dagger\} = \delta_{ss'} \delta_{pq}, \qquad \{d_{p,s}, d_{q,s'}^\dagger\} = \delta_{ss'} \delta_{pq},$$

$$\{b_{p,s}, b_{q,s'}\} = 0, \qquad \{d_{p,s}, d_{q,s'}\} = 0, \tag{19.15}$$

$$\{b_{p,s}, d_{q,s}\} = 0, \qquad \{b_{p,s}, d_{q,s'}^\dagger\} = 0$$

(further relevant identities are given by the Hermitian conjugation of (19.15)). Briefly, one may say that $\{b, b^\dagger\}$ and $\{d, d^\dagger\}$ are non-trivial (being expressed in terms of the notorious Kronecker symbols), and all remaining anticommutators vanish. Needless to say, if we decide to work in the infinite space, i.e. use a continuous variable $\vec{p}$ like in (19.11), (19.12), $\delta_{pq}$ in (19.15) is replaced with $\delta^{(3)}(\vec{p} - \vec{q})$.

The impact of the anticommutators (19.15) on the expressions (19.14) is immediately clear. In particular, using the relation $d_{p,s} d_{p,s}^\dagger + d_{p,s}^\dagger d_{p,s} = 1$, one has

$$H = \sum_{\vec{p}} \sum_s p_0 \left( b_{p,s}^\dagger b_{p,s} + d_{p,s}^\dagger d_{p,s} - 1 \right), \tag{19.16}$$

and, discarding the (negative) infinite constant emerging from (19.16), one is left with

$$H = \sum_{\vec{p}} \sum_s p_0 \left( b_{p,s}^\dagger b_{p,s} + d_{p,s}^\dagger d_{p,s} \right). \tag{19.17}$$

Such a redefinition guarantees positivity of the Hamiltonian (19.17); now it remains to be clarified whether $b^\dagger$, $b$, etc. may play indeed the desired role of creation and annihilation operators. Fortunately, it is so. The crucial point is that even though the basic relations (19.15) involve anticommutators, for the composite commutators of the type $[b^\dagger b, b^\dagger]$, etc. we recover the same structure as before, in the case of a scalar field. To see this, one employs the general identity

$$[AB, C] = A\{B, C\} - \{A, C\}B. \tag{19.18}$$

Using (19.15), one then gets, for example,

$$[b_{p,s}^\dagger b_{p,s}, b_{q,s'}^\dagger] = b_{p,s}^\dagger \{b_{p,s}, b_{q,s'}^\dagger\} - \{b_{p,s}^\dagger, b_{q,s'}^\dagger\} b_{p,s}$$
$$= \delta_{pq}\delta_{ss'} b_{p,s}^\dagger . \tag{19.19}$$

Similarly,

$$[b_{p,s}^\dagger b_{p,s}, b_{q,s'}] = b_{p,s}^\dagger \{b_{p,s}, b_{q,s'}\} - \{b_{p,s}^\dagger, b_{q,s'}\} b_{p,s}$$
$$= -\delta_{pq}\delta_{ss'} b_{p,s} . \tag{19.20}$$

Of course, analogous relations are obtained for the $d$-operators. The above results then imply readily

$$[H, b_{p,s}^\dagger] = p_0 b_{p,s}^\dagger ,$$
$$[H, b_{p,s}] = -p_0 b_{p,s} , \tag{19.21}$$

and similarly for $d_{p,s}^\dagger$, $d_{p,s}$. For the operator $\vec{P}$ one obviously gets analogous identities, with $p_0$ replaced by $\vec{p}$.

As we already know from the previous chapters, the relations like (19.21) represent the true basis for the particle interpretation of a quantized field. So, we have achieved our goal, at least a substantial part of it: $b^\dagger$ and $d^\dagger$ can be interpreted as creation operators, and $b$, $d$ are their annihilation counterparts. It remains to be clarified how can one distinguish the "$d$-particles" from "$b$-particles" (note that using just the information concerning the energy and momentum, both types of particles obviously carry the same mass). Having in mind our experience with the complex scalar field, it is natural to expect that we might uncover here a particle-antiparticle connection. To verify such an educated guess, we will consider the charge corresponding to the conserved current $J^\mu(x) = \bar{\psi}(x)\gamma^\mu\psi(x)$ derived in Chapter 15 (cf. (15.60)), i.e.

$$Q = \int d^3x\, \bar{\psi}(x)\gamma_0\psi(x) = \int d^3x\, \psi^\dagger(x)\psi(x) . \tag{19.22}$$

Substituting the expressions (19.6), (19.7) into (19.22) and performing basically the same manipulations as before, one obtains the result

$$Q = \int d^3p \sum_s \left[ b^\dagger(p,s)b(p,s) + d(p,s)d^\dagger(p,s) \right] \tag{19.23}$$

(note that such a result is in fact a natural counterpart of our earlier formula (10.21) obtained within the framework of relativistic quantum mechanics). For the field placed inside a box the formula for $Q$ reads, of course,

$$Q = \sum_{\vec{p}} \sum_s \left( b_{p,s}^\dagger b_{p,s} + d_{p,s} d_{p,s}^\dagger \right) . \tag{19.24}$$

So, $Q$ is a positive operator; however, by means of the anticommutation relations (19.15) it may be recast in the normal ordered form, namely

$$Q = \sum_{\vec{p}} \sum_s \left( b_{p,s}^\dagger b_{p,s} - d_{p,s}^\dagger d_{p,s} \right) \tag{19.25}$$

124

(needless to say, when passing from (19.24) to (19.25) a positive infinite constant is subtracted). The expression (19.25) is precisely what we need. Using (19.15) one gets

$$[Q, b^\dagger_{p,s}] = b^\dagger_{p,s},$$
$$[Q, d^\dagger_{p,s}] = -d^\dagger_{p,s}, \tag{19.26}$$

so that the one-particle states $b^\dagger_{p,s}|0\rangle$ and $d^\dagger_{p,s}|0\rangle$ carry opposite eigenvalues of $Q$. In this way, the description of states of the quantized Dirac field, incorporating particles and antiparticles, is implemented. Conventionally, we will associate particles with the $b$-operators and the $d$-operators will represent antiparticles.

As regards the structure of the corresponding Fock space, there is a very important point that must be mentioned here. The anticommutation relations

$$\{b^\dagger_{p,s}, b^\dagger_{q,s'}\} = 0, \qquad \{d^\dagger_{p,s}, d^\dagger_{q,s'}\} = 0$$

mean, in particular, that

$$\left(b^\dagger_{p,s}\right)^2 = 0, \qquad \left(d^\dagger_{p,s}\right)^2 = 0. \tag{19.27}$$

This in turn leads to enormous simplification of the possible contents of the particle states: the identities (19.27) imply, obviously, that the same state cannot be occupied by more than one particle (antiparticle). In other words, the occupation numbers, characterizing the state vectors in the Fock space of the quantized Dirac field, can take on just the values 0 and 1. Of course, such a statement is the famous **Pauli exclusion principle**. Thus, the particles and antiparticles corresponding to a quantum Dirac field are **fermions** (i.e. they obey the Fermi–Dirac statistics). This is another example of the spin-statistics connection, since there is a good reason to believe that we are working here with spin-$\frac{1}{2}$ particles (note that a clear hint to this point is provided by the form of the Dirac field angular momentum shown in (15.52)). In fact, the problem of the spin labels of the particles we have described up to now in terms of the creation operators $b^\dagger_{p,s}$ and $d^\dagger_{p,s}$ would deserve a more detailed treatment and we defer it to the Appendix D. Throughout the subsequent chapters we will assume, *bona fide*, that the spin state of a particle in question is described by means of the spin four-vector $s$, in much the same way as in the framework of relativistic quantum mechanics.

# Chapter 20

# Quantization of massive vector field

---

Following the sequence of relativistic wave equations for classical fields, the next example is the massive vector field satisfying the Proca equation that we have discussed earlier (cf. chapters 12 and 14). Let us stress again that while the basic equation for a four-vector function $A_\mu(x)$ that we are going to use is the same as in Chapter 12, its interpretation is totally different: in Chapter 12, we have dealt with a relativistic quantum mechanical wave function for a single spin-1 particle, but now we have in mind the classical field described in Chapter 14 (which, admittedly, is a rather abstract quantity).

The quantization of a massive vector (Proca) field is quite interesting for more than one reason. First of all, it is designed to describe massive spin-1 bosons (which, as we know, play a crucial role in weak interactions); in fact, in many applications of quantum electrodynamics one may also treat the photon as a massless limit of a massive vector boson. Moreover, the relevant quantization scheme follows basically the canonical paradigm used before for a scalar field, although one is dealing here with a set of dynamical variables subject to a non-trivial constraint. Anyway, such a procedure is conceptually simpler than a quantization of massless Maxwell field, where one has to cope with consequences of the electromagnetic gauge invariance.

So, let us see how the canonical quantization of the Proca field can be implemented. For simplicity, we are going to consider the case of a real field. According to our previous results, the Lagrangian density is given by (14.18) and the corresponding equation of motion reads

$$\partial_\mu F^{\mu\nu} + m^2 A^\nu = 0, \tag{20.1}$$

where $F_{\mu\nu} = \partial_\mu A_\nu - \partial_\nu A_\mu$. This, as we know, is equivalent to the pair of Klein–Gordon equation and a "Lorenz-like constraint", namely

$$\left(\Box + m^2\right) A^\mu = 0, \qquad \partial^\mu A_\mu = 0. \tag{20.2}$$

For a definition of the relevant canonical variables, the formula (14.24) is instrumental, so let us reproduce it here for reader's convenience:

$$\frac{\partial \mathscr{L}}{\partial(\partial_\rho A_\sigma)} = -F^{\rho\sigma}. \tag{20.3}$$

Supposing that the field components $A_\mu(x)$ play the role of generalized coordinates, the corresponding conjugate momenta can be defined, conventionally, by

$$A_\mu \longrightarrow \pi_\mu = \frac{\partial \mathscr{L}}{\partial(\partial_0 A_\mu)} = -F^{0\mu}. \tag{20.4}$$

It is seen that $\pi_0$ would be then identically zero, so it makes sense to use just $A_j(x)$, $j = 1, 2, 3$, as dynamical variables, and $A_0(x)$ is supposed to be calculated in terms of these, or the conjugate momenta $\pi_j(x)$. Thus, the canonical equal-time commutation relations may be written (in analogy with the scalar field case described in Chapter 16) as

$$[A_j(x), \pi_k(y)]_{\text{E.T.}} = i\delta_{jk}\delta^{(3)}(\vec{x} - \vec{y}),$$
$$[A_j(x), A_k(y)]_{\text{E.T.}} = 0, \qquad (20.5)$$
$$[\pi_j(x), \pi_k(y)]_{\text{E.T.}} = 0.$$

Note also that according to (20.4) one has

$$\pi_j = F_{0j} \quad (= \partial_0 A_j - \partial_j A_0). \qquad (20.6)$$

Now, the question is how to express $A_0$ in terms of the above canonical variables. For this purpose, one may invoke the equation of motion (20.1). One gets readily

$$A_0 = -\frac{1}{m^2}\partial_j F_{0j}, \qquad (20.7)$$

and this means, in view of (20.6),

$$A_0 = -\frac{1}{m^2}\partial_j \pi_j. \qquad (20.8)$$

As we will see shortly, the relation (20.8) is crucial for a successful implementation of the whole quantization procedure. In addition to (20.8), $\dot{A}_0$ can be immediately obtained from the constraint $\partial^\mu A_\mu = 0$:

$$\dot{A}_0 = \partial_j A_j. \qquad (20.9)$$

Of course, our ultimate goal is the particle interpretation of the quantized field, based on the appropriate creation and annihilation operators. To this end, one may proceed in analogy with the previous examples of the scalar and spinor field, i.e. write a solution of the equations (20.2) as a general superposition of plane waves and establish commutation relations of the corresponding expansion coefficients. We have examined the properties of plane-wave solutions of the Proca equation in Chapter 12, so it comes in handy now. The general expression for $A_\mu(x)$ that we need can be written as

$$A_\mu(x) = \int d^3k \, N_k \sum_{\lambda=1}^{3} \left[ a(k,\lambda)\varepsilon_\mu(k,\lambda) \, e^{-ikx} + a^\dagger(k,\lambda)\varepsilon_\mu^*(k,\lambda) \, e^{ikx} \right], \qquad (20.10)$$

where $\varepsilon_\mu(k,\lambda)$ are the three space-like polarization vectors satisfying the conditions (12.8), (12.28), and $N_k = (2\pi)^{-3/2}(2k_0)^{-1/2}$ is the usual conventional normalization factor; of course, $k_0 = (\vec{k}^2 + m^2)^{1/2}$. The next key step is deriving the commutation relations for the operator coefficients $a$, $a^\dagger$. Basically, the procedure is the same as in the case of the scalar field discussed in Chapter 16, but the practical calculation is algebraically much more complicated. First of all, one must express the operators $a$, $a^\dagger$ in terms of the original canonical variables $A_j$, $\pi_j$ appearing in the set of commutators (20.5). For this purpose, it is convenient to introduce the combinations

$$a_\mu(k) = \sum_{\lambda=1}^{3} a(k,\lambda)\varepsilon_\mu(k,\lambda),$$
$$a_\mu^\dagger(k) = \sum_{\lambda=1}^{3} a^\dagger(k,\lambda)\varepsilon_\mu^*(k,\lambda). \qquad (20.11)$$

Then the expansion (20.10) is recast as

$$A_\mu(x) = \int d^3k\, N_k \left[ a_\mu(k)\, e^{-ikx} + a_\mu^\dagger(k)\, e^{ikx} \right], \qquad (20.12)$$

and the operators $a_\mu(k)$, $a_\mu^\dagger(k)$ can be expressed with the help of the relations analogous to (16.9), namely

$$a_\mu(k) = i \int d^3x\, f_k^*(x) \overset{\leftrightarrow}{\partial_0} A_\mu(x),$$
$$a_\mu^\dagger(k) = -i \int d^3x\, f_k(x) \overset{\leftrightarrow}{\partial_0} A_\mu(x), \qquad (20.13)$$

where $f_k(x)$, $f_k^*(x)$ are given by (16.7). Now it is clear that for the evaluation of commutators involving $a_\mu(k)$ and $a_\mu^\dagger(k)$ one needs to know all possible equal-time commutators of $A_\mu$ and $\dot{A}_\mu$. This is why the relation (20.8) is so important: using it, along with (20.6) and (20.9), all commutators in question can be reduced to the basic relations (20.5). The corresponding calculation is quite boring, long and tedious; according to a little known literary character, it might be called "the horrors of the unborn" (in Czech: "hrůzy nezrozeného").[11] Anyway, let us at least summarize the results for the commutators in question. An awkward feature of the formulae shown below is that they cannot be presented in a covariant form (this, of course, is due to the intrinsically non-covariant way of distinguishing between $A_0$ and $A_j$, $j = 1, 2, 3$). So, there are basically three types of commutators:

a) Type $[A_\mu(x), A_\nu(y)]_{\text{E.T.}}$

$$[A_0(x), A_j(y)]_{\text{E.T.}} = -\frac{i}{m^2} \frac{\partial}{\partial x^j} \delta^{(3)}(\vec{x} - \vec{y}),$$
$$[A_j(x), A_k(y)]_{\text{E.T.}} = 0,$$
$$[A_0(x), A_0(y)]_{\text{E.T.}} = 0. \qquad (20.14)$$

b) Type $[\dot{A}_\mu(x), \dot{A}_\nu(y)]_{\text{E.T.}}$

$$[\dot{A}_0(x), \dot{A}_j(y)]_{\text{E.T.}} = i\frac{\partial}{\partial x^j} \delta^{(3)}(\vec{x} - \vec{y}) - \frac{i}{m^2} \frac{\partial}{\partial x^j} \Delta \delta^{(3)}(\vec{x} - \vec{y}),$$
$$[\dot{A}_j(x), \dot{A}_k(y)]_{\text{E.T.}} = 0,$$
$$[\dot{A}_0(x), \dot{A}_0(y)]_{\text{E.T.}} = 0. \qquad (20.15)$$

c) Type $[A_\mu(x), \dot{A}_\nu(y)]_{\text{E.T.}}$

$$[A_0(x), \dot{A}_0(y)]_{\text{E.T.}} = -\frac{i}{m^2} \Delta \delta^{(3)}(\vec{x} - \vec{y}),$$
$$[A_0(x), \dot{A}_j(y)]_{\text{E.T.}} = 0,$$
$$[A_j(x), \dot{A}_0(y)]_{\text{E.T.}} = 0,$$
$$[A_j(x), \dot{A}_k(y)]_{\text{E.T.}} = i\left( \delta_{jk} - \frac{1}{m^2} \frac{\partial}{\partial x^j} \frac{\partial}{\partial x^k} \right) \delta^{(3)}(\vec{x} - \vec{y}). \qquad (20.16)$$

---

[11] By way of explanation: True connoisseurs of the novel "The good soldier Švejk" by J. Hašek may recognize in it a statement of the cook occultist Jurajda concerning a difficult and confusing situation.

Well, the above list appears to be a welter of cumbersome formulae, but when these are utilized in commutators of the operators $a_\mu(k)$, $a_\nu^\dagger(k)$ expressed by means of (20.13), the final result that emerges from such an "ugly-duckling form" is quite elegant; one gets

$$[a_\mu(k), a_\nu^\dagger(k')] = \left(-g_{\mu\nu} + \frac{1}{m^2} k_\mu k_\nu\right) \delta^{(3)}(\vec{k} - \vec{k}'),$$

$$[a_\mu(k), a_\nu(k')] = 0.$$

(20.17)

From (20.17) it is then easy to obtain the corresponding commutators for $a(k, \lambda)$, $a^\dagger(k, \lambda)$. Taking into account the definition (20.11) and employing the orthogonality relation (12.28) one gets readily

$$a(k, \lambda) = -\varepsilon^{*\mu}(k, \lambda) a_\mu(k),$$

(20.18)

so that

$$[a(k, \lambda), a^\dagger(k', \lambda')] = [-\varepsilon^{*\mu}(k, \lambda) a_\mu(k), -\varepsilon^\nu(k', \lambda') a_\nu^\dagger(k')]$$

$$= \varepsilon^{*\mu}(k, \lambda) \varepsilon^\nu(k', \lambda') \left(-g_{\mu\nu} + \frac{1}{m^2} k_\mu k_\nu\right) \delta^{(3)}(\vec{k} - \vec{k}')$$

$$= -\varepsilon^*(k, \lambda) \cdot \varepsilon(k, \lambda') \delta^{(3)}(\vec{k} - \vec{k}')$$

$$= \delta_{\lambda\lambda'} \delta^{(3)}(\vec{k} - \vec{k}'),$$

(20.19)

where we have used again the basic properties of the polarization vectors. Obviously, one also gets immediately

$$[a(k, \lambda), a(k', \lambda')] = 0.$$

(20.20)

Thus, the results (20.19), (20.20) suggest that we are on the right track for the interpretation of $a(k, \lambda)$ and $a^\dagger(k, \lambda)$ as an annihilation and creation operator, respectively. Of course, as before, a necessary further step towards this goal is the calculation of the field energy and momentum. For this purpose, it is convenient to employ an alternative Lagrangian for the Proca field, which is equivalent to the original form (14.18); it reads

$$\widetilde{\mathscr{L}} = -\frac{1}{2}(\partial_\mu A_\nu)(\partial^\mu A^\nu) + \frac{1}{2}(\partial_\mu A^\mu)^2 + \frac{1}{2} m^2 A_\mu A^\mu.$$

(20.21)

The equivalence of (20.21) and (14.18) is verified easily. To derive the corresponding equation of motion, one gets first

$$\frac{\partial \widetilde{\mathscr{L}}}{\partial(\partial_\rho A_\sigma)} = -\partial^\rho A^\sigma + g^{\rho\sigma} \partial \cdot A,$$

(20.22)

and utilizing it in the general relation (14.10), one obtains readily the Proca equation (20.1). In addition to that, one may check that the Lagrangians (14.18) and (20.21) differ by a four-divergence. Indeed, the form (14.18) may be recast as

$$\mathscr{L} = -\frac{1}{2}(\partial_\mu A_\nu)(\partial^\mu A^\nu) + \frac{1}{2}(\partial_\mu A_\nu)(\partial^\nu A^\mu) + \frac{1}{2} m^2 A_\mu A^\mu,$$

(20.23)

and one can prove easily that the following identity holds:

$$(\partial_\mu A^\mu)^2 - (\partial_\mu A_\nu)(\partial^\nu A^\mu) = \partial_\mu \left(A^\mu \overset{\leftrightarrow}{\partial_\nu} A^\nu\right).$$

(20.24)

Thus, $\mathscr{L}$ and $\widetilde{\mathscr{L}}$ differ by a term proportional to the right-hand side of (20.24). To evaluate the energy (the field Hamiltonian), one has to employ the formula for $\mathscr{T}_{\text{can.}}^{00}$ (cf. (15.19)) expressed

129

in terms of $\widetilde{\mathcal{L}}$, substitute into it the solution of the equation of motion given by (20.10) and integrate over $\mathrm{d}^3 x$. The calculation is somewhat tedious, but it proceeds basically in the same manner as e.g. for the scalar field; of course, it is algebraically slightly more complicated (but, in fact, one gets along with the orthogonality relation (12.28) for the polarization vectors). The reader is encouraged to carry out the calculation in detail; in any case it is a useful revision exercise. As one may have rightly expected, the final result is

$$H = \frac{1}{2} \int \mathrm{d}^3 k \sum_{\lambda=1}^{3} k_0 \left[ a^\dagger(k, \lambda) a(k, \lambda) + a(k, \lambda) a^\dagger(k, \lambda) \right], \qquad (20.25)$$

and this may be subsequently recast in the normal-ordered form. The evaluation of the momentum operator proceeds similarly and it is easy to guess the result, which can be written as an obvious analogue of our previous formulae. An analysis of the angular momentum would be technically more complicated, so for the time being we will simply take for granted that the label $\lambda = 1, 2, 3$ corresponds to a definite spin (helicity) state of the spin-1 particle, in analogy with our earlier treatment within the framework of relativistic quantum mechanics.

Finally, let us add that the above discussion can be generalized in a straightforward manner to the case of a complex (non-Hermitian) Proca field; in such a case, the pair of the operators $a$, $a^\dagger$ appearing in (20.10) is replaced with $b$, $d^\dagger$ in much the same way as for the complex scalar field described in Chapter 18.

# Chapter 21

# Interactions of classical and quantum fields

---

The quantization of free fields that we have pursued in preceding chapters is certainly quite remarkable and elegant way of a description of relativistic particles. Particularly valuable is the connection of particle states with the creation and annihilation operators satisfying appropriate commutation or anticommutation relations. On the other hand, a set of totally free particles is a rather dull stuff, since the real life is obviously based on interactions. So, a next step forward is a treatment of dynamical processes involving particles, in terms of the field interactions.

Let us start with classical fields. A formal description of an envisaged interaction is, conceptually, quite simple: the idea is to have a coupling of fields that influence mutually each other (it is also possible to imagine a situation where a field influences itself; in such a case one may speak of a self-interaction). An instructive example is a system of coupled spinor (Dirac) and scalar (Klein–Gordon) fields. Full Lagrangian for such a system can be written e.g. as

$$\mathcal{L} = \frac{1}{2}\partial_\mu \varphi \, \partial^\mu \varphi - \frac{1}{2}M^2\varphi^2 + i\bar{\psi}\gamma^\mu\partial_\mu\psi - m\bar{\psi}\psi + g\bar{\psi}\psi\varphi \,. \tag{21.1}$$

The notation used in (21.1) is standard, so it is clear that the first four monomials constitute the familiar free Lagrangians discussed earlier. The last term represents an interaction; the dimensionless parameter $g$ is therefore traditionally called the "coupling constant". (Note that the type of the interaction shown in (21.1) is called, for historical reasons, the Yukawa coupling.) The role of the interaction term appearing in (21.1) becomes quite transparent when the corresponding equations of motion are written down. Using the general relations (14.10), one gets easily the Lagrange equations for the considered system; these read

$$\left(\Box + M^2\right)\varphi = g\bar{\psi}\psi \,,$$
$$i\gamma^\mu\partial_\mu\psi - m\psi + g\varphi\psi = 0 \,. \tag{21.2}$$

From (21.2) it is clear that the fields $\varphi$ and $\psi$ influence each other owing to the presence of the interaction; of course, the two equations become decoupled for $g = 0$.

Another example is the interaction of Dirac field with the electromagnetic Maxwell field; the corresponding Lagrangian has the form

$$\mathcal{L} = i\bar{\psi}\gamma^\mu\partial_\mu\psi - m\bar{\psi}\psi - \frac{1}{4}F_{\mu\nu}F^{\mu\nu} + e\bar{\psi}\gamma^\mu\psi A_\mu \,, \tag{21.3}$$

where, of course, $F_{\mu\nu} = \partial_\mu A_\nu - \partial_\nu A_\mu$. Again, in (21.3) one may distinguish the free part of $\mathcal{L}$ and an interaction, which now has the structure "current × field" (recall that $\bar{\psi}\gamma^\mu\psi$ is a conserved vector current in Dirac theory). It is worth noting that (21.3) may be recast as

$$\mathcal{L} = -\frac{1}{4}F_{\mu\nu}F^{\mu\nu} + i\bar{\psi}\gamma^\mu(\partial_\mu - ieA_\mu)\psi - m\bar{\psi}\psi \,. \tag{21.4}$$

The point is that the presence of the **covariant derivative** $D_\mu = \partial_\mu - ieA_\mu$ guarantees the invariance of the considered Maxwell–Dirac Lagrangian under **gauge transformations**

$$\psi'(x) = e^{i\omega(x)}\psi(x),$$
$$\overline{\psi}'(x) = e^{-i\omega(x)}\overline{\psi}(x),$$
$$A'_\mu(x) = A_\mu(x) + \frac{1}{e}\partial_\mu\omega(x),$$
(21.5)

where $\omega(x)$ is a real function. In other words, the expression (21.4) exhibits manifestly the **local U(1) symmetry**, which is a familiar feature of classical electrodynamics. The form (21.3) or (21.4) can be easily extended to incorporate also a mass term for the vector field $A_\mu$ (i.e. replace Maxwell with Proca); then, of course, the gauge invariance under (21.5) is lost, one is left just with global U(1) symmetry (corresponding to $\omega(x) \equiv \omega = \text{const.}$). Anyway, the current conservation is maintained in such a case. Adding the mass term $\frac{1}{2}M^2 A_\mu A^\mu$ to (21.3), one gets the equations of motion

$$\partial_\mu F^{\mu\nu} + M^2 A^\nu = -e\overline{\psi}\gamma^\nu\psi,$$
$$\left[i\gamma^\mu\left(\partial_\mu - ieA_\mu\right) - m\right]\psi = 0,$$
(21.6)

and thus it is seen again how the fields $\psi$ and $A_\mu$ are intertwined within the couple (21.6).

Finally, let us mention a simple example of a **self-interaction** of real scalar field. The Lagrangian in question reads

$$\mathscr{L} = \frac{1}{2}\partial_\mu\varphi\,\partial^\mu\varphi - \frac{1}{2}m^2\varphi^2 - \frac{1}{4}\lambda\varphi^4.$$
(21.7)

For definiteness, we assume that the coupling constant $\lambda$ is positive, so as to get a positive energy density (please check it). Note that the quartic term appearing in (21.7) occurs also in the standard model (SM) of particle physics, where $\varphi$ may be identified with the Higgs field (note, however, that this sector of SM has not been fully tested yet). From (21.7) one obtains easily the corresponding equation of motion:

$$\left(\Box + m^2\right)\varphi = -\lambda\varphi^3.$$
(21.8)

So, the "source term" on the right-hand side of Eq. (21.8) (cf. also (21.6)) is made of the field $\varphi$ itself and this makes the concept of self-interaction manifest.

The above considerations give a clear instruction for proceeding to the QFT case. One may utilize interaction terms written as products of quantum field operators, and figure out a viable procedure of calculating transition amplitudes for decay and scattering processes. Recall that, as we have already noted in Chapter 13, the formalism of quantized fields based on creation and annihilation operators should lead naturally to a description of physical processes, in which a transmutation of particles occurs, and also their number may be changed when passing from the initial to a final state (in the Fock space). Well, the problem is that we have only been able to quantize the free fields, but still one would like to utilize this valuable knowledge for calculations of dynamical processes involving interacting particles. It is not difficult to guess that a practicable scheme could be provided by an appropriate form of **perturbation theory** (assuming, *bona fide*, that the particle interactions in question are not too strong). To develop the pertinent technique, we are going to make now a brief digression concerning the general formalism for the description of time evolution in quantum theory (for most of the readers it will be probably just a revision of a part of knowledge gained in an earlier course of quantum mechanics).

To begin with, let us recall the two best known "pictures" (or representations) of time evolution. In the Schrödinger picture, the state vectors are time-dependent and satisfy the familiar equation

$$i \frac{\partial}{\partial t} |\Psi(t)\rangle = H |\Psi(t)\rangle , \qquad (21.9)$$

with $H$ being the relevant Hamiltonian, while an operator $A = A_S$ representing a standard observable is time-independent. In the Heisenberg picture it is the other way round, so that the observables are viewed as time-dependent dynamical variables

$$A_H(t) = e^{iHt} A_S e^{-iHt} \qquad (21.10)$$

(where one has to take into account that the label H has a different meaning in the l.h.s. and r.h.s. of Eq. (21.10)). One should notice that the value of a general matrix element of an observable is not changed upon transition between the two pictures; it holds

$$\langle \Phi_S(t) | A_S | \Psi_S(t) \rangle = \langle \Phi_H | A_H(t) | \Psi_H \rangle , \qquad (21.11)$$

where we have marked explicitly that the state vectors in the Heisenberg picture are constant in time.

An "interpolation" between the two above-mentioned schemes is the so-called **interaction picture** (sometimes also called **Dirac picture**). It is defined as follows. Suppose that the full Hamiltonian can be conveniently split into two terms,

$$H = H_0 + H_{int} , \qquad (21.12)$$

where $H_0$ is an appropriate "solvable" part (i.e. such that its spectrum and eigenstates can be computed explicitly); in particular, it may be the Hamiltonian for free particles. $H_{int}$ then denotes, conventionally, the "interaction term" (referring to the free Hamiltonian $H_0$). Typically, $H_{int}$ incorporates a small parameter that may warrant the use of perturbation expansion in powers of $H_{int}$. Let us see how such a scheme is implemented technically. The basic defining relation determines the passage from a time-independent operator $A$ in Schrödinger picture to $A_I(t)$ in the interaction picture:

$$A_I(t) = e^{iH_0t} A e^{-iH_0t} . \qquad (21.13)$$

Next, we require that all pictures of time evolution give the same results for matrix elements of the relevant operators; in such a way one can arrive at the corresponding law for the time-dependence of a state vector $|\Psi_I(t)\rangle$. So, one may impose e.g. the condition

$$\langle \Phi | A_H(t) | \Psi \rangle = \langle \Phi_I(t) | A_I(t) | \Psi_I(t) \rangle \qquad (21.14)$$

relating the Heisenberg and interaction pictures. Using (21.10) and (21.13), the identity (21.14) yields

$$|\Psi_I(t)\rangle = e^{iH_0t} e^{-iHt} |\Psi\rangle . \qquad (21.15)$$

Now we are in a position to derive an evolution equation for $|\Psi_I(t)\rangle$. From (21.15) one gets

$$\begin{aligned} \frac{\partial}{\partial t} |\Psi_I(t)\rangle &= \left( iH_0 e^{iH_0t} e^{-iHt} + e^{iH_0t} e^{-iHt} (-iH) \right) |\Psi\rangle \\ &= \left( i e^{iH_0t} H_0 e^{-iHt} - i e^{iH_0t} H e^{-iHt} \right) |\Psi\rangle \\ &= -i e^{iH_0t} (H - H_0) e^{-iHt} |\Psi\rangle . \end{aligned} \qquad (21.16)$$

However, according to (21.12), $H - H_0 = H_{int}$, so that the equation (21.16) may be recast conveniently as

$$i\frac{\partial}{\partial t}|\Psi_I(t)\rangle = e^{iH_0 t}\, H_{int}\, e^{-iH_0 t}\, e^{iH_0 t}\, e^{-iHt}|\Psi\rangle\,. \tag{21.17}$$

Taking into account the basic definitions (21.13) and (21.15), one thus gets finally

$$i\frac{\partial}{\partial t}|\Psi_I(t)\rangle = H_I^{int}(t)|\Psi_I(t)\rangle\,, \tag{21.18}$$

with

$$H_I^{int}(t) = e^{iH_0 t}\, H_{int}\, e^{-iH_0 t}\,. \tag{21.19}$$

In this way, one arrives at the desired equation, which determines the time evolution of a state vector just through the interaction part of the Hamiltonian (and the name of the considered picture is thereby justified). As any quantum theory practitioner knows, an equation like (21.18) may be solved formally by means of an evolution operator, which transforms a state vector at an initial time $t_0$ to its value at any other time $t$. So, one has

$$|\Psi_I(t)\rangle = U_I(t, t_0)|\Psi_I(t_0)\rangle\,, \tag{21.20}$$

and from (21.18) one obtains readily the differential equation for $U_I(t, t_0)$:

$$\frac{\partial}{\partial t}U_I(t, t_0) = -iH_I^{int}(t)U_I(t, t_0)\,. \tag{21.21}$$

Of course, an appropriate initial condition for the solution of Eq. (21.21) is

$$U_I(t_0, t_0) = \mathbb{1}\,. \tag{21.22}$$

The equation (21.21) may be recast in the form of an integral equation, namely

$$U_I(t, t_0) = \mathbb{1} - i\int_{t_0}^{t} dt'\, H_I^{int}(t')U_I(t', t_0)\,, \tag{21.23}$$

which incorporates automatically the initial condition (21.22). The main virtue of Eq. (21.23) is that it can be solved formally by means of iterations that lead eventually to an operator power expansion. Indeed, the first step amounts to

$$U_I(t, t_0) = \mathbb{1} - i\int_{t_0}^{t} dt'\, H_I^{int}(t')\left[\mathbb{1} - i\int_{t_0}^{t'} dt''\, H_I^{int}(t'')U_I(t'', t_0)\right], \tag{21.24}$$

and in such a way one may proceed further. Thus we end up with the result

$$U_I(t, t_0) = \mathbb{1} + (-i)\int_{t_0}^{t} H_I^{int}(t_1)\, dt_1 + (-i)^2\int_{t_0}^{t} dt_1\int_{t_0}^{t_1} dt_2\, H_I^{int}(t_1)H_I^{int}(t_2) + \dots$$

$$+ (-i)^n\int_{t_0}^{t} dt_1\int_{t_0}^{t_1} dt_2\cdots\int_{t_0}^{t_{n-1}} dt_n\, H_I^{int}(t_1)H_I^{int}(t_2)\cdots H_I^{int}(t_n) + \dots\,, \tag{21.25}$$

which is called the **Dyson series**.[12] It has obviously the form of a perturbation expansion, if $H_{int}$ may be viewed as a "small correction" (in whatever sense) to $H_0$ in the decomposition (21.12). One may notice that the individual terms in (21.25) exhibit a specific time ordering $t_1 > t_2 > \cdots t_n$ for the factors $H_1^{int}(t_j)$ in the integrands, which is naturally enforced by the structure of the original equation (21.23). It is useful to rewrite the expression (21.25) with the help of the formal operation of **chronological ordering**, or briefly **T-ordering**. For its general definition, let us start with two operator factors $A(t_1)$, $B(t_2)$. One defines

$$\begin{aligned} T\big(A(t_1)B(t_2)\big) &= A(t_1)B(t_2) \qquad \text{for } t_1 > t_2\,, \\ &= B(t_2)A(t_1) \qquad \text{for } t_2 > t_1\,. \end{aligned} \tag{21.26}$$

Then it is not difficult to realize that the third term in the expansion (21.25) may be recast as

$$\int_{t_0}^{t} dt_1 \int_{t_0}^{t_1} dt_2\, H_1(t_1)H_1(t_2) = \frac{1}{2!} \int_{t_0}^{t} \int_{t_0}^{t} dt_1\, dt_2\, T\big(H_1(t_1)H_1(t_2)\big)\,, \tag{21.27}$$

where we have omitted the label "int" for brevity. Indeed, the double integral on the right-hand side of (21.27) is equal

$$I = I_1 + I_2 = \int_{t_0}^{t} dt_1 \int_{t_0}^{t_1} dt_2\, H_1(t_1)H_1(t_2) + \int_{t_0}^{t} dt_1 \int_{t_1}^{t} dt_2\, H_1(t_2)H_1(t_1)\,. \tag{21.28}$$

However, in the second term in (21.28) one may interchange the order of integrations and get

$$I_2 = \int_{t_0}^{t} dt_2 \int_{t_0}^{t_2} dt_1\, H_1(t_2)H_1(t_1)\,. \tag{21.29}$$

Then the integration variables can be renamed, $t_1 \leftrightarrow t_2$, and thus it is immediately seen that $I_1 = I_2$. The factor $1/2!$ is thereby compensated and (21.27) is proven. Notice that no assumption concerning commutation properties of the operators $H_1(t)$ for different times was needed; the derivation of (21.27) relies just on standard rules for interchanging the order of integrations in a double integral.

It is straightforward to generalize the definition of T-ordering (T-product of operators) for an arbitrary number of factors. The operators appearing in the considered chain are ordered in such a way that the corresponding time arguments decrease from left to right. Obviously, for a product of $n$ factors there are $n!$ possible time orderings and one may repeat the previous argumentation based on interchanging the order of integrations in a multiple integral. In such a manner, the $n$-fold integral over the hypercube $(t_0, t) \times \cdots \times (t_0, t)$ is shown to consist of $n!$ identical contributions equal the $(n+1)$-th term in (21.25). This means that the formula (21.25) may be recast as

$$U_1(t, t_0) = \mathbb{1} + \sum_{n=1}^{\infty} \frac{(-i)^n}{n!} \int_{t_0}^{t} \cdots \int_{t_0}^{t} dt_1 \cdots dt_n\, T\big(H_1^{int}(t_1) \cdots H_1^{int}(t_n)\big)\,. \tag{21.30}$$

For obvious reason, the Dyson expansion (21.30) is usually written in a compact form as

$$U_1(t, t_0) = \text{Texp}\left(-i \int_{t_0}^{t} H_1^{int}(t')\, dt'\right) \tag{21.31}$$

---

[12]Freeman J. Dyson (1923-2020) was one of the founding fathers of modern QED.

(but one should still keep in mind that the definition of the symbol Texp is given by (21.30); the expression (21.31) is just a shorthand notation for the exponential-like series (21.30)).

So much for a digression concerning general methods of quantum theory that are instrumental in our subsequent discussion. Let us now come back to the field theory. The basic quantity we are working with is a Lagrangian density $\mathscr{L}$. From $\mathscr{L}$ one can proceed to the energy density $\mathscr{H}$, which is then integrated over the 3-dimensional space to get the full Hamiltonian; it remains to be clarified how the perturbative technique described above is implemented in a field theory model. For an illustration, we will consider the simple case of the $\varphi^4$ self-interaction shown in (21.7). Using the general formula (15.16), one gets readily

$$\mathscr{H} = \mathscr{T}^{00} = \frac{1}{2}\partial_0\varphi\,\partial_0\varphi + \frac{1}{2}\vec{\nabla}\varphi\,\vec{\nabla}\varphi + \frac{1}{2}m^2\varphi^2 + \frac{1}{4}\lambda\varphi^4. \tag{21.32}$$

Thus, it is natural to split $\mathscr{H}$ as

$$\mathscr{H} = \mathscr{H}_0 + \mathscr{H}_{\text{int}}, \tag{21.33}$$

with

$$\mathscr{H}_{\text{int}} = \frac{1}{4}\lambda\varphi^4. \tag{21.34}$$

One may notice that

$$\mathscr{H}_{\text{int}} = -\mathscr{L}_{\text{int}}, \tag{21.35}$$

which is an obvious consequence of the simple structure of the Lagrangian density (21.7). In fact, the relation (21.35) is valid for many other field theory models; for example, it is easy to find out that it is the case for the Yukawa interaction described by (21.1). Nevertheless, there are exceptions to this rule; we will discuss one prominent example in Chapter 29.

As regards the general formula (21.31), it is often used for a situation, in which one considers a transition from "distant past" to "distant future"; formally, one then sets $t_0 = -\infty$, $t = +\infty$ (an idealized description, of course). Typically, this is pertinent for the treatment of scattering or decay processes. Thus, it is not surprising that the corresponding evolution operator has a special name: it is the famous **S-matrix** ("$S$" from "scattering" in English or "Streuung" in German). So, one has, in the interaction picture,

$$S = U_{\text{I}}(+\infty, -\infty). \tag{21.36}$$

In field theory, the interaction Hamiltonian appearing in the formula (21.31) is the space integral of the density $\mathscr{H}_{\text{int}}$. This in turn means that in the expression for $S$ one ends up with the integral of $\mathscr{H}_{\text{int}}$ over the whole spacetime. Taking also into account the relation (21.35) (with the aforementioned caveat), the $S$-matrix operator may be written compactly as

$$S = \text{Texp}\left(i\int \mathscr{L}_{\text{I}}^{\text{int}}(x)\,\mathrm{d}^4x\right). \tag{21.37}$$

Now, the question is how to work it out practically. To this end, the key observation is that according to the definition of the interaction picture (cf. (21.13)), the time evolution of operators is controlled by the free Hamiltonian; this means that $\mathscr{L}^{\text{int}}$ (in its form copied from classical theory) is made of free fields, which we know explicitly. This is the main asset of the interaction picture in the context of field theory: within such a framework, the individual terms of the perturbation expansion based on the Dyson series

$$S = \mathbb{1} + i\int \mathscr{L}_{\text{I}}^{\text{int}}(x)\,\mathrm{d}^4x + \frac{i^2}{2!}\iint \mathrm{d}^4x\,\mathrm{d}^4y\,\mathrm{T}\Big(\mathscr{L}_{\text{I}}^{\text{int}}(x)\mathscr{L}_{\text{I}}^{\text{int}}(y)\Big) + \ldots \tag{21.38}$$

can be evaluated with the help of the results that we have obtained for free fields in previous chapters. Some instructive examples will be discussed in the sequel.

# Chapter 22

# Examples of $S$-matrix elements. Some simple Feynman diagrams

The definitions (21.20) and (21.36) mean that

$$|\Psi(+\infty)\rangle = S|\Psi(-\infty)\rangle\,, \tag{22.1}$$

where $|\Psi(\pm\infty)\rangle$ represent states of the considered system in distant past and distant future. For the sake of brevity, we will denote the initial state vector as $|i\rangle$; Eq. (22.1) thus tells us how this is evolved during a sufficiently long interval of time under an interaction embodied in the $S$-matrix operator. The probability amplitude for a definite process (scattering or whatever) $|i\rangle \rightarrow |f\rangle$ is thus given by the scalar product

$$\langle f|\Psi(+\infty)\rangle = \langle f|S|i\rangle\,. \tag{22.2}$$

In what follows, we are going to work out some examples of $S$-matrix elements $\langle f|S|i\rangle = S_{fi}$ in the lowest order of Dyson perturbation series (21.38).

Let us start with the model of Yukawa coupling (21.1). We will assume that $M > 2m$; then one may rightly expect that the interaction in question is responsible, among others, for the decay of the scalar boson $\varphi$ into the fermion-antifermion pair. To see how it may proceed, we will utilize the 1st order term in the expansion (21.38) (an exhilarating feature of such an otherwise quite boring calculation is that it touches some truly modern physics, namely a decay of the enigmatic Higgs boson; an attentive reader should appreciate it). In general, a basic input to the evaluation of an $S$-matrix element $S_{fi}$ is the description of the initial and final states: quite naturally, one may assume that these consist of free particles. To put it more eloquently, one starts with a set of free particles in a definite state prepared before the interaction and these (or a transmuted set) emerge from the interaction region (and are detected) as free particles again. Thus, the states $|i\rangle$ and $|f\rangle$ are described as the appropriate vectors belonging to the relevant Fock space, i.e. they are generated by means of the action of the corresponding creation operators on the vacuum state. In the considered case of the decay process $\varphi \rightarrow f\bar{f}$ one thus has

$$\begin{aligned} |i\rangle &= a^\dagger(q)|0\rangle\,, \\ |f\rangle &= b^\dagger(k,s)d^\dagger(p,r)|0\rangle\,, \end{aligned} \tag{22.3}$$

where we have brought back the familiar notation introduced in chapters 16 and 19. In the first order of the Dyson expansion (21.38) one then gets

$$S_{fi}^{(1)} = i \int \mathrm{d}^4x \, \langle f|\mathscr{L}_{\mathrm{int}}(x)|i\rangle\,, \tag{22.4}$$

where $\mathscr{L}_{\text{int}}$ is given by (21.1) (from now on, we are omitting the label I for the interaction picture). So, let us examine the matrix element of $\mathscr{L}_{\text{int}}(x)$. In order not to get lost in a welter of symbols, it is useful to write the structure of the field operators $\varphi, \psi$ and $\bar{\psi}$ as

$$\varphi = a \,\&\, a^\dagger, \qquad \psi = b \,\&\, d^\dagger, \qquad \bar{\psi} = b^\dagger \,\&\, d. \tag{22.5}$$

This is, hopefully, a comprehensible shortcut notation for the contents of the operators in question in terms of the relevant creation and annihilation operators. Of course, we refer here to the original formulae (16.6), (19.6) and (19.7). The matrix element $\langle f|\bar{\psi}\psi\varphi|i\rangle$ is then reduced, basically, to the vacuum expectation value (v.e.v.) that looks like

$$\langle 0|d\,b\,(b^\dagger \,\&\, d)\,(b \,\&\, d^\dagger)\,(a \,\&\, a^\dagger)\,a^\dagger|0\rangle. \tag{22.6}$$

Obviously, such an expression consists of eight separate contributions involving different chains of annihilation and creation operators. The good news is that there is only one term yielding a nontrivial result. Indeed, taking into account the basic property of an annihilation operator, namely $a|0\rangle = 0$ (equivalently, $\langle 0|a^\dagger = 0$), etc. (both for bosons and fermions), it is clear that a non-zero v.e.v. may be obtained just for a chain, in which annihilation and creation operators of the same kind are paired completely; the point is that the commutator or anticommutator of such a pair gives a Kronecker delta (or a delta function) and this, of course, survives in v.e.v. On the other hand, if there are some unpaired operators, their contribution to the v.e.v. (22.6) vanishes due to the action of an extra annihilation operator on the vacuum. Thus, the only non-zero contribution to (22.6) has the form

$$\langle 0|d\,b\,b^\dagger\,d^\dagger\,a\,a^\dagger|0\rangle, \tag{22.7}$$

while the remaining seven terms drop out; for example

$$\langle 0|d\,b\,b^\dagger\,b\,a\,a^\dagger|0\rangle = 0, \tag{22.8}$$

etc.

Now we are in a position to carry out the evaluation of the matrix element $\langle f|\mathscr{L}_{\text{int}}|i\rangle$ in detail. As we know (cf. Chapter 17), the basic formulae for the field operators $\varphi, \psi, \bar{\psi}$ can be expressed in two possible "dialects" ("continuous" or "discrete"). For simplicity, we are going to utilize now the discrete formalism and suppress, for brevity, the spin labels of the fermion operators (these can be easily retrieved at the end of the calculation). So, using the formula (17.2) as well as the discrete counterparts of (19.6) and (19.7), one may write

$$\langle f|\mathscr{L}_{\text{int}}(x)|i\rangle = g\langle 0|d_p b_k \sum_{l_1} N_{l_1} \left(b_{l_1}^\dagger\,\bar{u}(l_1)\,e^{il_1 x} + \dots\right)$$
$$\times \sum_{l_2} N_{l_2} \left(\dots + d_{l_2}^\dagger v(l_2)\,e^{il_2 x}\right) \sum_{l_3} N_{l_3} \left(a_{l_3}\,e^{-il_3 x} + \dots\right) a_q^\dagger|0\rangle, \tag{22.9}$$

where we have marked explicitly just the relevant parts of the field operators, indicated by the above discussion around the expression (22.7). As a next step, one employs the commutation and anticommutation relations, moving the annihilation operators to the right so as to let them act on the vacuum state. In this way, one is eventually left just with the corresponding Kronecker deltas and the v.e.v. of the operator chain in question becomes

$$\langle 0|d_p\,b_k\,b_{l_1}^\dagger\,d_{l_2}^\dagger\,a_{l_3}\,a_q^\dagger|0\rangle = \delta_{kl_1}\delta_{pl_2}\delta_{ql_3}. \tag{22.10}$$

138

Using the result (22.10) in (22.9), one thus gets

$$\langle f|\mathcal{L}_{\text{int}}(x)|i\rangle = gN_kN_pN_q\bar{u}(k,s)v(p,r)\,e^{i(k+p-q)x}\,, \qquad (22.11)$$

where we have restored also the spin labels, according to (22.3). It is quite clear that if we used in (22.9) the continuous formalism (i.e. integrals instead of infinite sums), the structure of the result would be the same, only the normalization factors in (22.11) would be different (recall that such a change amounts to the replacement $V^{-1/2} \to (2\pi)^{-3/2}$ in the corresponding formula).

So, let us now switch to the continuous values of the particle momenta, which correspond to the infinite 3-dimensional coordinate space. The spacetime integration in (22.4) then leads to

$$S_{fi}^{(1)} = igN_kN_pN_q\bar{u}(k,s)v(p,r)(2\pi)^4\delta^{(4)}(k+p-q)\,. \qquad (22.12)$$

Conventionally, we will write (22.12) in the form

$$S_{fi}^{(1)} = N_kN_pN_q\,i\mathcal{M}_{fi}^{(1)}(2\pi)^4\delta^{(4)}(k+p-q)\,, \qquad (22.13)$$

with

$$\mathcal{M}_{fi}^{(1)} = g\bar{u}(k,s)v(p,r)\,. \qquad (22.14)$$

The structure of the $S$-matrix element embodied in the formula (22.13) is quite general, as we will see repeatedly throughout this text. As expected, there is the four-dimensional delta function representing the energy–momentum conservation, and a product of normalization factors for particles involved in the considered process. These ingredients are universal; on the other hand, the quantity (22.14) is specific for a given type of interaction, i.e. it carries an information about the true dynamics of the process in question. It is also seen that its form is Lorentz invariant; such a quantity $\mathcal{M}_{fi}$ is thus usually called the **invariant matrix element**, or **invariant amplitude** (in the common parlance, it is just a "matrix element", whenever it cannot lead to confusion). The expression (22.14) for $\mathcal{M}_{fi}^{(1)}$ can be represented graphically by a simple Feynman diagram shown in Fig. 22.1. It exhibits the first few "canonical" rules that will be encountered in all

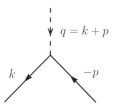

**Fig. 22.1:** Lowest-order Feynman graph for the process $\varphi \to f\bar{f}$. The outgoing fermion (e.g. electron) is depicted as an outgoing solid line, while the outgoing antifermion (e.g. positron) is represented by an incoming line carrying four-momentum with inverted sign.

forthcoming calculations. In particular, there is four-momentum conservation in the vertex and the contribution of the pair of fermion lines is read off conveniently by running against the direction of the arrows; for this purpose, the outgoing antiparticle is represented by an incoming line with inverted four-momentum.

Before leaving the above simple example, some more remarks are in order. It is gratifying that the decay process $\varphi \to f\bar{f}$ is so easily and naturally described by means of the QFT formalism; it is an obvious manifestation of its strength, when one is dealing with a process, in which the number as well as the nature of the particles involved is changing. Apart from the conservation of the energy and momentum, one gets automatically also the charge conservation in the considered decay of a neutral particle. It is instructive to check that charge non-conserving processes like $\varphi \to ff$, or $\varphi \to \bar{f}\,\bar{f}$ are impossible within the model (21.1) we are using; the reader is encouraged to find the relevant argument for such a statement.

Let us now consider an appropriate example of a scattering process. Obviously, restricting ourselves to the 1st order of Dyson expansion, within the model (21.1) we are not able to describe a process of fermion scattering, simply because the monomial $\bar{\psi}\psi\varphi$ does not contain a sufficient number of the relevant creation and annihilation operators. For this purpose, one would have to proceed to the 2nd order, but we will defer such a discussion to Chapter 26. Instead, one may contemplate another field theory model, which would not be too far from reality, but still be suitable for a description of purely fermionic processes in the lowest order. A good example is a toy model of a direct four-fermion interaction described by the interaction Lagrangian

$$\mathscr{L}_{\text{int}} = G\left(\bar{\psi}_1\psi_2\right)\left(\bar{\psi}_2\psi_1\right). \tag{22.15}$$

Here $\psi_1$ and $\psi_2$ represent two different Dirac fields; for definiteness, one may identify the label 1 with neutrino and 2 with electron. $G$ is a coupling constant, which, unlike the case of the Yukawa interaction, is not dimensionless: it is easy to find out that it has a dimension $(\text{mass})^{-2}$, since $\mathscr{L}_{\text{int}}$ has dimension $(\text{mass})^4$ and the dimension of each Dirac field is $(\text{mass})^{3/2}$ (the observant reader is supposed to guess that all these statements are based on the simple fact that the action functional is a dimensionless quantity in the natural system of units with $\hbar = 1$). As regards a relation of (22.15) to the "real world", let us note that it is a greatly simplified version of the effective electron-neutrino weak interaction (in a more realistic scheme corresponding to the standard model there would be some $\gamma$-matrices inside the field products). Let us also recall that an interaction of the above type has been used first by Enrico Fermi in the early 1930s for a description of the radioactive beta decay; the model introduced by Fermi involved fields of neutron, proton, electron and neutrino and it was one of the earliest successful applications of quantum field theory.

Anyway, the phenomenology is not our primary interest here; rather we will focus on the basic calculational technique. So, let us consider the process of elastic neutrino–electron scattering, $\nu(k) + e(p) \to \nu(k') + e(p')$, where we have marked explicitly the corresponding four-momenta. In such a case, the initial and final state may be represented as

$$|i\rangle = b_1^\dagger(k,r)b_2^\dagger(p,s)|0\rangle\,, \qquad |f\rangle = b_1^\dagger(k',r')b_2^\dagger(p',s')|0\rangle\,. \tag{22.16}$$

Utilizing the experience gained in the preceding example (cf. (22.5) through (22.9)), one may now proceed faster (in musical terms, the tempo can be "allegro" rather than "largo"). To evaluate the corresponding $S$-matrix element $S_{fi}^{(1)}$ given by the general formula (22.4), one has to work out first the v.e.v.

$$\langle 0|b_2(p')b_1(k')\left(\bar{\psi}_1\psi_2\right)\left(\bar{\psi}_2\psi_1\right)b_1^\dagger(k)b_2^\dagger(p)|0\rangle \tag{22.17}$$

(we have suppressed again the spin labels). Employing the "structural symbols" like (22.5), one can see that the expression (22.17) consists of 16 separate terms, but only one of them is non-zero, namely that involving the complete pairing of creation and annihilation operators of the same kind. These pairings are visualized in (22.17). Then, using the complete expressions for the field operators in (22.17) and carrying out some simple manipulations analogous to what we have done previously (cf. (22.11) through (22.14)), one arrives at the result

$$S_{fi}^{(1)} = N_k N_p N_{k'} N_{p'}\, i\mathscr{M}_{fi}^{(1)}(2\pi)^4\delta^{(4)}(k' + p' - k - p)\,, \tag{22.18}$$

where

$$\mathcal{M}_{fi}^{(1)} = G\left[\bar{u}(p',s')u(k,r)\right]\left[\bar{u}(k',r')u(p,s)\right] \tag{22.19}$$

(hopefully, an attentive reader would not have any problem to reproduce independently the above results).

The algebraic expression (22.19) can be represented graphically by the simple Feynman diagram shown in Fig. 22.2. One should notice that in addition to the Feynman rules uncovered

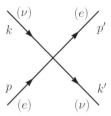

**Fig. 22.2:** The lowest-order Feynman diagram for the elastic scattering $v + e \rightarrow v + e$. One has to read its contribution separately along the upper and lower pair of lines, to copy the structure of the interaction Lagrangian.

in the previous example of the decay process, here one encounters also incoming particle lines, representing the plane-wave bispinor amplitudes $u(k,r)$ and $u(p,s)$.

In fact, the elastic scattering is not the only physical process that can be described (in the lowest order) by means of the interaction Lagrangian (22.15). Another possibility is e.g. the electron–positron annihilation $e^- e^+ \rightarrow v\bar{v}$. Sticking to the above notation of relevant operators, one may write the initial and final state as

$$\begin{aligned}|i\rangle &= b_2^\dagger(k,r)d_2^\dagger(p,s)|0\rangle\,, \\ |f\rangle &= b_1^\dagger(k',r')d_1^\dagger(p',s')|0\rangle\,. \end{aligned} \tag{22.20}$$

This means that the corresponding $S$-matrix element involves v.e.v. of an operator chain, which reads, following the paradigm (22.16),

$$\langle 0|d_1(p')b_1(k')\left(\bar{\psi}_1\psi_2\right)\left(\bar{\psi}_2\psi_1\right)b_2^\dagger(k)d_2^\dagger(p)|0\rangle\,. \tag{22.21}$$

Again, 15 out of the 16 possible terms descending from (22.21) are zero. It is easy to figure out what is the complete pairing of creation and annihilation operators yielding a non-vanishing result, so this is left to the reader as a simple one-minute exercise. Thus, recalling that the $b$-operators are associated with spinors $u$ or $\bar{u}$, while the $d$-operators are accompanied by $v$ or $\bar{v}$, the result for $S_{fi}^{(1)}$ reads, eventually,

$$S_{fi}^{(1)} = N_k N_p N_{k'} N_{p'}\, i\mathcal{M}_{fi}^{(1)}(2\pi)^4\delta^{(4)}(k' + p' - k - p)\,, \tag{22.22}$$

with

$$\mathcal{M}_{fi}^{(1)} = G\left[\bar{u}(k',r')u(k,r)\right]\left[\bar{v}(p,s)v(p',s')\right]\,. \tag{22.23}$$

So, in (22.22) one may observe (as expected) the recurrent general form of an $S$-matrix element, but the formula (22.23) exhibits a new Feynman rule: the incoming antiparticle (here positron) corresponds to $\bar{v}$ (while, as before, an outgoing antiparticle (antineutrino) is represented by $v$). The expression (22.23) may be now depicted by the Feynman diagram displayed in Fig. 22.3.

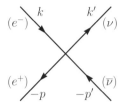

**Fig. 22.3:** The lowest-order Feynman diagram for the annihilation process $e^- e^+ \rightarrow \nu \bar{\nu}$. Its contribution (22.23) incorporates all relevant Feynman rules for the external lines representing fermions and antifermions.

In this way, we have arrived at a set of Feynman rules for Dirac fermions and these will be utilized in many forthcoming calculations. Another remarkable finding is that the content of the modest Lagrangian (22.15) is in fact quite rich: it embodies both elastic scattering and annihilation processes (needless to say, apart from those already discussed, the scattering $\bar{\nu} e \rightarrow \bar{\nu} e$, or the annihilation $\nu \bar{\nu} \rightarrow e^- e^+$, etc. are also possible within the considered model). This is a salient feature of QFT models involving various types of fields; moreover (and not surprisingly), the number of calculable processes increases in higher orders of the $S$-matrix perturbation expansion.

# Chapter 23

# Decay rates and cross sections

According to the results of the preceding chapter, we are now able to compute (at least in the lowest perturbative order) some quantum transition amplitudes, either for a particle decay, or a scattering process. To get measurable quantities, one has to calculate the corresponding probabilities. In general, such a transition probability is given by the square of an $S$-matrix element $|S_{fi}|^2$; but, as we will see, for obtaining a truly measurable quantity one must, roughly speaking, normalize such a probability in an appropriate manner.

Let us start with a decay process. Up to now, we have discussed explicitly just a two-body decay, but here we will consider a general case of a particle decaying into an arbitrary number $n$ of other particles. We will assume, *bona fide*, that the corresponding $S$-matrix element is factorized as

$$S_{fi} = N_i N_{f_1} \cdots N_{f_n} \, i \mathcal{M}_{fi} (2\pi)^4 \delta^{(4)}(p_{f_1} + \cdots + p_{f_n} - p_i). \tag{23.1}$$

Of course, the Ansatz (23.1) is inspired by our previous results (cf. (22.13), (22.18), (22.22)). Although momentarily we have no rigorous theorem for justifying it, the formula (23.1) is, hopefully, quite plausible, because of the arguments indicated in the preceding chapter. Note also that the relation (23.1) is supposed to be valid generally, without referring to perturbation expansion.

As we have noted above, $|S_{fi}|^2$ is the probability of the transition $|i\rangle \to |f\rangle$, but for a connection with an experimental observation one should rather consider a probability of the transition to a definite region of the space of final-state momenta. Thus, one may start with examining the differential decay probability corresponding to final states involving momenta in a close neighbourhood of a given value $\vec{p}_f = \vec{p}_{f_1} + \ldots + \vec{p}_{f_n}$. To proceed, we will streamline our notation a bit; the four-momenta of the decay products will be denoted as $p_1, \ldots, p_n$ and for the four-momentum of the decaying particle we choose the symbol $P$. For obtaining the number of possible states (momenta) in a close vicinity of a value $\vec{p}$, it is useful to employ the "discrete dialect" introduced in Chapter 17 (which means that the relevant fields are quantized within a finite spatial box with the volume $V = L^3$). Taking into account the formula (17.1), one can see that in the interval $(\vec{p}, \vec{p} + \Delta\vec{p})$ one has $\Delta N(\vec{p})$ states, where

$$\Delta N(\vec{p}) = \frac{\Delta p_1}{\dfrac{2\pi}{L}} \cdot \frac{\Delta p_2}{\dfrac{2\pi}{L}} \cdot \frac{\Delta p_3}{\dfrac{2\pi}{L}} = \frac{V \Delta p_1 \Delta p_2 \Delta p_3}{(2\pi)^3}. \tag{23.2}$$

So, passing from differences to differentials, the relation (23.2) becomes

$$\Delta N(\vec{p}) = \frac{V \, \mathrm{d}^3 p}{(2\pi)^3}. \tag{23.3}$$

Note that such an expression is traditionally called the **phase-space element**. Now it is clear that, including the relevant phase-space elements, the differential probability for the considered decay may be written as

$$d\mathscr{P}_{i \to f} = |S_{fi}|^2 \frac{V \, d^3 p_1}{(2\pi)^3} \cdots \frac{V \, d^3 p_n}{(2\pi)^3} \,. \tag{23.4}$$

Substituting the expression (23.1) into (23.4), one gets

$$d\mathscr{P}_{i \to f} = N_i^2 N_1^2 \cdots N_n^2 |\mathscr{M}_{fi}|^2 \left[ (2\pi)^4 \delta^{(4)}(P_f - P) \right]^2 \frac{V \, d^3 p_1}{(2\pi)^3} \cdots \frac{V \, d^3 p_n}{(2\pi)^3} \,, \tag{23.5}$$

where $P_f = p_1 + \ldots + p_n$ and the normalization factors $N_i, N_1, \ldots, N_n$ are given by the formula (17.3). The main difficulty is now the square of the delta function; this, in fact, is a kind of a "mathematician's nightmare". To cope with it, one may proceed as follows. First, since for an ordinary function $f(x)$ one has $f(x)\delta(x) = f(0)\delta(x)$, we will set, denoting $\Delta = P_f - P$,

$$\left[ \delta^{(4)}(\Delta) \right]^2 = \delta^{(4)}(0)\delta^{(4)}(\Delta) \,. \tag{23.6}$$

Thus, the challenge now consists in making sense of $\delta^{(4)}(0)$. This is an ill-defined infinity, but one may regularize it by means of a clever trick: utilizing the integral representation

$$(2\pi)^4 \delta^{(4)}(\Delta) = \int d^4 x \, e^{i\Delta \cdot x} \tag{23.7}$$

and restricting the infinite spacetime to the spatial box with a volume $V$, along with a finite (but large) time interval $(-T/2, T/2)$, for $\Delta = 0$ one gets on the right-hand side of (23.7) the value $VT$. Thus, our regularization prescription reads

$$(2\pi)^4 \delta^{(4)}(0) \longrightarrow VT \,. \tag{23.8}$$

Of course, from the point of view of pure rigorous mathematics, the replacement (23.8) is a dirty trick, but it is commonly used in physics literature, since, as we will see immediately, it is really instrumental in obtaining the relevant result for the description of a general decay process (and for scattering processes as well). At the end of this chapter, we will show how the short cut way leading to (23.8) can be (slightly) improved.

So, let us substitute the regularization recipe (23.8) into the expression (23.5), using also (23.6) and the standard formulae for the normalization factors according to (17.3). One thus gets

$$d\mathscr{P}_{i \to f} = \frac{1}{2E_i V} \frac{1}{2E_1 V} \cdots \frac{1}{2E_n V} |\mathscr{M}_{fi}|^2 \frac{V \, d^3 p_1}{(2\pi)^3} \cdots \frac{V \, d^3 p_n}{(2\pi)^3} VT (2\pi)^4 \delta^{(4)}(P_f - P) \,, \tag{23.9}$$

and it becomes obvious that the factors of the space volume $V$ (which is an auxiliary parameter) are cancelled. The quantity $d\mathscr{P}_{i \to f}$ is proportional to $T$; it means that one is led naturally to the definition of the **decay probability per unit time**, the quantity commonly called the **decay rate**. To distinguish it from $d\mathscr{P}_{i \to f}$, we will introduce a special notation

$$dw_{fi} = \frac{1}{T} d\mathscr{P}_{i \to f} \,. \tag{23.10}$$

From (23.9) one then gets

$$dw_{fi} = \frac{1}{2E_i} |\mathscr{M}_{fi}|^2 \frac{d^3 p_1}{(2\pi)^3 2E_1} \cdots \frac{d^3 p_n}{(2\pi)^3 2E_n} (2\pi)^4 \delta^{(4)}(p_1 + \cdots + p_n - P) \,, \tag{23.11}$$

144

and this is a basic formula that can be found in all handbooks of particle physics; with such a knowledge at hand, the reader may "join the crowd" and start calculating any fancy decay process occurring in the universe (provided that the matrix element $\mathcal{M}_{fi}$ is available). An attentive reader may have also noticed that the derivation of the formula (23.10) has been quite straightforward, due to the trick embodied in (23.8). Indeed, the above procedure is designed to be user-friendly; it can be reproduced easily e.g. when writing with your finger in the sand of a beach on a desert island.

A particularly simple case, which we have already touched in the preceding chapter, is a two-body decay. In such a case, the general formula for the differential decay rate may be further elaborated on; one can carry out explicitly an integration over the phase space and obtain another useful formula that is frequently employed in applications. So, let us consider such a decay process for the "parent" particle (with a mass $M$) at rest, i.e. the initial four-momentum is $P = (M, \vec{0})$. Then the delta function for the momentum conservation in (23.11) becomes $\delta^{(3)}(\vec{p}_1 + \vec{p}_2)$ and a first integration (e.g. over $\vec{p}_2$) is basically trivial; the decay products ("daughter" particles with masses $m_1, m_2$) have momenta $\vec{p}_1 = -\vec{p}_2 \equiv \vec{p}$ and one is left with

$$dw_{fi} = \frac{1}{2M} |\mathcal{M}_{fi}|^2 \frac{d^3p}{16\pi^2} \frac{1}{E_1 E_2} \delta(E_1 + E_2 - M), \tag{23.12}$$

where, of course,

$$E_1 = \sqrt{\vec{p}^2 + m_1^2}, \qquad E_2 = \sqrt{\vec{p}^2 + m_2^2}. \tag{23.13}$$

The next step is the integration involving the one-dimensional delta function in (23.12), through which the energy conservation is implemented. The integral to be evaluated is

$$d(\text{LIPS}_2) = \frac{1}{16\pi^2} \int \frac{d^3p}{E_1 E_2} \delta(E_1 + E_2 - M), \tag{23.14}$$

where LIPS is an acronym for "Lorentz-invariant phase space". To this end, one may employ the familiar formula for a composite delta function, namely

$$\delta[f(x)] = \frac{1}{|f'(x_0)|} \delta(x - x_0), \tag{23.15}$$

where $f(x_0) = 0$. In the present case, denoting $|\vec{p}| = x$, we have $\delta[f(x)]$ with

$$f(x) = \sqrt{x^2 + m_1^2} + \sqrt{x^2 + m_2^2} - M, \tag{23.16}$$

so that

$$f'(x) = \frac{x}{\sqrt{x^2 + m_1^2}} + \frac{x}{\sqrt{x^2 + m_2^2}}. \tag{23.17}$$

For the zero of $f(x)$, i.e. for $x_0$ such that

$$\sqrt{x_0^2 + m_1^2} + \sqrt{x_0^2 + m_2^2} = M, \tag{23.18}$$

one then gets readily

$$f'(x_0) = \frac{x_0 M}{\sqrt{x_0^2 + m_1^2}\sqrt{x_0^2 + m_2^2}}. \tag{23.19}$$

In the integral (23.14) one may introduce spherical variables, so that $d^3 p = |\vec{p}|^2 \, d|\vec{p}| \, d\Omega = x^2 \, dx \, d\Omega$, where $d\Omega$ is the element of solid angle corresponding to the direction of $\vec{p}$. Employing (23.15) and (23.19) we thus get, after a simple manipulation,

$$d(\text{LIPS}_2) = \frac{d\Omega}{16\pi^2} \frac{x_0}{M} . \tag{23.20}$$

Now, what is $x_0$? According to (23.18), it is the magnitude of the momentum of a decay product in the rest system of the parent particle, in other words, in the c. m. system of the daughter particles. So, we will denote $x_0$ as $|\vec{p}_{\text{c.m.}}|$ later on. The value of $x_0 = |\vec{p}_{\text{c.m.}}|$ is obtained easily from the kinematical relation (23.18). Indeed, this can be recast as

$$x_0^2 + m_1^2 = \left( M - \sqrt{x_0^2 + m_2^2} \right)^2 , \tag{23.21}$$

and one thus gets immediately

$$2M \sqrt{x_0^2 + m_2^2} = M^2 + m_2^2 - m_1^2 . \tag{23.22}$$

The solution of Eq. (23.22) may be written as

$$x_0 = \left[ \frac{\lambda(M^2, m_1^2, m_2^2)}{4M^2} \right]^{1/2} , \tag{23.23}$$

where

$$\lambda(x, y, z) = x^2 + y^2 + z^2 - 2xy - 2xz - 2yz \tag{23.24}$$

is the so-called Källén's function;[13] its alternative name is "triangle function" since it is closely related to the area $A$ of a triangle, expressed in terms of its sides (cf. also the famous Heron's formula). Explicitly, it holds

$$A = \frac{1}{4} \sqrt{-\lambda(a^2, b^2, c^2)}$$

if $a$, $b$, $c$ are the triangle side lengths. Note that another useful formula for the function $\lambda$ in (23.23), which can be obtained easily from (23.22), reads

$$\lambda(M^2, m_1^2, m_2^2) = \left[ M^2 - (m_1 + m_2)^2 \right] \left[ M^2 - (m_1 - m_2)^2 \right] . \tag{23.25}$$

So, returning to the relations (23.12) and (23.14), one has a general formula for the differential decay rate

$$dw_{fi} = \frac{1}{2M} |\mathcal{M}_{fi}|^2 \, d(\text{LIPS}_2) , \tag{23.26}$$

with the element of phase space given by (23.20), i.e.

$$d(\text{LIPS}_2) = \frac{d\Omega}{16\pi^2} \frac{|\vec{p}_{\text{c.m.}}|}{M} . \tag{23.27}$$

In a situation where the matrix element squared in (23.26) is independent of the spherical angles involved in $d\Omega$ (this happens e.g. for unpolarized particles), the relation (23.26) may be integrated

---

[13]Gunnar Källén (1926-1968) was an eminent Swedish theoretical physicist who made many important contributions to QFT and particle physics. He died tragically in an airplane accident; he was flying his own plane from Malmö to CERN and it crashed during the emergency landing in Hannover.

trivially and one gets a useful formula for the **integral decay rate** (called also the **decay width**), namely

$$\Gamma = \frac{1}{2M} |\mathscr{M}_{fi}|^2 \text{LIPS}_2, \tag{23.28}$$

where

$$\text{LIPS}_2 = \frac{1}{4\pi} \frac{|\vec{p}_{\text{c.m.}}|}{M}. \tag{23.29}$$

Using the above kinematical formulae (23.23) and (23.25), one may observe that there are some notable cases, in which the formula (23.29) becomes particularly simple. So, for $m_1 = m_2 = m$ one has

$$\text{LIPS}_2 = \frac{1}{8\pi} \sqrt{1 - \frac{4m^2}{M^2}}. \tag{23.30}$$

Further, for $m_1, m_2 \ll M$ one gets an extremely simple approximate formula

$$\text{LIPS}_2 \Big|_{m_1, m_2 \ll M} \doteq \frac{1}{8\pi}. \tag{23.31}$$

So much for decay processes. Let us now consider a general process of the scattering type

$$1 + 2 \to 3 + 4 + \cdots + n. \tag{23.32}$$

In analogy with our previous discussion we will employ an Ansatz for the corresponding $S$-matrix element, which now reads

$$S_{fi} = \delta_{fi} + N_1 N_2 N_3 N_4 \cdots N_n \, i \mathscr{M}_{fi} (2\pi)^4 \delta^{(4)} (P_f - P_i), \tag{23.33}$$

where the first term reflects the possibility of a trivial event with $f = i$, $P_f = p_3 + \ldots + p_n$ and $P_i = p_1 + p_2$. Considering just a non-trivial situation, in which $f \neq i$, the transition probability becomes, after the manipulations described above for the case of a decay,

$$d\mathscr{P}_{i \to f} = \frac{1}{2VE_1} \frac{1}{2VE_2} \frac{1}{2VE_3} \cdots \frac{1}{2VE_n} |\mathscr{M}_{fi}|^2 \frac{V \, d^3 p_3}{(2\pi)^3} \cdots \frac{V \, d^3 p_n}{(2\pi)^3} (2\pi)^4 \delta^{(4)} (P_f - P_i) VT. \tag{23.34}$$

The relevant quantity that can be derived from the transition probability is the **cross section**. As in the case of decay processes, one may first define a **differential cross section**. This is given by the number of events per unit time corresponding to the elements of the phase space shown in (23.34), divided by the current density of the colliding particles. Concerning the latter quantity, if one considers a target particle at rest (e.g. that labelled as 2), this would be

$$j_{\text{inc}} = \frac{N}{V} |\vec{v}_1| \qquad \text{for} \quad \vec{v}_2 = 0, \tag{23.35}$$

where $N$ is the total number of the incident particles (1) and $\vec{v}_1$ is their velocity (thus, $j_{\text{inc}}$ is the number of incident particles passing per unit time through unit area). For colliding beams of particles 1 and 2 (with parallel velocities), the relation (23.35) is generalized to

$$j_{\text{inc}} = \frac{N}{V} |\vec{v}_1 - \vec{v}_2| \tag{23.36}$$

(let us stress that $|\vec{v}_1 - \vec{v}_2|$ is not the relative velocity of the particles 1 and 2; it just defines an effective volume to be multiplied by the particle density $N/V$). Thus, one defines the differential cross section as

$$d\sigma = \frac{\text{number of events per unit time}}{\text{incident current density}} = \frac{N \cdot \frac{1}{T} d\mathscr{P}_{i \to f}}{\frac{N}{V} |\vec{v}_1 - \vec{v}_2|}. \tag{23.37}$$

147

Using the expression (23.34) for the transition probability, one gets from (23.37)

$$d\sigma = \frac{1}{|\vec{v}_1 - \vec{v}_2|} \frac{1}{2E_1} \frac{1}{2E_2} |\mathscr{M}_{fi}|^2 \frac{d^3 p_3}{(2\pi)^3 2E_3} \cdots \frac{d^3 p_n}{(2\pi)^3 2E_n} (2\pi)^4 \delta^{(4)} (P_f - P_i).$$ (23.38)

It is certainly reassuring that the auxiliary parameters $V$ and $T$ have dropped out (but this is something that we have in fact "ordered" in advance). Note that the cross section $d\sigma$ has, by its definition, the dimension of an area; in the natural system of units this is (mass)$^{-2}$, or, if you want, (energy)$^{-2}$. Let us also remark that the factor $|\vec{v}_1 - \vec{v}_2|^{-1}$ may be recast by means of the kinematical identity

$$|\vec{v}_1 - \vec{v}_2| = \frac{1}{E_1 E_2} \left[ (p_1 \cdot p_2)^2 - m_1^2 m_2^2 \right]^{1/2}$$ (23.39)

(proving it is left to the reader as an instructive exercise; please don't forget that $\vec{v}_1$ and $\vec{v}_2$ are supposed to be directed along the same line).

Now we are going to examine in more detail a very important special case, namely a **binary process** $1 + 2 \rightarrow 3 + 4$. Similarly as for the two-body particle decay, we will be able to carry out a substantial part of the integration over the phase space and obtain a practical formula for the angular distribution of the final-state particles. To this end, it is most convenient to work in the c. m. system; then $\vec{p}_2 = -\vec{p}_1$, and the momentum conservation yields $\vec{p}_4 = -\vec{p}_3$. Thus, we will use the notation $\vec{p}_1 = \vec{p}_{\text{c.m.}}$ and $\vec{p}_3 = \vec{p}'_{\text{c.m.}}$ in what follows. Returning to the general formula (23.38), let us first work out the factor multiplying the matrix element squared. One gets

$$\frac{1}{|\vec{v}_1 - \vec{v}_2|} \frac{1}{2E_1} \frac{1}{2E_2} = \frac{1}{4|\vec{p}_{\text{c.m.}}| E_{\text{c.m.}}} = \frac{1}{4|\vec{p}_{\text{c.m.}}| s^{1/2}},$$ (23.40)

where $E_{\text{c.m.}} = E_1 + E_2$ and we have expressed $E_{\text{c.m.}}$ in terms of the familiar **Mandelstam kinematical invariant** $s = (p_1 + p_2)^2 = E_{\text{c.m.}}^2$. Thus, the starting point of our calculation becomes

$$d\sigma = \frac{1}{4|\vec{p}_{\text{c.m.}}| s^{1/2}} |\mathscr{M}_{fi}|^2 \frac{d^3 p_3}{(2\pi)^3 2E_3} \frac{d^3 p_4}{(2\pi)^3 2E_4} (2\pi)^4 \delta^{(4)} (p_3 + p_4 - p_1 - p_2).$$ (23.41)

The integration over $d^3 p_4$ is trivial and one gets, after a simple manipulation,

$$d\sigma = \frac{1}{4|\vec{p}_{\text{c.m.}}| s^{1/2}} |\mathscr{M}_{fi}|^2 \frac{d\Omega_{\text{c.m.}} \, dx \, x^2}{16\pi^2 E_3 E_4} \delta(E_3 + E_4 - s^{1/2}),$$ (23.42)

where we have denoted $x = |\vec{p}_3| \, (= |\vec{p}'_{\text{c.m.}}|)$, and, of course, the final energies $E_3$, $E_4$ are given by $E_3 = (x^2 + m_3^2)^{1/2}$, $E_4 = (x^2 + m_4^2)^{1/2}$. Thus, we may utilize our previous results for a two-body decay (obviously, $s^{1/2}$ now plays basically the same role as the mass $M$ in (23.12)); in particular, integrating (23.42) over $x$ and taking into account the formula (23.27) for d(LIPS$_2$), one gets finally

$$\frac{d\sigma}{d\Omega_{\text{c.m.}}} = \frac{1}{64\pi^2} \frac{1}{s} \frac{|\vec{p}'_{\text{c.m.}}|}{|\vec{p}_{\text{c.m.}}|} |\mathscr{M}_{fi}|^2,$$ (23.43)

where, according to (23.23), it holds

$$|\vec{p}_{\text{c.m.}}| = \left[ \frac{\lambda(s, m_1^2, m_2^2)}{4s} \right]^{1/2}, \qquad |\vec{p}'_{\text{c.m.}}| = \left[ \frac{\lambda(s, m_3^2, m_4^2)}{4s} \right]^{1/2}.$$ (23.44)

The relation (23.43) is a key formula for the angular distribution of final-state particles (in the c. m. system) in a general binary process and it is used most frequently in applications. Thus, a reader who has mastered it, earns an enhanced status within the crowd of particle physicists and QFT practitioners. In a similar manner, one may derive a corresponding result for the laboratory frame, where the target particle 2 would be at rest; an explicit discussion of such a case is deferred to the Chapter 33. Note also that in some specific situations the formula (23.43) gets simplified considerably; in particular, for elastic scattering $1 + 2 \rightarrow 1 + 2$ one obviously has $|\vec{p}'_{\text{c.m.}}| = |\vec{p}_{\text{c.m.}}|$ and the formula for the differential cross section thus reads

$$\frac{d\sigma}{d\Omega_{\text{c.m.}}}\bigg|_{\text{elast.}} = \frac{1}{64\pi^2}\frac{1}{s}|\mathcal{M}_{fi}|^2 \,. \tag{23.45}$$

Similarly, for any binary process involving massless particles one also gets the result (23.45) (recall that the zero-mass approximation is relevant in the high-energy limit, i.e. for $s \gg m_j^2$, $j = 1, \ldots, 4$). Finally, it may be instructive to make a remark on dimensions of the quantities we are working with. Taking into account that the dimension of a cross section is $(\text{length})^2$, i.e. $(\text{mass})^{-2}$ in natural units, one may deduce easily from the formula (23.38) that an $n$-body matrix element $\mathcal{M}_{fi}$ has a dimension $(\text{mass})^{4-n}$. Thus, for a binary process it is a dimensionless quantity; this is confirmed immediately by the simple formula (23.43) (or (23.45)). It means that the structure of the relation (23.45) is particularly easy to remember: the square of $\mathcal{M}_{fi}$ is an obvious ingredient, and to get a cross section with the proper dimension, the factor $1/s$ is mandatory. If one is able to recall even the numerical factor $(64\pi^2)^{-1}$, it is an extra bonus (surprisingly, students frequently remember just such a detail).

In concluding this chapter, let us return to a technical point concerning the regularization rule (23.8). As we have stressed, its "derivation" has been a mathematical "hocus-pocus", so now we will try harder to improve it a bit. To this end, we are going to revisit the integral (23.7). Evaluating it in the "restricted spacetime" consisting of a finite spatial box with a volume $V = L^3$ and a finite time interval $(-T/2, T/2)$ one gets

$$I(V,T) = \int_{(V,T)} d^4x \, e^{i\Delta \cdot x} = \int_{-T/2}^{T/2} dx_0 \, e^{i\Delta_0 x_0} \int_{(V)} d^3x \, e^{-i\vec{\Delta}\cdot\vec{x}} = \frac{\sin\dfrac{\Delta_0}{2}T}{\dfrac{\Delta_0}{2}} \cdot V\delta_{\vec{\Delta},\vec{0}} \,. \tag{23.46}$$

Note that in the evaluation of the integral over $d^3x$ we have taken into account the fact that $\vec{\Delta}$ is a difference of momenta with discrete values given by (17.1). Now, the square of the expression (23.46) becomes

$$I^2(V,T) = \frac{\left(\sin\dfrac{\Delta_0}{2}T\right)^2}{\left(\dfrac{\Delta_0}{2}\right)^2} \cdot V^2\delta_{\vec{\Delta},\vec{0}} \,, \tag{23.47}$$

where we have used the obvious identity $\delta_{\vec{\Delta},\vec{0}}^2 = \delta_{\vec{\Delta},\vec{0}}$ (due to the fact that the Kronecker delta takes on just values 0 or 1). However, it holds

$$\lim_{T\to\infty} \frac{\sin^2 xT}{Tx^2} = \pi\delta(x) \tag{23.48}$$

(this is a rigorous statement) and thus one may write

$$I^2(V,T)\bigg|_{T\to\infty} = T\pi\delta\left(\frac{\Delta_0}{2}\right)V^2\delta_{\vec{\Delta},\vec{0}} \,. \tag{23.49}$$

149

At the same time, one relies on an (intuitive) rule that

$$V\delta_{\vec{\Delta},\vec{0}} \longrightarrow (2\pi)^3 \delta^{(3)}(\vec{\Delta}) \tag{23.50}$$

for $V \to \infty$. In this way, one eventually arrives at

$$I^2(V,T) \longrightarrow T \cdot 2\pi\delta(\Delta_0)V(2\pi)^3\delta^{(3)}(\vec{\Delta}) = VT(2\pi)^4\delta^{(4)}(\Delta) \tag{23.51}$$

for $V,T \to \infty$, and this corresponds precisely to the earlier replacement recipe (23.8).

A scrupulous reader has certainly noticed that the above argumentation is not fully rigorous; the main asset is the identity (23.48), but (23.50) is again hand-waving. Obviously, the main difficulty consists, from the very beginning, in working with the energy–momentum delta function. A rigorous treatment would be possible if one used a representation of the initial and final states by means of wave packets, smearing somewhat the values of the particle momenta (see e.g. [6]). Our excuse for adopting a more intuitive short cut approach is that in most textbooks the other authors proceed in the same way, so that it has become, as Germans say, "salonfähig", at least in a QFT environment.

# Chapter 24

# Sample lowest-order calculations for physical processes

Now we are equipped with some basic general formulae that may be employed for the evaluation of decay rates and scattering cross sections. In fact, the crucial dynamical part of such a quantity is the matrix element squared, $|\mathcal{M}_{fi}|^2$, which represents a specific contribution of the considered particular QFT model. In the present chapter we will work out several typical examples within the lowest order of the Dyson perturbation expansion for various field theory models.

Let us start with the process of fermion-antifermion decay of a scalar boson, discussed previously in Chapter 22. Using the model of Yukawa interaction (21.1), we have arrived at the simple result (22.14) for the matrix element in the first perturbative order. From the computational point of view, the simplest case corresponds to the situation, in which the final-state particles are unpolarized. It means that the conditions of a corresponding experiment (thought or real) are supposed to be set in such a way that the considered decay process is inclusive with respect to particle spins; consequently, in a pertinent calculation the final-state spins (more precisely, spin projections, such as e.g. helicities) are summed over. So, let us see how it works for the scalar boson decay in question. First, one gets

$$\begin{aligned} |\mathcal{M}|^2 = \mathcal{M}\mathcal{M}^* &= g^2 \bar{u}(k,s)v(p,r)v^\dagger(p,r)\gamma_0 u(k,s) \\ &= g^2 \bar{u}(k,s)v(p,r)\bar{v}(p,r)u(k,s) \,. \end{aligned} \tag{24.1}$$

Just to be sure, a trivial remark is perhaps in order here. The complex conjugation $\mathcal{M}^*$ is calculated, equivalently, as the Hermitian conjugation $\mathcal{M}^\dagger$; obviously, this is most convenient, since the number $\mathcal{M}$ is a matrix product. Needless to say, in (24.1) we have also utilized the familiar relation $\gamma_0^\dagger = \gamma_0$.

Now comes a crucial trick that will be instrumental in most of the forthcoming calculations. The matrix product in (24.1) is a number, so that it is equal to its trace; however, such a trace is invariant under a cyclic permutation and thus one has

$$\begin{aligned} |\mathcal{M}|^2 &= g^2 \operatorname{Tr}\big[\bar{u}(k,s)v(p,r)\bar{v}(p,r)u(k,s)\big] \\ &= g^2 \operatorname{Tr}\big[u(k,s)\bar{u}(k,s)v(p,r)\bar{v}(p,r)\big]. \end{aligned} \tag{24.2}$$

In this way, the original expression (24.1) has been recast as the trace of a product of $4 \times 4$ matrices and it is clear why such a seemingly simple-minded trick is so powerful: the reader should recall that the expressions like $u(k,s)\bar{u}(k,s)$ or $v(p,r)\bar{v}(p,r)$ (and their sums over the spin labels) are known explicitly (see chapters 6 and 7) and, moreover, there are many practical formulae for traces of products of Dirac matrices (see Chapter 3 and Appendix C). Thus, the reader may appreciate now that the promise made in Chapter 3, based on a quotation from A. P. Chekhov, is thereby fulfilled. The procedure outlined above is usually called simply the **trace technique**.[14]

So, how about the expression (24.2)? In the envisaged case of unpolarized particles in the final state one has, using (6.24) and (6.30),

$$\sum_{\text{pol.}} |\mathcal{M}|^2 = g^2 \sum_{r,s} \text{Tr}[u(k,s)\bar{u}(k,s)v(p,r)\bar{v}(p,r)]$$

$$= g^2 \text{Tr}\big[(\not{k}+m)(\not{p}-m)\big] \tag{24.3}$$

(note that here and in what follows we denote, conventionally, the sum over spins by the label "pol."). Employing some elementary formulae for Dirac traces, the result for (24.3) becomes

$$\sum_{\text{pol.}} |\mathcal{M}|^2 = 4g^2(k \cdot p - m^2). \tag{24.4}$$

The four-momentum conservation $q = k + p$ amounts to

$$k \cdot p = \frac{1}{2}(q^2 - k^2 - p^2) = \frac{1}{2}M^2 - m^2, \tag{24.5}$$

and thus we get finally

$$\sum_{\text{pol.}} |\mathcal{M}|^2 = 2g^2(M^2 - 4m^2). \tag{24.6}$$

Now we are in a position to evaluate the decay rate; according to the general formulae (23.28) and (23.30) one has

$$\Gamma(\varphi \to f\bar{f}) = \frac{g^2}{8\pi}M\left(1 - \frac{4m^2}{M^2}\right)^{3/2}. \tag{24.7}$$

Thus, any reader who has reached this point may be happy to learn how to compute, in the lowest order, at least one decay channel of the Higgs boson (the value of the relevant coupling constant $g$ can be found elsewhere, see e.g. ref. [22]).

In fact, we can do more. We certainly have sufficient technical means to evaluate a decay rate for polarized particles in the final state; of course, for such a purpose, the formulae (7.51) are instrumental. In the present context, the most instructive description of the spin states relies on the helicity formalism. So, let us study the "anatomy" of the result (24.6). It means that we would like to compute the matrix element squared for all combinations of the final state helicities, namely RR, LL, RL, LR, where R (right-handed) stands for the positive helicity and similarly L (left-handed) denotes a negative helicity. By the way, it is not difficult to guess the result *a priori*, on the basis of the angular momentum conservation. Indeed, staying in the rest frame of the parent particle, the decay products emerge with opposite momenta and thus the admissible combinations of helicities should be just RR or LL; for RL or LR configuration, the spin projections would be oriented in the same direction and this is incompatible with the zero angular momentum (spin) of the decaying scalar boson. Furthermore, an educated guess is that, for symmetry reasons, one should get $|\mathcal{M}_{\text{RR}}|^2 = |\mathcal{M}_{\text{LL}}|^2$. In this way, one might arrive at a complete solution of the above problem. Anyway, it will be instructive to perform an explicit

---

[14]Note that this elegant method has been invented by the well-known Dutch physicist Hendrik Casimir (1909-2000) in 1933, so it is sometimes called the Casimir's trick.

calculation and verify the expected results; at the same time, one thus may confirm that our formalism is sound indeed.

So, let us start with $|\mathcal{M}_{RR}|^2$. According to (24.2), one has

$$\frac{1}{g^2}|\mathcal{M}_{RR}|^2 = \mathrm{Tr}[u_R(k)\bar{u}_R(k)v_R(p)\bar{v}_R(p)]$$

$$= \frac{1}{4}\mathrm{Tr}\left[(\slashed{k}+m)\left(1+\gamma_5\slashed{s}_R(k)\right)(\slashed{p}-m)\left(1+\gamma_5\slashed{s}_R(p)\right)\right], \qquad (24.8)$$

where we have employed the formulae (8.12). Carrying out the matrix multiplication inside the trace, one gets 16 terms, but most of them do not contribute in the end. Taking into account the basic fact that the trace of the product of an odd number of $\gamma$-matrices vanishes, as well the identity $\mathrm{Tr}(\gamma_\mu\gamma_\nu\gamma_5) = 0$, the expression (24.8) is reduced to

$$\frac{1}{g^2}|\mathcal{M}_{RR}|^2 = \frac{1}{4}\mathrm{Tr}\left[\slashed{k}\slashed{p} - m^2 + \slashed{k}\gamma_5\slashed{s}_R(k)\slashed{p}\gamma_5\slashed{s}_R(p) - m^2\gamma_5\slashed{s}_R(k)\gamma_5\slashed{s}_R(p)\right]$$

$$= \frac{1}{4}\left[4k\cdot p - 4m^2 + \mathrm{Tr}(\slashed{k}\slashed{s}_R(k)\slashed{p}\slashed{s}_R(p)) + m^2\,\mathrm{Tr}(\slashed{s}_R(k)\slashed{s}_R(p))\right], \qquad (24.9)$$

where we have also utilized the anticommutation property of $\gamma_5$ and the identity $\gamma_5^2 = 1$. To work out the traces in (24.9), one may employ the orthogonality property of the spin four-vectors, $k\cdot s_R(k) = 0$, $p\cdot s_R(p) = 0$ and it is also useful to take into account that $k + p = q$, with $q = (M,\vec{0})$. Then we get readily

$$k\cdot s_R(p) = (q-p)\cdot s_R(p) = q\cdot s_R(p) = Ms_R^0(p) = M\frac{|\vec{p}|}{m}, \qquad (24.10)$$

and similarly

$$p\cdot s_R(k) = M\frac{|\vec{k}|}{m}. \qquad (24.11)$$

Note that to arrive at (24.10) and (24.11), we have used the explicit form of the helicity four-vector according to the formula (8.7). Putting all this together, the expression (24.9) becomes

$$\frac{1}{g^2}|\mathcal{M}_{RR}|^2 = k\cdot p - m^2 - (k\cdot p)(s_R(k)\cdot s_R(p)) + M^2\frac{|\vec{k}|^2}{m^2} + m^2 s_R(k)\cdot s_R(p), \qquad (24.12)$$

where we have also taken into account that $|\vec{p}| = |\vec{k}|$. The scalar product $s_R(k)\cdot s_R(p)$ is evaluated easily by means of the formula (8.7) (keeping in mind that $\vec{p} = -\vec{k}$); one gets

$$s_R(k)\cdot s_R(p) = \frac{|\vec{k}|^2}{m^2} + \frac{E^2}{m^2} = \frac{2E^2 - m^2}{m^2} = \frac{\frac{1}{2}M^2 - m^2}{m^2}. \qquad (24.13)$$

Now, substituting (24.13) and (24.5) (as well as $|\vec{k}|^2 = \frac{1}{4}M^2 - m^2$) into (24.12) we obtain, after a simple manipulation,

$$|\mathcal{M}_{RR}|^2 = g^2(M^2 - 4m^2). \qquad (24.14)$$

Note that this is just one half of (24.6), in accordance with our previous expectation. Let us also remark that from the intermediate result (24.9) one gets immediately $|\mathcal{M}_{LL}|^2 = |\mathcal{M}_{RR}|^2$, since $s_L = -s_R$. Consequently, $|\mathcal{M}_{RL}|^2 = |\mathcal{M}_{LR}|^2 = 0$, because

$$\sum_{\text{pol.}}|\mathcal{M}|^2 = |\mathcal{M}_{RR}|^2 + |\mathcal{M}_{LL}|^2 + |\mathcal{M}_{RL}|^2 + |\mathcal{M}_{LR}|^2, \qquad (24.15)$$

153

and the first two terms in (24.15) already saturate the value (24.6) for the left-hand side. Needless to say, using the relations derived above, it is easy to show directly that the combinations RL and LR give null results for the corresponding matrix elements.

As a next instructive example, we will consider the decay of a massive vector boson (i.e. spin-1 particle) into a fermion-antifermion pair within the model (21.3) involving the interaction Lagrangian

$$\mathscr{L}_{int} = g\bar{\psi}\gamma^{\mu}\psi A_{\mu} \, . \tag{24.16}$$

As we will see in Chapter 29, there is a subtle point concerning the validity of the relation (21.35), but in the first order of Dyson expansion it is safe, so we are going to proceed confidently, utilizing the basic formula (21.38).

For convenience, we will denote the vector boson as $V$. The derivation of the relevant matrix element for the decay $V \to f\bar{f}$ can be carried out in much the same way as in the preceding case of the Yukawa interaction. Indeed, taking into account the representation (20.10) of the Proca field, it is easy to realize that the only substantial difference in comparison with the scalar case is that here the annihilation and creation operators are accompanied by the polarization vectors $\varepsilon^{\mu}(k, \lambda)$. Thus, following the steps that in Chapter 22 led from (22.9) to (22.14), one gets

$$\mathscr{M}_{fi}^{(1)} = g\bar{u}(k, s)\gamma_{\mu}v(p, r)\varepsilon^{\mu}(q, \lambda) \, . \tag{24.17}$$

This is represented graphically by the Feynman diagram shown in Fig. 24.1.

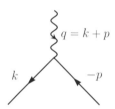

**Fig. 24.1:** First order Feynman diagram for the process $V \to f\bar{f}$. Feynman rules for the fermion lines are the same as before, the wavy line represents the initial vector boson and its contribution is given by the polarization vector $\varepsilon^{\mu}(q, \lambda)$.

Now, we would like to evaluate the contribution of the matrix element squared in the case of the decay of an unpolarized vector boson into the $f\bar{f}$ pair in an arbitrary admissible spin state. A novel issue here is a proper characterization of the state of the initial vector boson. From the point of view of the general formalism of quantum theory, such an unpolarized particle is considered to be in a mixed state described by a density matrix defined by equal probabilities for any spin state; in the present case there are three possible spin states (polarizations), so that all probabilities in question are equal to 1/3. This in turn means that $|\mathscr{M}_{fi}|^2$ should be summed over the vector boson polarizations $\lambda = 1, 2, 3$ and multiplied by 1/3. Such an operation is called, conventionally, **averaging over the initial spin states (polarizations)**. So, squaring the expression (24.17) we get first (dropping the obvious extra labels)

$$|\mathscr{M}|^2 = g^2 \left[\bar{u}(k, s)\gamma_{\mu}v(p, r)\right]\left[\bar{v}(p, r)\gamma_{\nu}u(k, s)\right]\varepsilon^{\mu}(q, \lambda)\varepsilon^{\nu *}(q, \lambda) \, , \tag{24.18}$$

where we have used the identities $\gamma_{\nu}^{\dagger} = \gamma_0\gamma_{\nu}\gamma_0$ and $(\gamma_0)^2 = 1$. Then, utilizing the trace technique, summing over the final spins and averaging over the initial polarizations, one obtains

$$\overline{|\mathscr{M}|^2} \equiv \frac{1}{3}\sum_{\text{pol.}}|\mathscr{M}|^2 = \frac{1}{3}g^2 \operatorname{Tr}\left[(\slashed{k} + m)\gamma_{\mu}(\slashed{p} - m)\gamma_{\nu}\right]\left(-g^{\mu\nu} + \frac{1}{M^2}q^{\mu}q^{\nu}\right), \tag{24.19}$$

154

where we have employed the formula (12.30) for the polarization sum for the massive vector boson. Note that the symbol $\overline{|\mathcal{M}|^2}$ used on the left-hand side of (24.19) is a standard notation for a spin-averaged matrix element squared that will be used from now on. The expression (24.19) can be worked out easily with the help of standard identities for traces of Dirac matrices. A welcome simplification is due to the fact that the term proportional to $1/M^2$ gives zero (an attentive reader may find out that such a striking result is not accidental: in fact, it can be traced back to the "current conservation" identity $\bar{u}(k,s)\slashed{q}v(p,r) = 0$). Thus, one is left with

$$\overline{|\mathcal{M}|^2} = -\frac{1}{3}g^2 \operatorname{Tr}\left[(\slashed{k} + m)\gamma_\mu(\slashed{p} - m)\gamma^\mu\right],$$

and another simple application of "diracology" yields the final result

$$\overline{|\mathcal{M}|^2} = \frac{1}{3}g^2\left(8k \cdot p + 16m^2\right) = \frac{4}{3}g^2(M^2 + 2m^2). \tag{24.20}$$

The corresponding decay rate then becomes

$$\Gamma(V \to f\bar{f}) = \frac{1}{2M}\overline{|\mathcal{M}|^2}\operatorname{LIPS}_2 = \frac{g^2}{12\pi}M\left(1 + \frac{2m^2}{M^2}\right)\sqrt{1 - \frac{4m^2}{M^2}}. \tag{24.21}$$

As for a physical background of the above technical study, note that we have calculated the decay rate of a hypothetical "heavy photon". More realistically, the result (24.21) corresponds to a simplified treatment of the neutral intermediate vector boson (Z) of the standard model of electroweak interactions. A knowledgeable reader may be aware of the fact that in the realistic case, the electroweak interaction of the Z boson involves, along with the vector current appearing in (24.16) also an axial-vector part $\bar{\psi}\gamma^\mu\gamma_5\psi$. Such a generalization of the above calculation is left to the reader as an instructive exercise.

As a last example to be discussed in this chapter, let us consider the process $e^- e^+ \to \nu\bar{\nu}$ that we have mentioned, within a simplified model, in Chapter 22. Here we would like to make its treatment more realistic; we are going to use for its description a theory of weak interactions that preceded the present-day standard model. The SM precursor we have in mind is called, historically, the $V - A$ theory.[15] The corresponding Lagrangian reads

$$\mathcal{L}_{\text{int}} = -\frac{G_F}{\sqrt{2}}\left[\bar{\psi}_1\gamma_\mu(1 - \gamma_5)\psi_2\right]\left[\bar{\psi}_2\gamma^\mu(1 - \gamma_5)\psi_1\right], \tag{24.22}$$

where the labelling of the Dirac fields is the same as in Chapter 22. The coupling constant is written here in the form generally accepted in particle physics; in particular, $G_F$ is the famous Fermi constant that can be found in any particle data tables (note that the overall minus sign is a pure convention). The most spectacular feature of the Lagrangian (24.22), in comparison with the toy model (22.15), is the algebraic structure involving $\gamma_\mu(1 - \gamma_5)$. This has been established on the phenomenological grounds and it also explains the $V - A$ name: the bilinear combinations of Dirac fields (currents) appearing in (24.22) can be read as vector ($V$) minus axial vector ($A$). So much for an excursion into the history of particle physics; our main concern now is a technical study of the relevant matrix element and the cross section for the above-mentioned process.

So, how about the matrix element $\mathcal{M}_{fi}$? Comparing the interaction Lagrangian (24.22) with (22.15), it is clear that one may rely on the basic structure of the expression (22.23) and

---

[15]Concerning its inventors, it is mostly attributed to Richard Feynman and Murray Gell-Mann, but there had been also a significant contribution of Robert Marshak and George Sudarshan. An interested reader can find some details of this story e.g. in [22].

just insert the matrix product $\gamma_\mu(1 - \gamma_5)$ between the Dirac spinors $\bar{u}, u$ and $\bar{v}, v$. We are thus led immediately to the result

$$\mathcal{M}_{fi} = -\frac{G_F}{\sqrt{2}}\left[\bar{u}(k', r')\gamma_\mu(1 - \gamma_5)u(k, r)\right]\left[\bar{v}(p, s)\gamma^\mu(1 - \gamma_5)v(p', s')\right]. \qquad (24.23)$$

For brevity, we will suppress the spin labels in what follows. The matrix element squared then becomes, after a simple manipulation,

$$|\mathcal{M}|^2 = \frac{1}{2}G_F^2\left[\bar{u}(k')\gamma_\rho(1 - \gamma_5)u(k)\right]\left[\bar{v}(p)\gamma^\rho(1 - \gamma_5)v(p')\right]$$
$$\times \left[\bar{u}(k)\gamma_\sigma(1 - \gamma_5)u(k')\right]\left[\bar{v}(p')\gamma^\sigma(1 - \gamma_5)v(p)\right]. \qquad (24.24)$$

To obtain the spin-averaged quantity $\overline{|\mathcal{M}|^2}$, one has to take into account that there are four possible spin combinations in the initial state, so that the averaging factor is $1/4$. Employing also the trace technique, we get

$$\overline{|\mathcal{M}|^2} = \frac{1}{8}G_F^2 \text{Tr}\left[(\slashed{k}' + m_\nu)\gamma_\rho(1 - \gamma_5)(\slashed{k} + m_e)\gamma_\sigma(1 - \gamma_5)\right]$$
$$\times \text{Tr}\left[(\slashed{p} - m_e)\gamma^\rho(1 - \gamma_5)(\slashed{p}' - m_\nu)\gamma^\sigma(1 - \gamma_5)\right]. \qquad (24.25)$$

To proceed further, one may neglect safely the neutrino mass, since this is tiny indeed; in any case, $m_\nu \ll m_e$. When this is done, it also becomes clear that the terms in (24.25) involving explicitly $m_e$ drop out because of $\text{Tr}(\text{odd \#}) = 0$ (note that we employ here the shorthand notation introduced in (3.17), see also (C.6), Appendix C). Taking into account these simplifications and using also the obvious identity $(1 - \gamma_5)^2 = 2(1 - \gamma_5)$, the expression (24.25) is reduced to

$$\overline{|\mathcal{M}|^2} = \frac{1}{2}G_F^2 \text{Tr}\left[\slashed{k}'\gamma_\rho\slashed{k}\gamma_\sigma(1 - \gamma_5)\right] \cdot \text{Tr}\left[\slashed{p}\gamma^\rho\slashed{p}'\gamma^\sigma(1 - \gamma_5)\right]. \qquad (24.26)$$

To work out the product of the traces in (24.26), it is very efficient to employ a set of formulae, which come in handy in many QFT calculations; these read

$$\text{Tr}(\slashed{a}\gamma_\mu\slashed{b}\gamma_\nu) \cdot \text{Tr}(\slashed{c}\gamma^\mu\slashed{d}\gamma^\nu) = 32\left[(a \cdot c)(b \cdot d) + (a \cdot d)(b \cdot c)\right], \qquad (24.27)$$
$$\text{Tr}(\slashed{a}\gamma_\mu\slashed{b}\gamma_\nu\gamma_5) \cdot \text{Tr}(\slashed{c}\gamma^\mu\slashed{d}\gamma^\nu\gamma_5) = 32\left[(a \cdot c)(b \cdot d) - (a \cdot d)(b \cdot c)\right], \qquad (24.28)$$
$$\text{Tr}(\slashed{a}\gamma_\mu\slashed{b}\gamma_\nu) \cdot \text{Tr}(\slashed{c}\gamma^\mu\slashed{d}\gamma^\nu\gamma_5) = 0. \qquad (24.29)$$

As a practical abbreviation, we will call these remarkably simple identities the "formulae 32". Their proof is straightforward, based on the standard formulae for products of Dirac matrices and the Levi-Civita pseudotensor $\varepsilon_{\mu\nu\rho\sigma}$. Anyway, it is somewhat lengthy, so we offer it to a diligent reader as an instructive exercise.

In the present case, using the above formulae in (24.26), one gets readily

$$\overline{|\mathcal{M}|^2} = 32G_F^2(k' \cdot p)(k \cdot p') \qquad (24.30)$$

(so, this is the right moment to appreciate the efficiency of the formulae 32, isn't it?). The kinematics of the considered process is such that $k + p = k' + p'$. Thus, the Mandelstam invariants may be defined conventionally as

$$s = (k + p)^2 = (k' + p')^2,$$
$$t = (k - k')^2 = (p - p')^2, \qquad (24.31)$$
$$u = (k - p')^2 = (k' - p)^2.$$

It means that the scalar products appearing in (24.32) can be recast as

$$k \cdot p' = k' \cdot p = -\frac{1}{2}(u - m_e^2), \tag{24.32}$$

and one gets finally

$$\overline{|\mathcal{M}|^2} = 8G_F^2(u - m_e^2)^2. \tag{24.33}$$

For an illustration (as well as for simplicity) let us see now what is the explicit form of the angular distribution of the final-state particles in the high-energy limit, i.e. for $s \gg m_e^2$. In such a case, $m_e^2$ in (24.33) may be neglected and it is not difficult to find out that

$$u = -\frac{1}{2}s\,(1 + \cos\vartheta_{\text{c.m.}}), \tag{24.34}$$

where $\vartheta_{\text{c.m.}}$ is the angle between the momenta of the incoming electron and outgoing neutrino. Thus, using the general formula (23.45) (valid for massless particles), we get

$$\frac{d\sigma}{d\Omega_{\text{c.m.}}} = \frac{G_F^2 s}{32\pi^2}\,(1 + \cos\vartheta_{\text{c.m.}})^2. \tag{24.35}$$

The expression (24.35) may be integrated easily over the scattering angle and one thus arrives at the result for the total cross section

$$\sigma(s)\Big|_{s \gg m_e^2} = \frac{1}{6\pi}G_F^2 s. \tag{24.36}$$

It is worth noting that a substantial part of the result (24.36) could have been guessed *a priori*, just on dimensional grounds. Indeed, the cross section has the dimension $(\text{mass})^{-2}$ and in the lowest order it is proportional to $G_F^2$. As we have already pointed out in Chapter 22, the dimension of $G_F$ is $(\text{mass})^{-2}$. In the high-energy limit, particle masses may be neglected, so the only other dimensionful quantity, apart from $G_F$, is the energy. Thus, obviously, to get a cross section $\sigma$ with the right dimension, one has to multiply $G_F^2$ by an energy squared; since $\sigma$ is Lorentz invariant, the relevant variable is $s$ $(= E_{\text{c.m.}}^2)$. Thus, one arrives at an estimate

$$\sigma(s)\Big|_{s \gg m_e^2} = \text{const.}\, G_F^2 s. \tag{24.37}$$

In other words, the detailed calculation that we have performed just supplements (24.37) with the numerical factor $1/6\pi$.

It may be also instructive to find a numerical value of the cross section (24.36) for a typical high energy, e.g. $E_{\text{c.m.}} = 1$ GeV, i.e. $s = 1$ GeV$^2$. According to particle data tables, the value of the Fermi constant amounts to $G_F \doteq 1.166 \times 10^{-5}$ GeV$^{-2}$. Thus, the cross section (24.36) comes out, in the natural units,

$$\sigma(s = 1\,\text{GeV}^2) \doteq 0.07 \times 10^{-10}\,\text{GeV}^{-2}. \tag{24.38}$$

To recast it in ordinary units, one employs the well-known conversion constant $\hbar c = 197$ MeV fm, where $1$ fm $= 10^{-13}$ cm. Thus, in natural units one has $1$ GeV$^{-1} \doteq 0.2$ fm. Putting all this together, one gets finally

$$\sigma(s = 1\,\text{GeV}^2) \doteq 2.8 \times 10^{-39}\,\text{cm}^2 = 2.8\,\text{fb (femtobarn)}, \tag{24.39}$$

157

where $1\,b = 10^{-24}\,cm^2$. Later on, we will see that for a typical electromagnetic process like $e^- e^+ \to \mu^- \mu^+$ (production of a muon pair in the electron–positron annihilation) one gets the result (within QED)

$$\sigma(e^- e^+ \to \mu^- \mu^+)\Big|_{s=1\,GeV^2} \doteq 8.7 \times 10^{-32}\,cm^2 \,,$$

i.e. a value that is seven orders of magnitude larger than (24.39) (such a huge difference justifies the attribute "weak" for the interaction responsible for the process $e^- e^+ \to \nu\bar{\nu}$). Well, an astute reader might object that the cross section (24.36) grows indefinitely for $s \to \infty$, so that the weak interaction might become strong at a sufficiently high energy. The standard model of electroweak interactions gives a clear answer to such a question, but this is another story that goes beyond the scope of these lectures (an interested reader is referred to [22]).

# Chapter 25

# Scattering in external Coulomb field. Mott formula

There is another example of a physical process that deserves a separate treatment. It is the scattering of a Dirac particle (e.g. electron) in an external static electromagnetic field; of particular interest is the case of the Coulomb field. For the description of such a situation one may employ the interaction Lagrangian

$$\mathcal{L}_{\text{int}} = e\bar{\psi}(x)\gamma^\mu\psi(x)A_\mu^{\text{clas.}}(x), \tag{25.1}$$

where $\psi(x)$ is the quantized Dirac field and $A_\mu^{\text{clas.}}(x)$ represents the external classical field. Of course, such a "hybrid" formalism is an idealized picture of a situation, in which the electron interacts with a heavy charged particle (e.g. proton or another atomic nucleus) that is the source of the pertinent electromagnetic field. We will come back to this issue later on, within the framework of quantum electrodynamics, but now we are going to proceed using (25.1).

It is not difficult to obtain the relevant matrix element for the considered scattering process. The states corresponding to the incident and scattered electron can be represented as

$$|i\rangle = b^\dagger(p, s)|0\rangle, \qquad |f\rangle = b^\dagger(p', s')|0\rangle, \tag{25.2}$$

and the 1st order $S$-matrix element is then

$$S_{fi}^{(1)} = ie \int d^4x \, \langle 0|b(p', s')\bar{\psi}(x)\gamma^\mu\psi(x)b^\dagger(p, s)|0\rangle A_\mu(x). \tag{25.3}$$

Note that here and in what follows we suppress the label "clas." on the electromagnetic field (four-potential) $A_\mu$. Moreover, we will assume that the field is static, i.e. $A_\mu(x) = A_\mu(\vec{x})$. The v.e.v. in the integrand in (25.3) is worked out easily, in analogy with the examples discussed previously; the non-trivial contribution corresponds to the pairings of $b(p', s')$ with $\bar{\psi}(x)$ and $b^\dagger(p, s)$ with $\psi(x)$. The rest of the calculation is routine and one ends up with

$$S_{fi}^{(1)} = ieN_pN_{p'}\bar{u}(p', s')\gamma^\mu u(p, s) \int d^4x \, e^{i(p'-p)x} A_\mu(\vec{x}). \tag{25.4}$$

The time integration is trivial and yields $2\pi\delta(p_0'-p_0)$, which means that the energy is conserved. Needless to say, this is what must happen in the presence of a static field. On the other hand, the three-momentum is not conserved, for the same reason (more precisely, the momentum magnitude is conserved, but its direction is changed, defining the scattering angle). Thus, we have

$$S_{fi}^{(1)} = N_pN_{p'} i\mathcal{M}_{fi}^{(1)} 2\pi\delta(E'-E), \tag{25.5}$$

159

with

$$\mathcal{M}_{fi}^{(1)} = e\,\bar{u}(p',s')\gamma^{\mu}u(p,s)\widetilde{A}_{\mu}(\vec{q}),\qquad(25.6)$$

where

$$\widetilde{A}_{\mu}(\vec{q}) = \int d^3x\, A_{\mu}(\vec{x})\,e^{-i\vec{q}\vec{x}}\qquad(25.7)$$

for

$$\vec{q} = \vec{p}' - \vec{p}.\qquad(25.8)$$

Such an expression can be represented graphically by means of a simple Feynman diagram shown in Fig. 25.1.

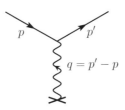

**Fig. 25.1:** Feynman diagram for the scattering in an external field. The contribution of the wavy line with the cross at the endpoint is the Fourier transform of the external field.

The evaluation of the scattering cross section proceeds basically along the same lines as before in Chapter 23, but there are some minor modifications due to the different kinematics of the considered process, embodied in (25.5). In analogy with (23.37), one may start with

$$d\sigma = \frac{\frac{1}{T}|S_{fi}|^2\,\dfrac{V\,d^3p_f}{(2\pi)^3}}{\frac{1}{V}|\vec{v}_i|}\qquad(25.9)$$

(for convenience, we have temporarily switched to the self-explanatory notation $p_i = p$, $p_f = p'$). The expression for $|S_{fi}|^2$ involves, among other things, the nasty singular factor $\delta(0)$; its pragmatic regularization now reads (cf. (23.8))

$$2\pi\delta(0) \longrightarrow \int_{-T/2}^{T/2} dx_0 = T.\qquad(25.10)$$

Thus, substituting all relevant ingredients into (25.9) one gets first, after a simple manipulation,

$$d\sigma = \frac{1}{16\pi^2}|\mathcal{M}_{fi}|^2\,\frac{d^3p_f}{|\vec{p}_i|E_f}\delta(E_f - E_i).\qquad(25.11)$$

Next, using spherical variables so that $d^3p_f = |\vec{p}_f|^2\,d|\vec{p}_f|\,d\Omega_f$ and taking into account that $|\vec{p}_f|\,d|\vec{p}_f| = E_f\,dE_f$, the integration over $dE_f$ is carried out readily and we obtain finally

$$\frac{d\sigma}{d\Omega} = \frac{1}{16\pi^2}|\mathcal{M}|^2\qquad(25.12)$$

(we have streamlined our notation as much as possible). Such a remarkably simple formula holds generally, for any static external field; now let us consider the particular case of the Coulomb field.

160

In such a case, the external field can be written as

$$A^\mu(\vec{x}) = \left(\frac{1}{4\pi}\frac{e}{r}, \vec{0}\right),\tag{25.13}$$

where $r = |\vec{x}|$. We have in mind here that the source carries the elementary charge, i.e. it may be viewed as a static proton. Note that the form (25.13) corresponds to the rationalized (Heaviside–Lorentz) electromagnetic units; in compliance with the natural system of units we are using, it then holds $e^2 = 4\pi\alpha$, where $\alpha \doteq 1/137$ is the famous fine-structure constant. For the evaluation of the Fourier transform of (25.13) one may utilize the well-known integral

$$\int d^3x \frac{1}{|\vec{x}|}e^{i\vec{q}\vec{x}} = \frac{4\pi}{\vec{q}^2}.\tag{25.14}$$

Thus,

$$\widetilde{A}^\mu(\vec{q}) = \left(\frac{e}{\vec{q}^2}, \vec{0}\right),\tag{25.15}$$

and the matrix element (25.6) becomes

$$\mathcal{M} = \frac{e^2}{\vec{q}^2}\bar{u}(p', s')\gamma_0 u(p, s).\tag{25.16}$$

We are going to compute now the differential cross section for unpolarized electrons. The spin-averaged matrix element squared is given by

$$\overline{|\mathcal{M}|^2} = \frac{1}{2}\left(\frac{e^2}{\vec{q}^2}\right)^2 \sum_{\text{pol.}} \left[\bar{u}(p', s')\gamma_0 u(p, s)\right]\left[\bar{u}(p, s)\gamma_0 u(p', s')\right]$$

$$= \frac{1}{2}\left(\frac{e^2}{\vec{q}^2}\right)^2 \text{Tr}\left[(p\!\!\!/' + m)\gamma_0(p\!\!\!/ + m)\gamma_0\right].\tag{25.17}$$

The trace in (25.17) is evaluated easily and one gets

$$\overline{|\mathcal{M}|^2} = \frac{2e^4}{(\vec{q}^2)^2}\left(2E^2 - p \cdot p' + m^2\right)$$

$$= \frac{2e^4}{(\vec{q}^2)^2}\left(E^2 + |\vec{p}|^2 \cos\vartheta + m^2\right)$$

$$= \frac{2e^4}{(\vec{q}^2)^2}\left[E^2 + m^2 + (E^2 - m^2)\cos\vartheta\right],\tag{25.18}$$

where $\vartheta$ is the scattering angle (between $\vec{p}'$ and $\vec{p}$). For $\vec{q}^2$ we have $\vec{q}^2 = (\vec{p}' - \vec{p})^2 = 2|\vec{p}|^2(1 - \cos\vartheta) = 4|\vec{p}|^2 \sin^2\frac{\vartheta}{2}$ and the cross section (25.12) becomes, finally,

$$\frac{d\sigma}{d\Omega} = \frac{\alpha^2}{4\beta^2|\vec{p}|^2}\frac{1}{\sin^4\frac{\vartheta}{2}}\left(1 - \beta^2 \sin^2\frac{\vartheta}{2}\right),\tag{25.19}$$

where $\beta$ is the particle velocity, $\beta = |\vec{p}|/E$. We have thus arrived at the celebrated **Mott formula**[16] announced in the title of the present chapter. It represents a relativistic generalization

---

[16]Sir Nevill Mott (1905-1996) published this formula in 1929. Later he conducted highly successful research in solid state physics and received the Nobel Prize in 1977 for his work on disordered systems such as amorphous semiconductors.

of the familiar Rutherford formula known from non-relativistic quantum and classical mechanics. Indeed, let us see what one gets from (25.19) in the low-energy approximation, i.e. for $|\vec{p}| \ll m$. In such a case, one has $\beta \ll 1$ and the second term in parentheses can be neglected. In the factor multiplying the angular distribution $1/\sin^4 \frac{\vartheta}{2}$ one may then set $4\beta^2 |\vec{p}|^2 \doteq 4\beta^2 \cdot m^2\beta^2 = 16E_{\text{kin}}^2$, where $E_{\text{kin}} = \frac{1}{2}m\beta^2$ is the non-relativistic kinetic energy. Putting all this together, (25.19) is reduced to

$$\frac{d\sigma}{d\Omega}\bigg|_{\beta \ll 1} = \left(\frac{\alpha}{4E_{\text{kin}}}\right)^2 \frac{1}{\sin^4 \frac{\vartheta}{2}}, \tag{25.20}$$

and this is just the Rutherford formula. Of course, in the opposite limit $\beta \to 1$ the angular distribution is substantially modified by the relativistic effect embodied in the factor $1 - \beta^2 \sin^2 \frac{\vartheta}{2}$; in particular, the backward scattering, corresponding to $\vartheta = 180°$, is suppressed.

There is another aspect of the Mott scattering that is worth mentioning, namely the behaviour of polarized particles. To elucidate this point, let us consider the case of the scattering of fully polarized electron, e.g. with positive helicity. One may define the degree of polarization of scattered electrons as

$$P = \frac{\dfrac{d\sigma_R}{d\Omega} - \dfrac{d\sigma_L}{d\Omega}}{\dfrac{d\sigma_R}{d\Omega} + \dfrac{d\sigma_L}{d\Omega}}, \tag{25.21}$$

where the indices $R$, $L$ mark the scattered electron with positive and negative helicity, respectively. The calculation based on (25.12) (and relying heavily on the trace techniques) is straightforward, but rather tedious, so we will give here just the final result. This reads

$$P = \frac{E^2 \cos^2 \dfrac{\vartheta}{2} - m^2 \sin^2 \dfrac{\vartheta}{2}}{E^2 \cos^2 \dfrac{\vartheta}{2} + m^2 \sin^2 \dfrac{\vartheta}{2}} = 1 - \frac{2m^2 \sin^2 \dfrac{\vartheta}{2}}{E^2 \cos^2 \dfrac{\vartheta}{2} + m^2 \sin^2 \dfrac{\vartheta}{2}}. \tag{25.22}$$

The derivation of this formula is left as a challenge for a truly diligent reader. One can thus see that there is a depolarization effect depending, in general, on the energy and the scattering angle. A salient feature of the result (25.22) is that in the ultrarelativistic limit $E \gg m$ the initial polarization is preserved, $P \to 1$. In fact, there is a deeper reason for such a behaviour. In the high-energy limit, the mass can be neglected and the helicity becomes equivalent to chirality (cf. Chapter 8); however, in the electromagnetic interaction the chirality is conserved, since e.g.

$$\bar{u}_L \gamma_\mu u_R = \bar{u}_L \frac{1 + \gamma_5}{2} \gamma_\mu \frac{1 + \gamma_5}{2} u_R = 0.$$

One may also notice immediately that in this aspect, namely the chirality conservation, the electromagnetic (vector-like) force differs substantially from the Yukawa interaction, since, in general, $\bar{u}_L u_R \neq 0$.

# Chapter 26

# Propagator of scalar field

In the preceding chapters we have discussed some elementary applications of perturbative QFT, employing just the first-order term of the Dyson expansion of the $S$-matrix. These lowest-order calculations are very simple indeed and the resulting Feynman diagrams consist of vertices and external lines only; nevertheless, one thus gets a lot of useful results.

Now, let us proceed to the 2nd order of Dyson expansion. This will lead us naturally to a new crucial ingredient of the diagram technique, namely to internal lines connecting the vertices; mathematically, such an internal line corresponds to the **Feynman propagator** of a quantized field.

As a motivation example, we are going to consider a model of Yukawa-like interaction

$$\mathcal{L}_{\text{int}} = g_1 \bar{\psi}_1 \psi_1 \varphi + g_2 \bar{\psi}_2 \psi_2 \varphi \,, \tag{26.1}$$

where $\psi_1$, $\psi_2$ are two different Dirac fields (corresponding, for definiteness, e.g. to electron and muon) and $\varphi$ is a real scalar field. (Note that a knowledgeable reader may recognize in (26.1) the interaction of the Higgs boson field with fermions within the current standard model of particle physics.) Let us consider the electron–muon scattering process

$$e(k) + \mu(p) \rightarrow e(k') + \mu(p') \,, \tag{26.2}$$

where we have also indicated the corresponding particle four-momenta. Obviously, to get a non-trivial contribution to such a process, one has to use (at least) the 2nd order of the $S$-matrix expansion, since one needs two electron operators, as well as two muon operators. Thus, we start with the $S$-matrix operator approximation

$$S^{(2)} = \frac{i^2}{2!} \iint \mathrm{d}^4 x \, \mathrm{d}^4 y \, \mathrm{T}\big(\mathcal{L}_{\text{int}}(x)\mathcal{L}_{\text{int}}(y)\big) \,, \tag{26.3}$$

and with the initial and final states defined as

$$
\begin{aligned}
|i\rangle &= b_1^\dagger(k)b_2^\dagger(p)|0\rangle \,, \\
|f\rangle &= b_1^\dagger(k')b_2^\dagger(p')|0\rangle \,,
\end{aligned} \tag{26.4}
$$

where the notation is self-explanatory; for brevity, we have suppressed the relevant spin labels. For convenience, let us denote the two parts of the interaction Lagrangian (26.1) as $\mathcal{L}_1$ and $\mathcal{L}_2$, respectively. It is clear that a non-vanishing contribution to the matrix element $S^{(2)}_{fi}$ can be obtained only from products $\mathcal{L}_1(x)\mathcal{L}_2(y)$ and $\mathcal{L}_2(x)\mathcal{L}_1(y)$ in (26.3). Taking into account that $T(A(x)B(y)) = T(B(y)A(x))$, as well as the simple fact that in the integral (26.3) one can perform the change of variables $x \leftrightarrow y$ if necessary, $S_{fi}^{(2)}$ can be written as

$$S_{fi}^{(2)} = i^2 g_1 g_2 \iint d^4x\, d^4y \, \langle f|T(\bar\psi_1(x)\psi_1(x)\varphi(x)\bar\psi_2(y)\psi_2(y)\varphi(y))|i\rangle, \qquad (26.5)$$

because the above-mentioned two terms under the T-product give the same contribution. To work out the last expression, we use the definitions (26.4) and the usual general representation of the T-product in terms of the Heaviside step functions,

$$T(A(x)B(y)) = \vartheta(x_0 - y_0)A(x)B(y) + \vartheta(y_0 - x_0)B(y)A(x). \qquad (26.6)$$

For the integrand in Eq. (26.5) one then has

$$\langle f|T(\bar\psi_1(x)\psi_1(x)\varphi(x)\bar\psi_2(y)\psi_2(y)\varphi(y))|i\rangle$$
$$= \vartheta(x_0 - y_0)\langle 0|b_2(p')b_1(k')\bar\psi_1(x)\psi_1(x)\varphi(x)\bar\psi_2(y)\psi_2(y)\varphi(y)b_1^\dagger(k)b_2^\dagger(p)|0\rangle \qquad (26.7)$$
$$+ \vartheta(y_0 - x_0)\langle 0|b_2(p')b_1(k')\bar\psi_2(y)\psi_2(y)\varphi(y)\bar\psi_1(x)\psi_1(x)\varphi(x)b_1^\dagger(k)b_2^\dagger(p)|0\rangle.$$

According to our previous experience, we know that the non-vanishing contributions to the vacuum expectation values like those in (26.7) are given by complete pairings of the annihilation and creation operators of the same kind. Of course, we exclude pairings between $b_1(k')$ and $b_1^\dagger(k)$, etc., since these correspond to kinematically trivial situation (no real scattering). Taking into account the operator structure of Dirac fields, one gets for the relevant pairings (determined by the basic anticommutators), in the usual notation

$$b_1(k')\bar\psi_1(x) = N_{k'}\bar u(k')e^{ik'x}, \qquad \psi_1(x)b_1^\dagger(k) = N_k u(k)e^{-ikx},$$
$$b_2(p')\bar\psi_2(y) = N_{p'}\bar u(p')e^{ip'y}, \qquad \psi_2(y)b_2^\dagger(p) = N_p u(p)e^{-ipy}. \qquad (26.8)$$

While the expressions (26.8) are factorized from the vacuum matrix elements in (26.7), the scalar fields "stay inside", i.e. one is left with their time-ordered product sandwiched between the vacuum states. Thus, the integrand in (26.5) becomes

$$\langle f|T(\bar\psi_1(x)\psi_1(x)\varphi(x)\bar\psi_2(y)\psi_2(y)\varphi(y))|i\rangle$$
$$= N_k N_p N_{k'} N_{p'} e^{-ikx} e^{ik'x} e^{-ipy} e^{ip'y} [\bar u(k')u(k)][\bar u(p')u(p)]\langle 0|T(\varphi(x)\varphi(y))|0\rangle. \qquad (26.9)$$

So, now it is in order to examine the new object involving scalar fields in detail. First of all, let us introduce an appropriate (standard) notation, namely

$$i\mathscr{D}_F(x - y) \equiv \langle 0|T(\varphi(x)\varphi(y))|0\rangle. \qquad (26.10)$$

The function $\mathscr{D}_F(x - y)$ is called the **Feynman propagator** and we have already used the fact that it is a function of the difference $x - y$, not $x$ and $y$ separately. This is easy to prove, as we are going to find out shortly. It is certainly highly desirable to have an explicit representation of the function $\mathscr{D}_F$, so let us now evaluate the vacuum expectation value in (26.10). The quantized (free) scalar field has the form

$$\varphi(x) = \int d^3k \, N_k \left[a(k)e^{-ikx} + a^\dagger(k)e^{ikx}\right]. \qquad (26.11)$$

Just to be sure, let us recall that in Eq. (26.11), $kx \equiv k \cdot x = k_0 x_0 - \vec{k} \cdot \vec{x}$, with $k_0 = \sqrt{|\vec{k}|^2 + m^2}$. Denoting the annihilation and creation parts of the field $\varphi(x)$ as $\varphi^-(x)$ and $\varphi^+(x)$, respectively, it holds, clearly,

$$
\begin{aligned}
\langle 0|T(\varphi(x)\varphi(y))|0\rangle &= \vartheta(x_0 - y_0)\langle 0|\varphi^-(x)\varphi^+(y)|0\rangle \\
&+ \vartheta(y_0 - x_0)\langle 0|\varphi^-(y)\varphi^+(x)|0\rangle \, .
\end{aligned}
\tag{26.12}
$$

So, using the explicit expressions appearing in (26.11), one gets

$$
\begin{aligned}
\langle 0|T(\varphi(x)\varphi(y))|0\rangle &= \vartheta(x_0 - y_0) \iint d^3k\, d^3l\, N_k N_l\, \langle 0|a(k)a^\dagger(l)|0\rangle e^{-ikx+ily} \\
&+ \vartheta(y_0 - x_0) \iint d^3k\, d^3l\, N_k N_l\, \langle 0|a(l)a^\dagger(k)|0\rangle e^{ikx-ily} \, .
\end{aligned}
\tag{26.13}
$$

However, one finds out easily that

$$
\langle 0|a(k)a^\dagger(l)|0\rangle = \langle 0|a(l)a^\dagger(k)|0\rangle = \delta^{(3)}(\vec{k} - \vec{l}) \, ,
$$

as a simple consequence of the canonical commutation relations. So, we have finally

$$
\begin{aligned}
\langle 0|T(\varphi(x)\varphi(y))|0\rangle &= \vartheta(x_0 - y_0) \int d^3k\, N_k^2\, e^{-ik(x-y)} \\
&+ \vartheta(y_0 - x_0) \int d^3k\, N_k^2\, e^{ik(x-y)} \, .
\end{aligned}
\tag{26.14}
$$

The last result makes it obvious that the above statement about the dependence of the function $\mathscr{D}_F$ on $x - y$ is valid; one may notice that technically, it is due to the commutation relation $[a(k), a^\dagger(l)] = \delta^{(3)}(\vec{k} - \vec{l})$.

A remark is in order here. In fact, there is another explanation of the $x - y$ dependence manifested now in (26.14), which is more general and in a sense more elegant than the explicit calculation. Indeed, one may employ the relation

$$
\varphi(x + a) = e^{iP \cdot a} \varphi(x) e^{-iP \cdot a}
\tag{26.15}
$$

(see (16.33)), where $a$ is a constant spacetime shift and $P$ is the operator of four-momentum of the quantized field $\varphi(x)$, i.e.

$$
P^0 = \int d^3k\, k^0\, a^\dagger(k)a(k) \, , \quad \vec{P} = \int d^3k\, \vec{k}\, a^\dagger(k)a(k) \, .
\tag{26.16}
$$

Thus,

$$
\langle 0|\varphi(x + a)\varphi(y + a)|0\rangle = \langle 0|e^{iP \cdot a}\varphi(x)\varphi(y)e^{-iP \cdot a}|0\rangle \, ,
$$

but $e^{-iP \cdot a}|0\rangle = |0\rangle$, since $P^\mu|0\rangle = 0$ due to the normal ordering in the expressions (26.16). In this way, we see that

$$
\langle 0|\varphi(x + a)\varphi(y + a)|0\rangle = \langle 0|\varphi(x)\varphi(y)|0\rangle \, ,
$$

and this in turn means that the vacuum expectation value in (26.10) is invariant under spacetime translations; this, of course, is tantamount to the statement that $\mathscr{D}_F$ depends just on $x - y$.

Having in mind our ultimate goal of constructing an appropriate momentum-space Feynman diagram representing the scattering amplitude contained in (26.5), one should endeavour to get the four-dimensional Fourier transform of the function $\mathscr{D}_{\mathrm{F}}(x - y)$. To this end, one can utilize the known formula for the Fourier transformation of the $\vartheta$-function, namely

$$\vartheta(t) = \frac{1}{2\pi i} \int\limits_{-\infty}^{\infty} \frac{e^{i\omega t}}{\omega - i\epsilon} \, d\omega, \tag{26.17}$$

where $\epsilon > 0$ is an arbitrarily small constant. Using this in the expression (26.14) and taking into account that $N_k^2 = (2\pi)^{-3}(2k_0)^{-1}$ one has, after a trivial manipulation,

$$\langle 0|T(\varphi(x)\varphi(y))|0\rangle$$
$$= \frac{1}{i} \int \frac{d^3k \, d\omega}{(2\pi)^4 \, 2k_0} \left[ \frac{e^{i(\omega - k_0)(x_0 - y_0)}}{\omega - i\epsilon} e^{i\vec{k}\cdot(\vec{x} - \vec{y})} + \frac{e^{i(\omega - k_0)(y_0 - x_0)}}{\omega - i\epsilon} e^{-i\vec{k}\cdot(\vec{x} - \vec{y})} \right]. \tag{26.18}$$

The integration variable $\omega$ can be shifted by means of the substitution $\omega - k_0 = \omega'$. Further, in the second term in (26.18), one may perform the change $\omega' \to -\omega'$ and in the first term one substitutes $\vec{k} \to -\vec{k}$. After these simple manipulations one has (renaming $\omega'$ as $\omega$)

$$\langle 0|T(\varphi(x)\varphi(y))|0\rangle$$
$$= \frac{1}{i} \int \frac{d^3k \, d\omega}{(2\pi)^4 \, 2k_0} e^{i\omega(x_0 - y_0) - i\vec{k}\cdot(\vec{x} - \vec{y})} \left( \frac{1}{\omega + k_0 - i\epsilon} + \frac{1}{-\omega + k_0 - i\epsilon} \right). \tag{26.19}$$

The sum of the fractions in the integrand in (26.19) becomes

$$\frac{1}{\omega + k_0 - i\epsilon} - \frac{1}{\omega - k_0 + i\epsilon} = \frac{-2k_0}{\omega^2 - k_0^2 + 2ik_0\epsilon},$$

where we have discarded irrelevant terms involving the auxiliary parameter $\epsilon \to 0^+$. As for the term $2ik_0\epsilon$ in the denominator, this is in fact equivalent to $i\epsilon$, since $k_0 = \sqrt{|\vec{k}|^2 + m^2} > m$ and $\epsilon$ is taken to be infinitesimal, but otherwise arbitrary. Thus, (26.19) can be recast as

$$\langle 0|T(\varphi(x)\varphi(y))|0\rangle = i \int \frac{d^3k \, d\omega}{(2\pi)^4} e^{i\omega(x_0 - y_0) - i\vec{k}\cdot(\vec{x} - \vec{y})} \frac{1}{\omega^2 - k_0^2 + i\epsilon}. \tag{26.20}$$

Now, taking into account that $k_0^2 = |\vec{k}|^2 + m^2$, and introducing the notation $q = (\omega, \vec{k})$, i.e. $\omega = q_0$ and $\vec{q} = \vec{k}$, the result (26.20) can be written finally as

$$\langle 0|T(\varphi(x)\varphi(y))|0\rangle = i \int \frac{d^4q}{(2\pi)^4} \frac{e^{iq(x - y)}}{q^2 - m^2 + i\epsilon}, \tag{26.21}$$

which is the desired four-dimensional Fourier representation. It also shows the utility of the conventional factor of $i$ in the definition (26.10); for the function $\mathscr{D}_{\mathrm{F}}(x - y)$ we have

$$\mathscr{D}_{\mathrm{F}}(x - y) = \int \frac{d^4q}{(2\pi)^4} \frac{e^{iq(x - y)}}{q^2 - m^2 + i\epsilon}. \tag{26.22}$$

Thus, we see that the Fourier transform of the function $\mathscr{D}_{\mathrm{F}}(x - y)$ is remarkably simple; denoting it as $D_{\mathrm{F}}(q)$, one has

$$D_{\mathrm{F}}(q) = \frac{1}{q^2 - m^2 + i\epsilon} \tag{26.23}$$

166

(on the other hand, the function $\mathscr{D}_F(x)$ is quite complicated for $m \neq 0$, but we will not need its explicit form). The simple form (26.23) is not accidental; the point is that the function $\mathscr{D}_F$ defined by (26.10) is in fact a **Green's function** of the Klein–Gordon equation. We are going to prove it now *ab initio* (i.e. without any reference to our preceding calculation).

So, our task is to find out how the d'Alembert operator acts on the T-product in (26.10). In our calculation we will employ the canonical equal-time commutation relations (ETCR)

$$\left[\varphi(x), \varphi(y)\right]_{\text{E.T.}} = 0\,, \tag{26.24}$$

$$\left[\varphi(x), \dot{\varphi}(y)\right]_{\text{E.T.}} = i\delta^{(3)}(\vec{x} - \vec{y})\,, \tag{26.25}$$

and the familiar Klein–Gordon equation

$$(\Box + m^2)\varphi(x) = 0\,. \tag{26.26}$$

Let us start with the time derivative $\partial_0 = \partial/\partial x_0$. To proceed, the readers should recall any good math course, where they learned that $\partial_0 \vartheta(x_0 - y_0) = \delta(x_0 - y_0)$. Then one has

$$
\begin{aligned}
\partial_0 T\big(\varphi(x)\varphi(y)\big) &= \partial_0\big[\vartheta(x_0 - y_0)\varphi(x)\varphi(y) + \vartheta(y_0 - x_0)\varphi(y)\varphi(x)\big] \\
&= \delta(x_0 - y_0)\varphi(x)\varphi(y) + \vartheta(x_0 - y_0)\dot{\varphi}(x)\varphi(y) \\
&\quad - \delta(x_0 - y_0)\varphi(y)\varphi(x) + \vartheta(y_0 - x_0)\varphi(y)\dot{\varphi}(x) \\
&= \delta(x_0 - y_0)\big[\varphi(x), \varphi(y)\big]_{\text{E.T.}} \\
&\quad + \vartheta(x_0 - y_0)\dot{\varphi}(x)\varphi(y) + \vartheta(y_0 - x_0)\varphi(y)\dot{\varphi}(x)\,.
\end{aligned} \tag{26.27}
$$

Now, taking into account (26.24), the second derivative becomes

$$
\begin{aligned}
\partial_{00} T\big(\varphi(x)\varphi(y)\big) &= \delta(x_0 - y_0)\dot{\varphi}(x)\varphi(y) + \vartheta(x_0 - y_0)\ddot{\varphi}(x)\varphi(y) \\
&\quad - \delta(x_0 - y_0)\varphi(y)\dot{\varphi}(x) + \vartheta(y_0 - x_0)\varphi(y)\ddot{\varphi}(x)\,,
\end{aligned} \tag{26.28}
$$

and using (26.25) one is then left with

$$\partial_{00} T\big(\varphi(x)\varphi(y)\big) = \vartheta(x_0 - y_0)\ddot{\varphi}(x)\varphi(y) + \vartheta(y_0 - x_0)\varphi(y)\ddot{\varphi}(x) - i\delta^{(4)}(x - y)\,. \tag{26.29}$$

The differentiation with respect to spatial coordinates is much simpler, since the $\vartheta$-functions then do not get in the way. So, for the action of the Laplacian one gets immediately

$$\Delta_x T\big(\varphi(x)\varphi(y)\big) = \vartheta(x_0 - y_0)\,\Delta\varphi(x)\varphi(y) + \vartheta(y_0 - x_0)\varphi(y)\,\Delta\varphi(x)\,. \tag{26.30}$$

Putting together the results (26.29) and (26.30), the Klein–Gordon equation (26.26) may be utilized and one obtains, finally,

$$(\Box_x + m^2)T\big(\varphi(x)\varphi(y)\big) = -i\delta^{(4)}(x - y)\,.$$

This in turn means that the function $\mathscr{D}_F(x - y)$ defined by (26.10) satisfies the differential equation

$$(\Box_x + m^2)\mathscr{D}_F(x - y) = -\delta^{(4)}(x - y)\,, \tag{26.31}$$

and this is indeed the advertised equation for the Green's function. Now it also becomes obvious that the Fourier transform $D_F(q)$ must have the form $(q^2 - m^2)^{-1}$, at least for $q^2 \neq m^2$: the point is that the d'Alembert operator $\Box$ is turned into $-q^2$ upon Fourier transformation and the delta function on the right-hand side of Eq. (26.31) becomes 1. We know that the equation (26.31) has infinitely many solutions, consisting of a particular solution of the inhomogeneous

Klein–Gordon equation and the general solution of its homogeneous counterpart. The specific form of the Feynman propagator (26.23) involving $i\epsilon$ (which serves as a regularization of the pole singularity for $q^2 = m^2$) is the choice that corresponds to the so-called causal Green's function. We will not discuss this concept here, since it is not necessary for our description of momentum-space Feynman diagrams (for some details concerning Green's functions in relativistic quantum theory see e.g. the books [2], [6] or [16]).

With the results (26.22) and (26.23) at hand, let us now come back to the $S$-matrix element (26.5) and the expression (26.9). Putting all things together, one has

$$S^{(2)}_{fi} = i^2 \cdot i \, g_1 g_2 N_k N_p N_{k'} N_{p'} \left[\bar{u}(k')u(k)\right]\left[\bar{u}(p')u(p)\right]$$

$$\times \iint d^4x \, d^4y \, e^{-ikx} e^{ik'x} e^{-ipy} e^{ip'y} \int \frac{d^4q}{(2\pi)^4} \, D_F(q) \, e^{iq(x-y)} . \quad (26.32)$$

The integration over $x$ and $y$ is trivial; one gets

$$\iint d^4x \, d^4y \, e^{i(-k+k'+q)x} e^{i(-p+p'-q)y} = (2\pi)^4 \delta^{(4)}(k'-k+q)(2\pi)^4 \delta^{(4)}(p'-p-q) . \quad (26.33)$$

The ensuing integration over $q$ then leads to the expected delta function that embodies the overall four-momentum conservation; at the same time, the intermediate delta functions in (26.33) mean that the momentum-space propagator $D_F(q)$ is to be taken here for $q = p' - p$, or $k - k'$ (which is the same). Thus, we have arrived at the result for $S^{(2)}_{fi}$ that can be written in the by now familiar form

$$S^{(2)}_{fi} = N_k N_p N_{k'} N_{p'} \, i \mathcal{M}^{(2)}_{fi} (2\pi)^4 \delta^{(4)}(k'+p'-k-p) , \quad (26.34)$$

with

$$i \mathcal{M}^{(2)}_{fi} = i^3 g_1 g_2 \left[\bar{u}(k')u(k)\right] D_F(q) \left[\bar{u}(p')u(p)\right] . \quad (26.35)$$

The last expression suggests naturally a graphical representation in the form of the Feynman diagram in Fig. 26.1. Note that when drawing such a diagram, the four-momentum conservation

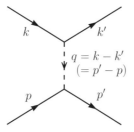

**Fig. 26.1:** Electron–muon scattering at the lowest order.

holds at each vertex and the rules for the external lines are the same as before, in the case of 1st order diagrams. Let us also notice that the direction of the internal line, indicated in Fig. 26.1, is in fact not important, since $D_F(q)$ is an even function.

The usual verbal description of a diagram like Fig. 26.1 refers to an "exchange of the **virtual particle**"; one can also say that the interaction of fermions is mediated by the exchange of the scalar boson. Of course, the adjective "virtual" here means that, in general, $q^2 \neq m^2$; the would-be pole at $q^2 = m^2$ corresponds to the real "on-shell" particle. Fig. 26.1 and its intuitive verbal description also provides a natural justification of the term "propagator": one may continue the standard popular lore by saying that the incident electron emits a virtual scalar boson, which is captured by the incident muon, and the four-momenta of the initial-state particles are thereby changed; in such a thought ("gedanken") process the intermediate scalar boson propagates between the interaction vertices in the diagram in Fig. 26.1.

In any case, one should take the above-mentioned popular common parlance with a grain of salt and always keep in mind the simple fact that the internal line in a diagram like Fig. 26.1 is just a graphical representation of a characteristic mathematical function — the propagator $D_F(q)$.

Let us now turn to a practical aspect of the propagator contribution in Feynman diagram calculations. The question is what is in fact the role played by the still somewhat mysterious term $i\epsilon$ in the propagator denominator. Well, it depends. For instance, in the contribution of the diagram in Fig. 26.1 the term $i\epsilon$ can be safely neglected, since in the elastic scattering process in question one can never encounter the pole at $q^2 = m^2$; the point is that here one has always $q^2 \leq 0$ (proving this is left to the reader as an exercise in kinematics). On the other hand, the infinitesimal $i\epsilon$ term plays quite important role in the evaluation of closed-loop diagrams, as we will see in later chapters. Some more remarks concerning the problem of the potential pole singularity of the Feynman propagator are deferred to the Chapter 29.

In closing this chapter, a remark is in order here concerning the form of the propagator in coordinate space. As we have already mentioned before, the explicit expression for $\mathscr{D}_F(x)$ is quite complicated for $m \neq 0$; indeed, one needs all sorts of Bessel functions for its full description (see e.g. [16]). Nevertheless, for $m = 0$ the result is quite simple; in such a case one gets

$$\mathscr{D}_F(x) = -\frac{1}{4\pi}\delta(x^2) + \frac{i}{4\pi^2}\frac{1}{x^2} \,,$$

which can be recast as

$$\mathscr{D}_F(x) = \frac{i}{4\pi^2}\frac{1}{x^2 - i\epsilon} \,. \tag{26.36}$$

In fact, such a simple form is not so surprising, on dimensional grounds; by its definition, the scalar field propagator has obviously the dimension of inverse length squared, so in the absence of a mass scale, the only functional forms with such a dimension are just $\delta(x^2)$ and $1/x^2$. The above-mentioned formulae will not be needed in our subsequent treatment of perturbative QFT, but provide at least a nice illustration of the singular behaviour of massless field theories.

169

# Chapter 27

# Propagator of Dirac field

---

Taking into account the results of the preceding chapter, one may naturally expect that the concept of the propagator is relevant for any quantized field. The case we are going to consider now is the Dirac field. To have a clear motivation for an appropriate definition of the corresponding propagator, we will employ another variant of a Yukawa-type interaction, namely

$$\mathscr{L}_{\text{int}} = g\bar{\psi}_1\psi_2\varphi + g\bar{\psi}_2\psi_1\varphi^\dagger \,, \tag{27.1}$$

where $\psi_1$, $\psi_2$ are two different Dirac fields (corresponding to fermions $f_1$, $f_2$) and $\varphi$ is a complex Klein–Gordon field (the corresponding particles will be denoted as $\varphi^-$, $\varphi^+$). Note that the interaction Lagrangian (27.1) represents just a "toy model", since there is actually no realistic physical system that would be described in such a way; nevertheless, it is good enough to set the stage for our purpose.

Let us consider the process of pair production of charged scalars in the fermion–antifermion annihilation, i.e.

$$f_1 + \bar{f}_1 \rightarrow \varphi^- + \varphi^+ \,, \tag{27.2}$$

with four-momenta $k$, $l$ for $f_1$, $\bar{f}_1$ and $p$, $r$ for $\varphi^-$, $\varphi^+$. In the usual notation, the initial and final states are defined as

$$|i\rangle = b_1^\dagger(k)d_1^\dagger(l)|0\rangle \,, \quad |f\rangle = b_\varphi^\dagger(p)d_\varphi^\dagger(r)|0\rangle \,. \tag{27.3}$$

Clearly, there is a non-trivial contribution to such a process in the 2nd order of the Dyson expansion of the $S$-matrix. Following analogous arguments as in the preceding chapter, it is not difficult to realize that the relevant $S$-matrix element can be written immediately as

$$S_{fi}^{(2)} = i^2 g^2 \iint \mathrm{d}^4 x \, \mathrm{d}^4 y \, \langle f|\mathrm{T}\big(\bar{\psi}_1(x)\psi_2(x)\varphi(x)\bar{\psi}_2(y)\psi_1(y)\varphi^\dagger(y)\big)|i\rangle \,. \tag{27.4}$$

Let us examine the matrix element in the integrand. It reads

$$\langle f|\mathrm{T}\big(\bar{\psi}_1(x)\psi_2(x)\varphi(x)\bar{\psi}_2(y)\psi_1(y)\varphi^\dagger(y)\big)|i\rangle$$
$$= \vartheta(x_0 - y_0)\langle 0|b_\varphi(p)d_\varphi(r)\bar{\psi}_{1a}(x)\psi_{2a}(x)\varphi(x)\bar{\psi}_{2b}(y)\psi_{1b}(y)\varphi^\dagger(y)b_1^\dagger(k)d_1^\dagger(l)|0\rangle \tag{27.5}$$
$$+ \vartheta(y_0 - x_0)\langle 0|b_\varphi(p)d_\varphi(r)\bar{\psi}_{2b}(y)\psi_{1b}(y)\varphi^\dagger(y)\bar{\psi}_{1a}(x)\psi_{2a}(x)\varphi(x)b_1^\dagger(k)d_1^\dagger(l)|0\rangle \,,$$

where we have introduced explicitly the bispinor indices of the Dirac field operators with regard to the later appearance of $4 \times 4$ matrices. Now we can utilize our universal finding concerning the vacuum expectation values like those shown in (27.5): the field operators have to be paired completely, in the usual sense. Here it means that $\psi_{1b}(y)$ is paired with $b_1^\dagger(k)$, $\bar{\psi}_{1a}(x)$ is paired with $d_1^\dagger(l)$, and $b_\varphi(p)$, $d_\varphi(r)$ get paired with $\varphi^\dagger(y)$ and $\varphi(x)$, respectively. Looking carefully at the fermion pairings in the two terms in (27.5), one should notice that when moving, say, $\psi_{1b}(y)$ to its counterpart $b_1^\dagger(k)$ in the second term, the operator $\bar\psi_{1a}(x)$ gets in the way, so that there is one more anticommutation as compared with the first term (where $\psi_{1b}(y)$ and $b_1^\dagger(k)$ are neighbours). Thus, one may conclude that there is a relative minus sign in the contributions of the two terms in the expression (27.5). (It is instructive to get back for a moment to the preceding chapter, in order to see that there is no such thing in the evaluation of the expression (26.7).)

The operator pairings have the familiar form and so it is easy to see that the matrix element (27.5) can be recast as

$$
\begin{aligned}
&\langle f|T\big(\bar\psi_1(x)\psi_2(x)\varphi(x)\bar\psi_2(y)\psi_1(y)\varphi^\dagger(y)\big)|i\rangle \\
&= N_k N_l N_p N_r e^{-ilx} e^{-iky} e^{irx} e^{ipy} \\
&\quad \times \bar v_a(l)\big[\vartheta(x_0-y_0)\langle 0|\psi_{2a}(x)\bar\psi_{2b}(y)|0\rangle - \vartheta(y_0-x_0)\langle 0|\bar\psi_{2b}(y)\psi_{2a}(x)|0\rangle\big]u_b(k).
\end{aligned}
\tag{27.6}
$$

Now, it is natural to denote the expression in square brackets as the T-**product of Dirac field operators** and subsequently we can also define the **propagator of Dirac field** by means of the relation

$$
i\mathscr{S}_{Fab}(x-y) = \langle 0|T\big(\psi_a(x)\bar\psi_b(y)\big)|0\rangle,
\tag{27.7}
$$

with

$$
T\big(\psi_a(x)\bar\psi_b(y)\big) = \vartheta(x_0-y_0)\psi_a(x)\bar\psi_b(y) - \vartheta(y_0-x_0)\bar\psi_b(y)\psi_a(x).
\tag{27.8}
$$

We have omitted the label 2, since the relations (27.7), (27.8) will serve as general definitions from now on. Note also that the symbol $\mathscr{S}_F$ for the Feynman propagator of Dirac field refers to the spinor nature of the field in question. The dependence of the matrix function $\mathscr{S}_F$ on the difference $x-y$ can be proven analogously as in the case of the scalar field.

In this way, we have arrived at a generalization of the definition of chronological product: for fermion fields we have to take it with change of sign when passing from $x_0 > y_0$ to $x_0 < y_0$. This, of course, is not in contradiction with the original definition of the T-product à la Dyson: the Lagrangian density is a composite scalar and we already know that even for an elementary scalar field the original definition of the T-product is pertinent. So, the next step is an explicit evaluation of the propagator $\mathscr{S}_F(x-y)$. Employing the familiar representation of the Dirac field in terms of creation and annihilation operators, one has

$$
\begin{aligned}
&\langle 0|T\big(\psi_a(x)\bar\psi_b(y)\big)|0\rangle \\
&= \vartheta(x_0-y_0)\sum_{s,s'}\iint d^3k\,d^3k'\,N_k N_{k'} u_a(k,s)\bar u_b(k',s')\langle 0|b(k,s)b^\dagger(k',s')|0\rangle e^{-ikx+ik'y} \\
&\quad - \vartheta(y_0-x_0)\sum_{s,s'}\iint d^3k\,d^3k'\,N_k N_{k'} \bar v_b(k',s')v_a(k,s)\langle 0|d(k',s')d^\dagger(k,s)|0\rangle e^{ikx-ik'y}.
\end{aligned}
\tag{27.9}
$$

Using the standard anticommutation relations, the vacuum expectation values in Eq. (27.9) are equal to $\delta_{ss'}\delta^{(3)}(\vec k - \vec k')$ and one thus gets

$$
\begin{aligned}
\langle 0|T\big(\psi_a(x)\bar\psi_b(y)\big)|0\rangle &= \vartheta(x_0-y_0)\sum_s \int d^3k\,N_k^2 u_a(k,s)\bar u_b(k,s)e^{-ikx+iky} \\
&\quad - \vartheta(y_0-x_0)\sum_s \int d^3k\,N_k^2 v_a(k,s)\bar v_b(k,s)e^{ikx-iky} \\
&= \vartheta(x_0-y_0)\int d^3k\,\frac{1}{(2\pi)^3 2k_0}(\slashed{k}+m)_{ab}\,e^{-ik(x-y)} \\
&\quad - \vartheta(y_0-x_0)\int d^3k\,\frac{1}{(2\pi)^3 2k_0}(\slashed{k}-m)_{ab}\,e^{ik(x-y)},
\end{aligned}
\tag{27.10}
$$

where we have also utilized the "completeness relations" (6.24), (6.30) for $u(k, s)$ and $v(k, s)$.

Next, one employs the Fourier representation of the $\vartheta$-functions and Eq. (27.10) then becomes

$$\langle 0|T\big(\psi_a(x)\bar{\psi}_b(y)\big)|0\rangle = \frac{1}{i} \int \frac{d^3k\, d\omega}{(2\pi)^4\, 2k_0}\, (\slashed{k} + m)_{ab}\, \frac{e^{i(\omega - k_0)(x_0 - y_0)}}{\omega - i\epsilon}\, e^{i\vec{k}\cdot(\vec{x}-\vec{y})}$$
$$- \frac{1}{i} \int \frac{d^3k\, d\omega}{(2\pi)^4\, 2k_0}\, (\slashed{k} - m)_{ab}\, \frac{e^{i(\omega - k_0)(y_0 - x_0)}}{\omega - i\epsilon}\, e^{-i\vec{k}\cdot(\vec{x}-\vec{y})} \, . \tag{27.11}$$

Performing now the same substitutions as in the case of scalar field (cf. (26.18) and (26.19)), the expression (27.11) is recast as

$$\langle 0|T\big(\psi_a(x)\bar{\psi}_b(y)\big)|0\rangle = \frac{1}{i} \int \frac{d^3k\, d\omega}{(2\pi)^4\, 2k_0}\, (\widetilde{\slashed{k}} + m)_{ab}\, \frac{e^{i\omega(x_0 - y_0) - i\vec{k}\cdot(\vec{x}-\vec{y})}}{\omega + k_0 - i\epsilon}$$
$$- \frac{1}{i} \int \frac{d^3k\, d\omega}{(2\pi)^4\, 2k_0}\, (\slashed{k} - m)_{ab}\, \frac{e^{i\omega(x_0 - y_0) - i\vec{k}\cdot(\vec{x}-\vec{y})}}{-\omega + k_0 - i\epsilon} \, , \tag{27.12}$$

where $\widetilde{k} = (k_0, -\vec{k})$, i.e. $\widetilde{\slashed{k}} = k_0\gamma_0 + \vec{k}\cdot\vec{\gamma}$, while $\slashed{k} = k_0\gamma_0 - \vec{k}\cdot\vec{\gamma}$. The integrals in (27.12) can then be reorganized as follows (omitting from now on the matrix indices $a$, $b$):

$$\langle 0|T\big(\psi(x)\bar{\psi}(y)\big)|0\rangle$$
$$= \frac{1}{i} \int \frac{d^3k\, d\omega}{(2\pi)^4\, 2k_0}\, k_0\gamma_0 \left( \frac{1}{\omega + k_0 - i\epsilon} - \frac{1}{-\omega + k_0 - i\epsilon} \right) e^{i\omega(x_0 - y_0) - i\vec{k}\cdot(\vec{x}-\vec{y})}$$
$$+ \frac{1}{i} \int \frac{d^3k\, d\omega}{(2\pi)^4\, 2k_0}\, (\vec{k}\cdot\vec{\gamma} + m) \left( \frac{1}{\omega + k_0 - i\epsilon} + \frac{1}{-\omega + k_0 - i\epsilon} \right) e^{i\omega(x_0 - y_0) - i\vec{k}\cdot(\vec{x}-\vec{y})} \, . \tag{27.13}$$

Then, after some simple manipulations analogous to those carried out in the scalar case, one gets

$$\langle 0|T\big(\psi(x)\bar{\psi}(y)\big)|0\rangle = \frac{1}{i} \int \frac{d^3k\, d\omega}{(2\pi)^4}\, (\omega\gamma_0 - \vec{k}\cdot\vec{\gamma} - m)\, \frac{1}{\omega^2 - k_0^2 + i\epsilon}\, e^{i\omega(x_0 - y_0) - i\vec{k}\cdot(\vec{x}-\vec{y})} \, . \tag{27.14}$$

Thus, taking into account that $k_0^2 = |\vec{k}|^2 + m^2$, introducing the notation $q = (\omega, \vec{k})$, and performing finally the substitution $q \to -q$, (27.14) can be rewritten as

$$\langle 0|T\big(\psi(x)\bar{\psi}(y)\big)|0\rangle = i \int \frac{d^4q}{(2\pi)^4}\, \frac{\slashed{q} + m}{q^2 - m^2 + i\epsilon}\, e^{-iq(x-y)} \, . \tag{27.15}$$

According to the definition (27.7) one then has for the propagator of Dirac field

$$\mathscr{S}_F(x - y) = \int \frac{d^4q}{(2\pi)^4}\, \frac{\slashed{q} + m}{q^2 - m^2 + i\epsilon}\, e^{-iq(x-y)} \, . \tag{27.16}$$

So, in the momentum-space representation it reads simply

$$S_F(q) = \frac{\slashed{q} + m}{q^2 - m^2 + i\epsilon} \, . \tag{27.17}$$

A QFT newcomer should thus remember that $S_F(q)$ is a $4 \times 4$ matrix. Note that the last expression can also be written as (neglecting the infinitesimal term $i\epsilon$ for a moment)

$$S_F(q) = \frac{1}{\not q - m}, \tag{27.18}$$

where, of course, $1/(\not q - m) = (\not q - m)^{-1}$. It is easy to see that the expressions (27.18) and (27.17) are equivalent. Indeed, taking into account that $\not q \not q = q^2$, one has, obviously,

$$(\not q - m) \frac{\not q + m}{q^2 - m^2} = 1,$$

and that's it. In fact, (27.18) is the most common form used for the momentum-space propagator $S_F(q)$.

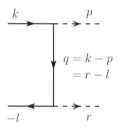

**Fig. 27.1:** The tree-level diagram for the process (27.2).

Let us now return to the $S$-matrix element for the annihilation process (27.2). Employing the expressions (27.4), (27.6), (27.7) and (27.16), one has

$$S^{(2)}_{fi} = i^2 \cdot i\, g^2 N_k N_l N_p N_r \iint d^4x \, d^4y \int \frac{d^4q}{(2\pi)^4} \, e^{-ilx-iky+irx+ipy} e^{-iq(x-y)} \, \bar u(k) S_F(q) v(l) \,. \tag{27.19}$$

The integrations over $x$ and $y$ lead to the product of delta functions

$$\delta^{(4)}(-l+r-q)\, \delta^{(4)}(-k+p+q)\,, \tag{27.20}$$

and the subsequent integration over $q$ produces the anticipated delta function $\delta^{(4)}(p+r-k-l)$ for the overall four-momentum conservation. The delta functions (27.20) show that the variable $q$ is effectively equal to $r-l$ or $k-p$ (which is the same). Thus, we get finally

$$S^{(2)}_{fi} = N_k N_l N_p N_r \, i\mathcal{M}^{(2)}_{fi} (2\pi)^4 \delta^{(4)}(p+r-k-l)\,,$$

with

$$i\mathcal{M}^{(2)}_{fi} = i^3 g^2 \bar v(l) S_F(q) u(k)\,, \tag{27.21}$$

and such an expression can be represented graphically as shown in Fig. 27.1.

In this case we see that the direction of the internal fermion line, corresponding to the Dirac field propagator, is important, since $S_F(q)$ is not an even function. So, in this way we have extended our catalogue of Feynman rules by including diagrams involving another type of the internal line. The other rules implemented in the contribution of the diagram in Fig. 27.1 are reproduced here.

Another remark is perhaps in order. The simple expression (27.18) indicates that the propagator of quantized Dirac field can also be interpreted as an appropriate Green's function of the Dirac equation. It is indeed so, but we will not discuss this theme here; an interested reader is encouraged to examine this problem independently.

# Chapter 28

# Propagator of massive vector field

Proceeding in the spirit of the previous two chapters, let us now consider a model of the interaction of massive vector (Proca) field with fermions, and an appropriate scattering process that will lead us directly to the propagator of such a vector field. We will see that the straightforward calculation will bring a new ingredient to our perturbative treatment of the relevant $S$-matrix elements.

So, let our model be defined by the interaction Lagrangian

$$\mathcal{L}_{\text{int}} = g\big(\bar{\psi}_1\gamma^\mu\psi_1 + \bar{\psi}_2\gamma^\mu\psi_2\big)A_\mu, \qquad (28.1)$$

where $\psi_1,\psi_2$ are two different Dirac fields (the corresponding fermions being denoted as $f_1,f_2$) and $A_\mu$ is a (real) massive vector field. Note that such a model can be understood as quantum electrodynamics (QED) with a spin-1 "massive photon". The process we are going to examine is the elastic scattering

$$f_1 + f_2 \to f_1 + f_2, \qquad (28.2)$$

where the particle four-momenta can be denoted e.g. as $k,p$ for the initial state and $k',p'$ for the final state. Thus, the initial and final state are defined as

$$\begin{aligned}
|i\rangle &= b_1^\dagger(k)b_2^\dagger(p)|0\rangle, \\
|f\rangle &= b_1^\dagger(k')b_2^\dagger(p')|0\rangle.
\end{aligned} \qquad (28.3)$$

We will start our calculation by assuming *bona fide* that the Dyson expansion in terms of powers of the interaction Lagrangian is valid (though we will see later on that this is a subtle point to be clarified separately). The first few steps are then almost identical with what we did before in the case of Yukawa-type interaction involving a scalar field. In the present case, when working out the corresponding $S$-matrix element, we encounter the vacuum expectation value of the T-product of vector fields

$$\langle 0|\text{T}\big(A_\mu(x)A_\nu(y)\big)|0\rangle = \vartheta(x_0 - y_0)\langle 0|A_\mu(x)A_\nu(y)|0\rangle + \vartheta(y_0 - x_0)\langle 0|A_\nu(y)A_\mu(x)|0\rangle, \quad (28.4)$$

that we would like to relate to the vector field propagator. For the explicit evaluation of the expression (28.4) we employ the standard representation of $A_\mu(x)$ in terms of the creation and annihilation operators, i.e. (see (20.10))

$$A_\mu(x) = \int \mathrm{d}^3k\, N_k \sum_{\lambda=1}^{3}\big[a(k,\lambda)\epsilon_\mu(k,\lambda)e^{-ikx} + a^\dagger(k,\lambda)\epsilon_\mu^*(k,\lambda)e^{ikx}\big]. \qquad (28.5)$$

Inserting (28.5) into (28.4) and using the canonical commutation relation (20.19)

$$\big[a(k,\lambda), a^\dagger(k',\lambda')\big] = \delta_{\lambda\lambda'}\delta^{(3)}(\vec{k} - \vec{k}'),$$

174

one gets, after a simple manipulation,

$$
\langle 0 | T\big(A_\mu(x) A_\nu(y)\big) | 0 \rangle = \vartheta(x_0 - y_0) \int d^3k \, N_k^2 \sum_{\lambda=1}^{3} \epsilon_\mu(k,\lambda)\epsilon_\nu^*(k,\lambda) e^{-ik(x-y)}
$$

$$
+ \vartheta(y_0 - x_0) \int d^3k \, N_k^2 \sum_{\lambda=1}^{3} \epsilon_\mu^*(k,\lambda)\epsilon_\nu(k,\lambda) e^{ik(x-y)} \tag{28.6}
$$

$$
= \int d^3k \, N_k^2 \, P_{\mu\nu}(k) \Big[ \vartheta(x_0 - y_0) e^{-ik(x-y)} + \vartheta(y_0 - x_0) e^{ik(x-y)} \Big],
$$

where $P_{\mu\nu}(k)$ is the well-known polarization sum for massive spin-1 boson (see (12.30))

$$
P_{\mu\nu}(k) = -g_{\mu\nu} + \frac{1}{m^2} k_\mu k_\nu \tag{28.7}
$$

(let us stress that $P_{\mu\nu}(k)$ should in fact be written as $P_{\mu\nu}(\vec{k})$, since only $\vec{k}$ are independent variables, and $k_0 = \sqrt{|\vec{k}|^2 + m^2}$). The following steps are routine; the reader may revisit the paradigm of the scalar field for details, if necessary. Upon the relevant manipulations one gets

$$
\langle 0 | T\big(A_\mu(x) A_\nu(y)\big) | 0 \rangle
$$

$$
= \frac{1}{i} \int \frac{d^3k \, d\omega}{(2\pi)^4 \, 2k_0} \left[ P_{\mu\nu}(-\vec{k}) \frac{1}{\omega + k_0 - i\epsilon} + P_{\mu\nu}(\vec{k}) \frac{1}{-\omega + k_0 - i\epsilon} \right] e^{i\omega(x_0 - y_0) - i\vec{k}\cdot(\vec{x} - \vec{y})}. \tag{28.8}
$$

Let us now work out the expression (28.8) step by step, for all possible combinations of the indices $\mu, \nu$.

1) First, let $(\mu, \nu) = (i, j)$, $i, j = 1, 2, 3$. Then (28.7) becomes

$$
P_{ij}(\vec{k}) = \delta_{ij} + \frac{1}{m^2} k_i k_j,
$$

so that obviously $P_{ij}(-\vec{k}) = P_{ij}(\vec{k})$. Repeating again the simple manipulations performed in the case of scalar field and using the notation $q = (\omega, \vec{k})$, one has

$$
\langle 0 | T\big(A_i(x) A_j(y)\big) | 0 \rangle = i \int \frac{d^4q}{(2\pi)^4} P_{ij}(q) \frac{1}{q^2 - m^2 + i\epsilon} e^{iq(x-y)}. \tag{28.9}
$$

2) Next, for $(\mu, \nu) = (0, j)$, $j = 1, 2, 3$, we have

$$
P_{0j}(\vec{k}) = \frac{1}{m^2} k_0 k_j,
$$

so that $P_{0j}(-\vec{k}) = -P_{0j}(\vec{k})$. Then the usual sequence of elementary manipulations yields first

$$
\langle 0 | T\big(A_0(x) A_j(y)\big) | 0 \rangle = i \int \frac{d^3k \, d\omega}{(2\pi)^4} \frac{1}{m^2} \omega k_j \frac{1}{\omega^2 - k_0^2 + i\epsilon} e^{i\omega(x_0 - y_0) - i\vec{k}\cdot(\vec{x} - \vec{y})}, \tag{28.10}
$$

and introducing our favourite four-component variable $q$, the expression (28.10) is recast as

$$
\langle 0 | T\big(A_0(x) A_j(y)\big) | 0 \rangle = i \int \frac{d^4q}{(2\pi)^4} P_{0j}(q) \frac{1}{q^2 - m^2 + i\epsilon} e^{iq(x-y)}. \tag{28.11}
$$

3) Finally, consider the combination $(\mu\nu) = (0,0)$. Then

$$P_{00}(\vec{k}) = -1 + \frac{1}{m^2} k_0^2 \,,$$

so that $P_{00}(-\vec{k}) = P_{00}(\vec{k})$. Proceeding as usual, from (28.8) one gets

$$\langle 0|T\big(A_0(x)A_0(y)\big)|0\rangle = i \int \frac{d^3k\, d\omega}{(2\pi)^4} \left(-1 + \frac{k_0^2}{m^2}\right) \frac{1}{\omega^2 - k_0^2 + i\epsilon} \, e^{i\omega(x_0-y_0) - i\vec{k}\cdot(\vec{x}-\vec{y})}. \quad (28.12)$$

Now we can see that there is a problem: in order to recover the $00$ component of the $P_{\mu\nu}(q)$ with $q = (\omega, \vec{k})$ in the integrand in (28.12), one would like to have there the factor $-1 + \omega^2/m^2$ rather than $-1 + k_0^2/m^2$! Well, the situation is serious, but not desperate. So as to get the desired term at play, we can rewrite Eq. (28.12) as

$$\langle 0|T\big(A_0(x)A_0(y)\big)|0\rangle = i \int \frac{d^3k\, d\omega}{(2\pi)^4} \left(-1 + \frac{\omega^2}{m^2} + \frac{k_0^2 - \omega^2}{m^2}\right) \frac{1}{\omega^2 - k_0^2 + i\epsilon} \, e^{i\omega(x_0-y_0) - i\vec{k}\cdot(\vec{x}-\vec{y})}.$$
$$(28.13)$$

The term in parentheses that is proportional to $k_0^2 - \omega^2$ is seen to cancel the denominator of the integrand. Introducing now the variable $q$, the integral (28.13) can be written down as

$$\langle 0|T\big(A_0(x)A_0(y)\big)|0\rangle = i \int \frac{d^4q}{(2\pi)^4} P_{00}(q) \frac{1}{q^2 - m^2 + i\epsilon} \, e^{iq(x-y)} - \frac{i}{m^2} \int \frac{d^4q}{(2\pi)^4} e^{iq(x-y)}$$

$$= i \int \frac{d^4q}{(2\pi)^4} \frac{P_{00}(q)}{q^2 - m^2 + i\epsilon} \, e^{iq(x-y)} - \frac{i}{m^2} \delta^{(4)}(x-y) \,. \quad (28.14)$$

Putting all things together, we may write our results in a unified form

$$\langle 0|T\big(A_\mu(x)A_\nu(y)\big)|0\rangle = i \int \frac{d^4q}{(2\pi)^4} \left(\frac{P_{\mu\nu}(q)}{q^2 - m^2 + i\epsilon} - \frac{1}{m^2} \delta_{\mu 0}\delta_{\nu 0}\right) e^{iq(x-y)}$$

$$= i\mathscr{D}_{\mu\nu}^{\text{covar.}}(x-y) - \frac{i}{m^2} \delta_{\mu 0}\delta_{\nu 0}\delta^{(4)}(x-y) \,. \quad (28.15)$$

Thus, while the first term in (28.15) is a good candidate for covariant Feynman propagator of the massive vector field, the second term is manifestly non-covariant (of course, we could write equivalently $g_{\mu 0}g_{\nu 0}$ instead of $\delta_{\mu 0}\delta_{\nu 0}$, but it does not help). On the other hand, the non-covariant contribution is extremely simple — it is a "contact term", proportional to the delta function. In fact, its emergence is closely related to our straightforward definition of the chronological product (28.4): due to the presence of time-dependent $\vartheta$-functions it is not manifestly covariant and the contribution of the whole expression at $x = y$ is not uniquely defined. In the next chapter we will uncover a way out of this difficulty; the good news is that when the dust settles, one is left with just the covariant part of the expression (28.15) and a simple structure of the perturbation expansion is thereby saved.

# Chapter 29

# Fate of non-covariant term in vector boson propagator

In order to solve the difficulty with the non-covariant term in the propagator of massive vector field (28.15), one has to return to the original form of the Dyson expansion of the $S$-matrix. It is a perturbation series in powers of the interaction Hamiltonian (21.31), in the interaction picture of time evolution. In field-theory models it is recast in terms of powers of the interaction Hamiltonian density $\mathscr{H}_{\text{int}}$. We know that in some simple cases, such as e.g. the direct four-fermion interaction, or a Yukawa-type coupling of scalar field, $\mathscr{H}_{\text{int}}$ is equal to the interaction Lagrangian density with minus sign. We will see that in the theory of massive vector field interacting with fermions (like e.g. (28.1)) such a simple relation between $\mathscr{H}_{\text{int}}$ and $\mathscr{L}_{\text{int}}$ may fail. The reason is that the component $A_0$ of the vector field is not an independent dynamical variable — rather it is to be understood as a solution of a constraint equation.

To examine the relation between $\mathscr{H}_{\text{int}}$ and $\mathscr{L}_{\text{int}}$ in detail, let us consider the full Lagrangian of the model in question, i.e.

$$\mathscr{L} = -\frac{1}{4}F_{\mu\nu}F^{\mu\nu} + \frac{1}{2}m^2 A_\mu A^\mu + J_\mu A^\mu + \dots , \tag{29.1}$$

with $F_{\mu\nu} = \partial_\mu A_\nu - \partial_\nu A_\mu$; we have denoted here

$$J_\mu = g\bar{\psi}_1 \gamma_\mu \psi_1 + g\bar{\psi}_2 \gamma_\mu \psi_2 . \tag{29.2}$$

The ellipsis in (29.1) means the Lagrangian of the free Dirac fields; these terms do not play any substantial role in our discussion. Now, the essential "strategic" point is that the independent dynamical variables for the description of the quantized vector field are the components $A_j$, $j = 1, 2, 3$; the corresponding canonical conjugate momenta are then (see (20.4))

$$\pi_j = \frac{\partial \mathscr{L}}{\partial(\partial_0 A_j)} = -F^{0j} = F_{0j} . \tag{29.3}$$

In the Hamiltonian density we will single out the part quadratic in $A_j$ and $\pi_j$ (including their derivatives) and the rest will be identified with $\mathscr{H}_{\text{int}}$.

So much for our strategy; now let us proceed to evaluate the relevant Hamiltonian density $\mathscr{H}$. This is defined, in general, as the component $\mathscr{T}^{00}$ of the energy–momentum tensor; so, we have

$$\mathscr{H} = \partial_0 A_j \frac{\partial \mathscr{L}}{\partial(\partial_0 A_j)} - \mathscr{L} + \dots \tag{29.4}$$

(here and in what follows, three dots always mean irrelevant terms originating in Dirac fields). Using (29.3), one gets

$$\mathcal{H} = \partial_0 A_j F_{0j} - \mathcal{L} + \dots$$
$$= (\partial_0 A_j - \partial_j A_0 + \partial_j A_0) F_{0j}$$
$$- \left( -\frac{1}{4} F_{0j} F^{0j} - \frac{1}{4} F_{j0} F^{j0} - \frac{1}{4} F_{jk} F^{jk} + \frac{1}{2} m^2 (A_0)^2 - \frac{1}{2} m^2 A_j A_j + J_0 A_0 - J_k A_k \right) + \dots$$

After some simple manipulations, and employing again (29.3), it is simplified to

$$\mathcal{H} = \frac{1}{2} \pi_j \pi_j + \partial_j A_0 F_{0j} + \frac{1}{4} F_{jk} F^{jk} - \frac{1}{2} m^2 (A_0)^2 + \frac{1}{2} m^2 A_j A_j - J_0 A_0 + J_k A_k + \dots \quad (29.5)$$

Now comes a crucial step: we are going to express $A_0$ in terms of canonical variables by using the equation of motion corresponding to the Lagrangian $\mathcal{L}$. It is easy to see that the relevant Euler–Lagrange equation reads

$$\partial_\mu F^{\mu\nu} + m^2 A^\nu + J^\nu = 0,$$

i.e. for $\nu = 0$ one gets

$$A_0 = -\frac{1}{m^2} (\partial_j F_{0j} + J_0)$$
$$= -\frac{1}{m^2} (\partial_j \pi_j + J_0). \quad (29.6)$$

Employing now the expression (29.6) in (29.5), one has

$$\mathcal{H} = \frac{1}{2} \pi_j \pi_j + \frac{1}{2} m^2 A_j A_j + \frac{1}{4} F_{jk} F^{jk}$$
$$- \frac{1}{m^2} \partial_j (\partial_k \pi_k + J_0) \pi_j$$
$$- \frac{1}{2m^2} \left( \partial_j \pi_j \partial_k \pi_k + 2 \partial_j \pi_j J_0 + (J_0)^2 \right)$$
$$+ \frac{1}{m^2} \partial_j \pi_j J_0 + \frac{1}{m^2} (J_0)^2 + J_k A_k + \dots \quad (29.7)$$

As a next step, one may discard terms that have the form of total derivative, since they do not contribute upon the integration of the density $\mathcal{H}$ over the 3-dimensional space. From (29.7) one thus gets an equivalent expression

$$\mathcal{H} = \frac{1}{2} \pi_j \pi_j + \frac{1}{2} m^2 A_j A_j + \frac{1}{4} F_{jk} F^{jk} + \frac{1}{2m^2} \partial_j \pi_j \partial_k \pi_k$$
$$+ \frac{1}{m^2} \partial_j \pi_j J_0 + \frac{1}{2m^2} (J_0)^2 + J_k A_k + \dots \quad (29.8)$$

The terms quadratic in canonical variables $A_j$ and $\pi_j$ (including their derivatives) constitute the free part of the Hamiltonian density $\mathcal{H}$ and the rest can be identified with the interaction part, as we have indicated above. Thus, we have (using again (29.3))

$$\mathcal{H}_{\text{int}} = \frac{1}{m^2} J_0 \partial_j F_{0j} + \frac{1}{2m^2} (J_0)^2 + J_k A_k. \quad (29.9)$$

178

However, in the Dyson expansion we use the interaction picture, i.e. the expression (29.9) should be evaluated by inserting there free fields. The equation of motion (29.6) for free fields (i.e. for $J_0 = 0$) amounts to

$$\frac{1}{m^2} \partial_j F_{0j}^{\text{free}} = -A_0^{\text{free}} . \tag{29.10}$$

So, denoting $\mathcal{H}_{\text{int}}$ in the interaction picture as $\mathcal{H}_{\text{int}}^{(I)}$, one has

$$\begin{aligned}
\mathcal{H}_{\text{int}}^{(I)} &= -J_0 A_0 + J_k A_k + \frac{1}{2m^2} (J_0)^2 \\
&= -J_\mu A^\mu + \frac{1}{2m^2} (J_0)^2 \\
&= -\mathcal{L}_{\text{int}}^{(I)} + \frac{1}{2m^2} (J_0)^2 .
\end{aligned} \tag{29.11}$$

We have thus obtained the result announced above. Indeed, one has $\mathcal{H}_{\text{int}} \neq -\mathcal{L}_{\text{int}}$, in particular,

$$\mathcal{H}_{\text{int}}^{(I)} = -\mathcal{L}_{\text{int}}^{(I)} + \frac{1}{2m^2} (J_0)^2 . \tag{29.12}$$

It remains to be clarified if the two "anomalous" effects, namely the non-covariant term in the propagator (28.15) and the extra term in $\mathcal{H}_{\text{int}}^{(I)}$ appearing in (29.12) could somehow cancel each other. To this end, let us consider the scattering process (28.2) within the model described by (29.1), (29.2). Now we know that one has to start correctly with the $S$-matrix operator written as

$$\begin{aligned}
S &= \mathrm{Texp}\left(-i \int \mathcal{H}_{\text{int}}(x) \, \mathrm{d}^4 x\right) \\
&= \mathbb{1} - i \int \mathcal{H}_{\text{int}}(x) \, \mathrm{d}^4 x + \frac{(-i)^2}{2!} \iint \mathrm{T}\big(\mathcal{H}_{\text{int}}(x) \mathcal{H}_{\text{int}}(y)\big) \, \mathrm{d}^4 x \, \mathrm{d}^4 y + \dots ,
\end{aligned} \tag{29.13}$$

where $\mathcal{H}_{\text{int}}$ is given by Eq. (29.12). In the Dyson expansion we should collect terms of the same order in the coupling constant $g$. In the considered case we are interested in contributions of the order $\mathcal{O}(g^2)$. There are obviously two such terms descending from Eq. (29.13), namely

$$S_1^{(2)} = -\frac{i}{2m^2} \int J_0^2(x) \, \mathrm{d}^4 x \tag{29.14}$$

and

$$S_2^{(2)} = \frac{i^2}{2!} \iint \mathrm{T}\big(\mathcal{L}_{\text{int}}(x) \mathcal{L}_{\text{int}}(y)\big) \, \mathrm{d}^4 x \, \mathrm{d}^4 y . \tag{29.15}$$

As regards (29.14), only its part containing the product $(\bar{\psi}_1 \gamma_0 \psi_1)(\bar{\psi}_2 \gamma_0 \psi_2)$ can contribute to the considered process; denoting it as $X_1$ for convenience, one has

$$X_1 = -i \frac{g^2}{m^2} \int (\bar{\psi}_1 \gamma_0 \psi_1)(\bar{\psi}_2 \gamma_0 \psi_2) \, \mathrm{d}^4 x . \tag{29.16}$$

Concerning (29.15), in the integrand one also has to take into account just the products combining the fermion fields 1 and 2 (according to our previous experience, the factor $1/2!$ is thus cancelled). Now, there are two types of contributions originating in (29.15). First, the "normal" term leading

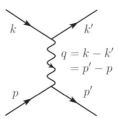

**Fig. 29.1:** 2nd order Feynman diagram involving the exchange of the massive vector boson.

ultimately to the standard 2nd order Feynman diagram involving the covariant part of the vector-field propagator (28.15). Second, there is the contribution of the non-covariant part of the propagator; denoting it as $X_2$, one gets

$$
\begin{aligned}
X_2 &= i^2 g^2 \iint \mathrm{d}^4 x \, \mathrm{d}^4 y \left( \bar{\psi}_1(x) \gamma_\mu \psi_1(x) \right) \left( \bar{\psi}_2(y) \gamma_\nu \psi_2(y) \right) \left( -\frac{i}{m^2} \right) \delta_{\mu 0} \delta_{\nu 0} \delta^{(4)}(x-y) \\
&= i \frac{g^2}{m^2} \int \left( \bar{\psi}_1(x) \gamma_0 \psi_1(x) \right) \left( \bar{\psi}_2(x) \gamma_0 \psi_2(x) \right) \mathrm{d}^4 x \,.
\end{aligned}
\tag{29.17}
$$

So, $X_1 + X_2 = 0$; this is the envisaged "miraculous" cancellation of the unwanted ingredients of our calculations. The "normal" term from $S_2^{(2)}$ can be processed in a standard routine way, and we are thus left with the 2nd order contribution to the considered scattering process that is represented by the Feynman diagram in Fig. 29.1. Its contribution reads

$$
i\mathcal{M} = i^3 g^2 \left[ \bar{u}(k') \gamma_\mu u(k) \right] D_{\mathrm{F}}^{\mu\nu}(q) \left[ \bar{u}(p') \gamma_\nu u(p) \right],
\tag{29.18}
$$

with

$$
D_{\mathrm{F}}^{\mu\nu}(q) = \frac{-g^{\mu\nu} + \dfrac{q^\mu q^\nu}{m^2}}{q^2 - m^2 + i\epsilon} \,.
\tag{29.19}
$$

Notice that the propagator (29.19) is an even function of $q$, so the direction of the wavy line in Fig. 29.1 is irrelevant. Finally, let us add that our discussion of the cancellation of the terms $X_1$ and $X_2$ is just an illustration of such a remarkable mechanism at the level of the lowest non-trivial order. More comments on this problem can be found e.g. in [10] or [13].

# Chapter 30

# Some applications:
# QED with massive photon

After the long and perhaps rather boring exposition of the preceding four chapters, we have now sufficient tools for considering some further physically interesting applications, in addition to those already discussed earlier, which have been restricted to the first order of Dyson expansion of the $S$-matrix.

In fact, with the propagators of Dirac field and massive vector field at hand, we have come very close to the quantum electrodynamics (QED). So, let us continue working with the model defined by (29.1), (29.2); since we envisage a road to QED, we will change the notation for the coupling constant, using $e$ instead of $g$. For definiteness, let us identify the fermion 1 with the electron and 2 will be the muon; the mass of the "heavy photon" is denoted simply as $m$ in what follows. Apart from the elastic scattering process $e + \mu \rightarrow e + \mu$ mentioned briefly in the preceding chapters, one can also consider the production of muon pair in the electron–positron annihilation,

$$e^-(k) + e^+(p) \rightarrow \mu^-(k') + \mu^+(p') . \tag{30.1}$$

Surely, the reader is already knowledgeable enough to draw the relevant Feynman diagram immediately; obviously, this is as shown in Fig. 30.1 (any "doubting Thomas" is encouraged to

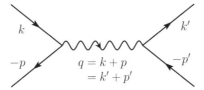

**Fig. 30.1:** Electron–positron annihilation into a muon pair at the leading order in QED with massive photon.

put a finger into the 2nd order of the Dyson expansion involving (29.1) and (29.2) so as to verify it). The matrix element corresponding to the diagram in Fig. 30.1 reads

$$i\mathcal{M} = i^3 e^2 \left[ \bar{v}(p) \gamma^\mu u(k) \right] \frac{-g_{\mu\nu} + \dfrac{q_\mu q_\nu}{m^2}}{q^2 - m^2 + i\epsilon} \left[ \bar{u}(k') \gamma^\nu v(p') \right] . \tag{30.2}$$

At first sight, the desired passage to the massless photon could be thwarted by the presence of the term proportional to $1/m^2$ in the propagator numerator. Fortunately, it is not so; in fact, the contribution of this term vanishes. Indeed, one has e.g.

$$\bar{v}(p)\gamma^\mu u(k)q_\mu = \bar{v}(p)\slashed{q}\, u(k) = \bar{v}(p)(\slashed{p} + \slashed{k})u(k)\,. \tag{30.3}$$

But the Dirac equation tells us that

$$\slashed{k}u(k) = m\, u(k)\,, \qquad \bar{v}(p)\slashed{p} = -m\,\bar{v}(p)\,, \tag{30.4}$$

and there you are. Then the limit $m \to 0$ in the denominator is safe; at the same time, $i\epsilon$ becomes irrelevant, since $q^2 = (k' + p')^2 \geq 4m_\mu^2$ for the given process (in other words, total energy in the c.m. system must be at least $2m_\mu$). Thus, for $m = 0$ the propagator in the expression (30.2) is effectively equal to $-g_{\mu\nu}/q^2$. As we will see in Chapter 32, this is precisely the simplest form of the photon propagator obtained within the covariant quantization of the electromagnetic field.

Let us now proceed to the evaluation of the cross section. For $m = 0$ we have from (30.2)

$$\mathscr{M} = e^2\big[\bar{v}(p)\gamma_\mu u(k)\big]\big[\bar{u}(k')\gamma^\mu v(p')\big]\frac{1}{q^2}\,. \tag{30.5}$$

We are going to consider first the case of unpolarized particles. Using the familiar trace technique, the spin-averaged square of the matrix element becomes, after some simple manipulations,

$$\overline{|\mathscr{M}|^2} = \frac{1}{4}\sum_{\text{pol.}}|\mathscr{M}|^2$$
$$= \frac{1}{4}\frac{e^4}{(q^2)^2}\,\mathrm{Tr}\big[(\slashed{p} - m_e)\gamma_\mu(\slashed{k} + m_e)\gamma_\nu\big]\cdot\mathrm{Tr}\big[(\slashed{k}' + m_\mu)\gamma^\mu(\slashed{p}' - m_\mu)\gamma^\nu\big]\,. \tag{30.6}$$

As we have already noted, the threshold energy for the considered process is $E_{\text{c.m.}} = 2m_\mu$. The electron mass is two orders of magnitude smaller than $m_\mu$, so $m_e$ can be safely neglected in (30.6). Employing the familiar identities for $\gamma$-matrices, in particular our "formulae 32" (C.24), the expression (30.6) is worked out easily; one gets, for $m_e = 0$,

$$\overline{|\mathscr{M}|^2} = 8\frac{e^4}{s^2}\big[m_\mu^2\, k\cdot p + (k\cdot p')(k'\cdot p) + (k\cdot k')(p\cdot p')\big]\,, \tag{30.7}$$

where we have also introduced the standard Mandelstam variable $s = q^2$.

It is instructive to consider now the high-energy limit, i.e. the collision energy such that $s \gg m_\mu^2$. Then one may set $m_\mu = 0$ wherever it occurs; this also means that the scalar products in (30.7) can be expressed simply in terms of the Mandelstam variables $t = (k - k')^2 = (p - p')^2$ and $u = (k - p')^2 = (k' - p)^2$. One thus ends up with an elegant result

$$\overline{|\mathscr{M}|^2}\Big|_{s \gg m_\mu^2} = 2e^4\frac{t^2 + u^2}{s^2}\,. \tag{30.8}$$

Now we can utilize the kinematical formulae for $t$ and $u$ that hold in the massless case, namely

$$t = -\frac{1}{2}s\,(1 - \cos\vartheta_{\text{c.m.}})\,,$$
$$u = -\frac{1}{2}s\,(1 + \cos\vartheta_{\text{c.m.}})\,, \tag{30.9}$$

where $\vartheta_{\text{c.m.}}$ is the angle between $\vec{k}$ and $\vec{k}'$ in the c.m. system. The expression (30.8) is thus recast as

$$\overline{|\mathscr{M}|^2}\Big|_{s \gg m_\mu^2} = e^4\,(1 + \cos^2\vartheta_{\text{c.m.}})\,. \tag{30.10}$$

182

Further, the differential cross section in terms of variables of c.m. system becomes

$$\frac{d\sigma}{d\Omega_{\text{c.m.}}}\bigg|_{s \gg m_\mu^2} = \frac{\alpha^2}{4s} \left(1 + \cos^2 \vartheta_{\text{c.m.}}\right), \tag{30.11}$$

where we have introduced the fine-structure constant $\alpha = e^2/(4\pi)$. Note that the angular distribution (30.11) is characteristic for an interaction mediated by a spin-1 particle (here the photon, either massless or massive). The reader is encouraged to perform an analogous calculation in a model of Yukawa-type interaction of a scalar field like that promoted in Chapter 26 (it is algebraically much simpler than within QED); it turns out that in such a case the angular distribution is trivial, i.e. isotropic.

Finally, one may integrate the expression (30.11) over the solid angle $\Omega_{\text{c.m.}}$; for the integral cross section we arrive at the well-known approximate QED formula

$$\sigma(s)\big|_{s \gg m_\mu^2} = \frac{4\pi\alpha^2}{3s}. \tag{30.12}$$

A remark is in order here. The most important part of the formula (30.12) may in fact be obtained simply by means of an "educated guess": In the 2nd order of perturbation expansion, the matrix element is proportional to $e^2$, i.e. $\alpha$; thus, the cross section must be proportional to $\alpha^2$, to the given order. Further, in high-energy limit, where all masses are neglected, the total cross section can only depend on the energy, i.e. on the Mandelstam invariant $s$. The cross section has the dimension of the length squared, i.e. inverse energy squared. So, it is clear that in high-energy limit the cross section is proportional to $\alpha^2/s$, since $\alpha$ is dimensionless. The explicit evaluation using the diagram in Fig. 30.1 thus makes this estimate more precise just by adding the factor $4\pi/3$. In this way we see that the simple guesswork described above gives us essentially a correct order-of-magnitude estimate of the cross section in question.

If the muon mass is not neglected, i.e. if one wants to get a formula for $\sigma(s)$ valid in the whole kinematical region $E_{\text{c.m.}} \geq 2m_\mu$, the relevant calculation is slightly more complicated than before and the result is (of course, still taking $m_e = 0$)

$$\sigma(s) = \frac{4\pi\alpha^2}{3s} \left(1 + \frac{2m_\mu^2}{s}\right) \sqrt{1 - \frac{4m_\mu^2}{s}}. \tag{30.13}$$

The derivation of the formula (30.13) is left as an instructive exercise for any diligent reader. Note that another useful exercise would be the evaluation of the relevant cross sections for various combinations of polarizations of the participating particles (i.e. combining left-handed and right-handed states). Such a calculation is quite easy in the high-energy limit, where one can utilize the known simple relation between the helicity and chirality (see Chapter 8).

Another physically interesting process is the annihilation of the electron–positron pair into a pair of photons. In such a case, there are two contributing diagrams, namely those in Fig. 30.2. These are quite often depicted also as shown in Fig. 30.3 in order to stress "crossing" of the external photon lines. We will make more explanatory comments on the origin of these diagrams in the forthcoming chapters, but in fact their genesis should be intuitively clear already now: in the 2nd order of Dyson expansion there appear two operators of the vector field, say $A_\mu(x)$ and $A_\nu(y)$, and there are obviously two possibilities of pairing them with the operators representing the final-state vector bosons, $a(p)$ and $a(r)$. For the time being, let us adopt a pragmatic position, employing the diagrams from Fig. 30.2 as they stand; note that the outgoing vector boson lines correspond to the polarization vectors $\epsilon^*$ (let us recall that for a vector boson

**Fig. 30.2:** Electron–positron annihilation into a pair of photons.

**Fig. 30.3:** Electron–positron annihilation into a pair of photons: an alternative way of graphical representation of the configuration of external photons.

in the initial state, the incoming line would represent the polarization vector $\epsilon$). Using the by now familiar Feynman rules, the sum of contributions of the diagrams in Fig. 30.2 reads

$$
\begin{aligned}
i\mathcal{M} = {}& i^3 e^2\, \bar{v}(l)\gamma^\mu \frac{1}{\slashed{q} - m_e}\gamma^\nu u(k)\epsilon_\mu^*(r,\lambda)\epsilon_\nu^*(p,\lambda') \\
& + i^3 e^2\, \bar{v}(l)\gamma^\nu \frac{1}{\slashed{Q} - m_e}\gamma^\mu u(k)\epsilon_\mu^*(r,\lambda)\epsilon_\nu^*(p,\lambda')\,,
\end{aligned}
\tag{30.14}
$$

where $\lambda, \lambda' = 1, 2, 3$ label the possible polarization (spin) states of the vector bosons (we still consider "heavy photons" with three possible polarizations). For later convenience, the matrix element $\mathcal{M}$ can be written as

$$
\mathcal{M} = -e^2 \mathcal{M}^{\mu\nu}\epsilon_\mu^*(r,\lambda)\epsilon_\nu^*(p,\lambda')\,,
\tag{30.15}
$$

where

$$
\begin{aligned}
\mathcal{M}^{\mu\nu} = {}& \bar{v}(l)\gamma^\mu \frac{1}{\slashed{q} - m_e}\gamma^\nu u(k) \\
& + \bar{v}(l)\gamma^\nu \frac{1}{\slashed{Q} - m_e}\gamma^\mu u(k)\,.
\end{aligned}
\tag{30.16}
$$

Let us now consider unpolarized vector bosons in the final state. For the matrix element squared, involving the summation over polarization states $\lambda, \lambda'$ we have, using (30.15),

$$
\begin{aligned}
\sum_{\lambda,\lambda'} |\mathcal{M}|^2 = {}& e^4\, \mathcal{M}^{\mu\nu}\epsilon_\mu^*(r,\lambda)\epsilon_\nu^*(p,\lambda')\, \mathcal{M}^{*\rho\sigma}\epsilon_\rho(r,\lambda)\epsilon_\sigma(p,\lambda') \\
= {}& e^4\, \mathcal{M}^{\mu\nu}\mathcal{M}^{*\rho\sigma}\left(-g_{\mu\rho} + \frac{1}{m^2} r_\mu r_\rho\right)\left(-g_{\nu\sigma} + \frac{1}{m^2} p_\nu p_\sigma\right),
\end{aligned}
\tag{30.17}
$$

184

where we have utilized the standard formula for the vector boson polarization sum. Now we face again the problem of the feasibility of the limit $m \to 0$. At first sight, there is the same difficulty as before concerning the heavy-photon propagator. In the present case, the solution is similar to that discovered above: we will be able to show that the terms proportional to $1/m^2$ in the polarization sums in (30.17) do not contribute at all. Indeed, it turns out that the following identities hold,

$$r_\mu \mathcal{M}^{\mu\nu} = 0 \,, \quad p_\nu \mathcal{M}^{\mu\nu} = 0 \,, \tag{30.18}$$

and this makes the above statement obvious. It is sufficient to prove the first identity (30.18), for the other one it goes in the same way. So, using (30.16), one gets

$$
\begin{aligned}
r_\mu \mathcal{M}^{\mu\nu} &= \bar{v}(l)\slashed{r}\frac{1}{\slashed{q}-m_e}\gamma^\nu u(k) + \bar{v}(l)\gamma^\nu\frac{1}{\slashed{Q}-m_e}\slashed{r}u(k) \\
&= \bar{v}(l)(\slashed{l}+\slashed{q})\frac{1}{\slashed{q}-m_e}\gamma^\nu u(k) + \bar{v}(l)\gamma^\nu\frac{1}{\slashed{Q}-m_e}(\slashed{k}-\slashed{Q})u(k) \\
&= \bar{v}(l)(-m_e+\slashed{q})\frac{1}{\slashed{q}-m_e}\gamma^\nu u(k) + \bar{v}(l)\gamma^\nu\frac{1}{\slashed{Q}-m_e}(m_e-\slashed{Q})u(k) \\
&= \bar{v}(l)\gamma^\nu u(k) - \bar{v}(l)\gamma^\nu u(k) \\
&= 0 \,.
\end{aligned}
$$

Note that the relations (30.18) in fact represent an example of the so-called **Ward identities**. A deeper aspect of the considered situation is that we are working here with a neutral vector field coupled to a conserved current. We will come across more examples of such a kind later on.

Thus, we see that the limit $m \to 0$ can be performed safely. In the forthcoming chapters, where we will discuss the quantization of the electromagnetic field, we will convince ourselves that the polarization sums for massless photons (i.e. massless from the very beginning) lead to the same result as our approach based on the limit $m \to 0$. So, in principle, we are now in a position to evaluate the two-photon annihilation of the electron–positron pair in detail. However, we will not pursue this theme now and rather defer it to later chapters devoted to an extensive treatment of QED. The above exposition was meant as an elucidation of a part of the intriguing problem of massless limit in the physics of massive vector bosons. Actually, the problem is more complex. For charged vector bosons the situation is different and the simple procedure described above does not work. It is in fact one of the crucial topics in the standard model of electroweak interactions, but this would be another tale (see e.g. [22]).

Let us now leave the theme of massless limit as a possible road towards QED, and consider an opposite case, where the vector boson mass is sufficiently large, in particular $m > 2m_\mu$. Then, revisiting the expression (30.2), it is clear that one may run into a potential singularity of the propagator for $q^2 = m^2$; a legitimate question is then what can be done about it (obviously, the infinitesimal term $i\epsilon$ does not provide a satisfactory answer). The solution goes beyond the tree-level diagrams we have been working with up to now. If one includes higher-order corrections to the propagator, these modify the propagator denominator in such a way that its resulting form is $q^2 - m^2 + im\Gamma$, where $\Gamma$ is the width of the unstable heavy vector boson corresponding to its decay into the muon pair. Such an expression for the corrected propagator is called the **Breit–Wigner form** and it is relevant in many situations. In particular, in the theory of electroweak interactions, it is appropriate for the description of processes of production of massive vector bosons, the most prominent example being the production of the neutral boson $Z$ that was studied in much detail at the electron–positron collider LEP at CERN during the last decade of 20th century.

# Chapter 31

# Quantization of electromagnetic field: covariant and non-covariant

As we have seen in the preceding chapter, in some situations one may describe photons (both real and virtual) by means of the massless limit of quantized Proca field. Nevertheless, one should also take up the task of a direct quantization of the electromagnetic Maxwell field. In a sense, it is the most intriguing case to be discussed in the present text and that's why we have postponed it up until this moment (though, historically, the photon had been the first example of a particle interpreted as a field quantum).

So, let us start with the familiar Lagrangian for the free Maxwell field

$$\mathcal{L} = -\frac{1}{4} F_{\mu\nu} F^{\mu\nu} , \tag{31.1}$$

with $F_{\mu\nu} = \partial_\mu A_\nu - \partial_\nu A_\mu$. The corresponding equation of motion reads $\partial_\mu F^{\mu\nu} = 0$, i.e.

$$\Box A^\mu - \partial^\mu (\partial \cdot A) = 0 , \tag{31.2}$$

where we have employed the usual shorthand notation $\partial \cdot A = \partial^\mu A_\mu$. An attempt to implement the straightforward procedure of canonical quantization runs into a difficulty that we have already encountered in Chapter 20 in the case of the Proca field: from the relation

$$\frac{\partial \mathcal{L}}{\partial(\partial_\rho A_\sigma)} = -F^{\rho\sigma} \tag{31.3}$$

(see Eq. (20.3)) one gets

$$\frac{\partial \mathcal{L}}{\partial \dot{A}_0} = 0 ,$$

and this means that $A_0$ certainly cannot play the role of an appropriate canonical variable. In any case, it is clear *a priori* that all four components of the four-potential $A_\mu$, $\mu = 0, 1, 2, 3$, cannot be taken as independent dynamical variables; one should recall that, classically, electromagnetic plane waves have two physical polarizations. Thus, there are just two independent degrees of freedom to be quantized. Of course, this is intimately related to the gauge symmetry, i.e. the invariance of the field strength $F_{\mu\nu}$ under the gauge transformation $A_\mu \to A_\mu + \partial_\mu f$, with $f$ being an essentially arbitrary function. In view of the gauge freedom, one may impose the Lorenz condition[17]

$$\partial_\mu A^\mu(x) = 0 . \tag{31.4}$$

---

[17]Note that in many textbooks and papers the relation (31.4) is called the "Lorentz condition". However, it is originally due to the Danish physicist Ludvig Lorenz (1829–1891), who had introduced it in 1867, when the more famous Hendrik Lorentz (1853–1928) was only 14 years old.

Let us recall that such a relation holds also for the massive vector (Proca) field, but there it emerges as a consequence of the equations of motion. For the Maxwell field, Eq. (31.4) is just a subsidiary condition imposed by hand. An obvious advantage of utilizing this standard constraint is that Eq. (31.2) is thereby reduced to the d'Alembert equation

$$\Box A^{\mu}(x) = 0.$$ (31.5)

In this manner, one unphysical degree of freedom is eliminated; as a next step in this reduction procedure, one may set e.g. $A_0 = 0$ (recall that in the classical theory, this is a usual way of description of the electromagnetic radiation far from its source). Thus, taking into account Eq. (31.4), our gauge conditions read

$$A_0 = 0, \quad \vec{\nabla} \cdot \vec{A} = 0.$$ (31.6)

This defines what is usually called, for obvious reasons, the **radiation gauge**. Note that such a way of gauge fixing is not Lorentz covariant, since the component $A_0$ is singled out and treated differently from $A_j$, $j = 1, 2, 3$. These latter components can be employed as the canonical variables for quantization, but one must keep in mind the constraint $\vec{\nabla} \cdot \vec{A} = 0$. Defining the conjugate momenta $\pi_j$ as

$$\pi_j = \frac{\partial \mathscr{L}}{\partial(\partial_0 A_j)},$$

one has, using (31.3) and (31.6),

$$\pi_j = F_{0j} = \partial_0 A_j.$$ (31.7)

It is immediately clear that the quantization based on the canonical commutation relations

$$\left[A_j(\vec{x}, t), \pi_k(\vec{y}, t)\right] \overset{?}{=} i\delta_{jk}\delta^{(3)}(\vec{x} - \vec{y})$$ (31.8)

would not work, since it is not compatible with the constraint $\vec{\nabla} \cdot \vec{A} = 0$. As we already know, the ultimate goal of any quantization procedure is an identification of creation and annihilation operators worth this name, i.e. endowed with standard properties regarding the field energy and momentum. So, we will proceed in the indicated direction, starting with an "educated guess" for the annihilation and creation operators (i.e. the other way round than e.g. in the case of the Klein–Gordon or Proca fields) and subsequently one may find out what would be a pertinent modification of the commutation relation (31.8). Notice that such an approach is similar to the procedure we have employed previously for the quantization of the Dirac field.

To this end, let us examine solutions of the d'Alembert equation (31.5) satisfying the gauge constraint (31.6). It is not difficult to see that such a general solution can be written as

$$A_j(x) = \int \mathrm{d}^3 k \, N_k \sum_{\lambda=1}^{2} \left[a(k, \lambda)\epsilon_j(k, \lambda)e^{-ikx} + a^{\dagger}(k, \lambda)\epsilon_j^*(k, \lambda)e^{ikx}\right],$$ (31.9)

where $k^2 = 0$, i.e. $k_0 = |\vec{k}|$ and $\epsilon_j(k, \lambda)$, $\lambda = 1, 2$, are two linearly independent polarization vectors such that (in compliance with $\vec{\nabla} \cdot \vec{A} = 0$)

$$\vec{k} \cdot \vec{\epsilon}(k, \lambda) = 0.$$ (31.10)

A convenient normalization condition is

$$\vec{\epsilon}(k, \lambda) \cdot \vec{\epsilon}^*(k, \lambda') = \delta_{\lambda\lambda'}.$$ (31.11)

187

Needless to say, $N_k$ is the conventional factor $N_k = (2\pi)^{-3/2}(2k_0)^{-1/2}$.

Before considering the envisaged commutation relations for the operator coefficients $a$, $a^\dagger$ it is useful to establish an important relation for the polarization vectors. The formula we have in mind is

$$P_{ij} \equiv \sum_{\lambda=1}^{2} \epsilon_i(k,\lambda)\epsilon_j^*(k,\lambda) = \delta_{ij} - \frac{k_i k_j}{|\vec{k}|^2}, \tag{31.12}$$

where $P_{ij}$ is the natural notation for the **polarization sum**. The identity (31.12) can be proved quite easily. One may start with the completeness relation for the orthogonal basis in the 3-dimensional Euclidean space, made of $\vec{\epsilon}(k,\lambda)$, $\lambda = 1, 2$, and $\vec{\epsilon}(k,3) = \vec{k}/|\vec{k}|$. This reads

$$\sum_{\lambda=1}^{2} \epsilon_i(k,\lambda)\epsilon_j^*(k,\lambda) + \frac{k_i}{|\vec{k}|}\frac{k_j}{|\vec{k}|} = \delta_{ij},$$

and the result (31.12) thus becomes immediately obvious.

It is also convenient to recast (31.12) in a "pseudo-covariant" form, namely

$$\sum_{\lambda=1}^{2} \epsilon_\mu(k,\lambda)\epsilon_\nu^*(k,\lambda) = -g_{\mu\nu} + \frac{1}{\eta \cdot k}(k_\mu\eta_\nu + \eta_\mu k_\nu) - \frac{1}{(\eta \cdot k)^2}k_\mu k_\nu, \tag{31.13}$$

where $\eta = (1,0,0,0)$. The proof goes as follows. We can use the completeness relation for a basis in the four-dimensional Minkowski space that is formed by $\epsilon^\mu(k,\lambda) = (0,\vec{\epsilon}(k,\lambda))$, $\lambda = 1, 2$, $\epsilon^\mu(k,3) = (0,\vec{k}/|\vec{k}|)$ and $\eta = (1,0,0,0)$. The first three $\epsilon$'s are space-like and $\eta$ is time-like. Thus, one may write

$$-\sum_{\lambda=1}^{2} \epsilon^\mu(k,\lambda)\epsilon^{*\nu}(k,\lambda) - \epsilon^\mu(k,3)\epsilon^{*\nu}(k,3) + \eta^\mu\eta^\nu = g^{\mu\nu}. \tag{31.14}$$

The vector $\epsilon^\mu(k,3)$ can be rewritten artificially as

$$\epsilon^\mu(k,3) = \frac{1}{k_0}k^\mu - \eta^\mu = \frac{1}{\eta \cdot k}k^\mu - \eta^\mu. \tag{31.15}$$

From (31.14) and (31.15) one then gets

$$\sum_{\lambda=1}^{2} \epsilon^\mu(k,\lambda)\epsilon^{*\nu}(k,\lambda) = -g^{\mu\nu} - \left(\frac{1}{\eta \cdot k}k^\mu - \eta^\mu\right)\left(\frac{1}{\eta \cdot k}k^\nu - \eta^\nu\right) + \eta^\mu\eta^\nu.$$

The term $\eta^\mu\eta^\nu$ gets cancelled and the identity (31.13) is thereby proved.

As we have indicated above, we will proceed by postulating commutation relations

$$\left[a(k,\lambda), a^\dagger(k',\lambda')\right] = \delta_{\lambda\lambda'}\delta^{(3)}(\vec{k} - \vec{k}'), \quad \left[a(k,\lambda), a(k',\lambda')\right] = 0 \tag{31.16}$$

for the annihilation and creation operators "in spe". In order to confirm their anticipated contents, one would like to show that the energy (Hamiltonian) has the proper form, i.e.

$$H = \frac{1}{2}\int d^3k\, k_0 \sum_{\lambda=1}^{2} \left[a^\dagger(k,\lambda)a(k,\lambda) + a(k,\lambda)a^\dagger(k,\lambda)\right]. \tag{31.17}$$

Then, in accordance with our previous experience, one may conclude that

$$\begin{aligned} [H, a^\dagger(k, \lambda)] &= k_0 a^\dagger(k, \lambda), \\ [H, a(k, \lambda)] &= -k_0 a(k, \lambda), \end{aligned} \tag{31.18}$$

and this guarantees the particle interpretation of the quantized field. So, how about the energy operator? Employing the standard formula for the component $\mathscr{T}^{00}$ of the energy–momentum tensor and the Lagrangian (31.1), one gets, after some simple manipulations (utilizing also the gauge condition (31.6)),

$$\begin{aligned} \mathscr{H} = \mathscr{T}^{00} &= \frac{\partial \mathscr{L}}{\partial(\partial_0 A_j)} \partial_0 A_j - \mathscr{L} \\ &= (\partial_0 A_j)(\partial_0 A_j) + \frac{1}{4} F_{\mu\nu} F^{\mu\nu} \\ &= \frac{1}{2}(\partial_0 A_j)(\partial_0 A_j) + \frac{1}{4} F_{jk} F^{jk} \\ &= \frac{1}{2}(\partial_0 A_j)(\partial_0 A_j) + \frac{1}{2}(\partial_j A_k)(\partial_j A_k) - \frac{1}{2}(\partial_j A_k)(\partial_k A_j). \end{aligned} \tag{31.19}$$

Then the energy becomes

$$H = \int d^3x \, \mathscr{H} = \frac{1}{2} \int d^3x \left[ (\partial_0 A_j)(\partial_0 A_j) - A_j \Delta A_j \right], \tag{31.20}$$

where we have used the integration by parts as well as the gauge condition $\partial_j A_j = 0$. Now one may substitute the solution (31.9) into (31.20) and arrive at the result (31.17). The calculation is somewhat tedious, but straightforward; so, the hard-working reader is encouraged to perform it as an instructive exercise.

With these results at hand, we are on the right track to photons as the quanta of the Maxwell field. We will not proceed further in such an analysis; let us only remark that one may expect rightly that the spin (helicity) states of photons are determined by the polarization vectors $\vec{\epsilon}(k, \lambda)$, similarly as in the case of the quantized Proca field. However, for photons only two helicity states are possible, corresponding to the transverse polarizations displayed in Eq. (31.10). So, a "massive photon" would differ from the massless one by the additional degree of freedom — the longitudinal polarization (zero helicity). A remark is in order here. One might wonder what is the fate of the would-be "canonical" commutation relation (31.8) within the quantization scheme that we have described above. The answer is that the right-hand side of (31.8) is replaced by a more complicated object, aptly dubbed "transverse delta function"; its explicit form can be found e.g. in the book [2].

Next, let us examine the propagator of the quantized Maxwell field. One may define, as usual,

$$i \mathscr{D}_F^{\mu\nu}(x - y) = \langle 0 | T(A^\mu(x) A^\nu(y)) | 0 \rangle. \tag{31.21}$$

Since $A_0 = 0$, this in fact means that only the components with $(\mu\nu) = (ij)$, $i, j = 1, 2, 3$, are non-zero. Following then the similar steps as in the case of massive vector (Proca) field, one gets first (cf. (28.8))

$$\begin{aligned} &\langle 0 | T(A_i(x) A_j(y)) | 0 \rangle \\ &= \frac{1}{i} \int \frac{d^3k \, d\omega}{(2\pi)^4} \frac{1}{2k_0} \left[ P_{ij}(-\vec{k}) \frac{1}{\omega + k_0 - i\epsilon} + P_{ij}(\vec{k}) \frac{1}{-\omega + k_0 - i\epsilon} \right] e^{i\omega(x_0 - y_0) - i\vec{k} \cdot (\vec{x} - \vec{y})}. \end{aligned}$$

Since $P_{ij}(\vec{k})$ is an even function (see (31.12)), this yields immediately, after some simple manipulations,

$$\langle 0|T(A_i(x)A_j(y))|0\rangle = i \int \frac{d^4q}{(2\pi)^4} P_{ij}(\vec{q}) \frac{1}{q^2 + i\epsilon} e^{iq(x-y)}, \qquad (31.22)$$

where we have denoted, as usual, $q = (\omega, \vec{k})$. Thus, we have

$$\mathscr{D}_F^{\mu\nu}(x - y) = \int \frac{d^4q}{(2\pi)^4} D_F^{\mu\nu}(q) e^{iq(x-y)},$$

where

$$D_F^{\mu\nu}(q) = \frac{P^{\mu\nu}(q)}{q^2 + i\epsilon}, \qquad (31.23)$$

with

$$P^{\mu\nu}(q) = \delta^{ij} - \frac{q^i q^j}{|\vec{q}|^2} \quad \text{for } (\mu, \nu) = (i, j),$$

$$P^{\mu\nu}(q) = 0 \quad \text{otherwise}. \qquad (31.24)$$

Let us also note that the above result is equivalent to the following "pseudo-covariant" form of $P_{\mu\nu}(q)$:

$$P_{\mu\nu}(q) = -g_{\mu\nu} + \frac{(\eta \cdot q)(\eta_\mu q_\nu + q_\mu \eta_\nu) - q_\mu q_\nu - q^2 \eta_\mu \eta_\nu}{(\eta \cdot q)^2 - q^2}, \qquad (31.25)$$

with $\eta = (1, 0, 0, 0)$ as before. The reader is encouraged to verify the equivalence of (31.24) and (31.25). Hint: One may proceed similarly as in the derivation of (31.13), but here one must take into account that $q^2 \neq 0$ in general. In particular, one has $|\vec{q}| = \sqrt{(\eta \cdot q)^2 - q^2}$.

Thus, one may notice that the resulting form of the photon propagator is rather complicated. Owing to the non-covariant character of the gauge condition (31.6), the expression (31.25) is non-covariant as well, as manifested by the presence of $\eta$, which is just a fixed set of numbers, not a four-vector (the attribute "pseudo-covariant" is therefore pertinent in the context). In particular, the term $\eta_\mu \eta_\nu$ in $P_{\mu\nu}(q)$ indicates that in a simple application, like e.g. describing the process $e^+ e^- \to \mu^+ \mu^-$ discussed earlier, one cannot get the same result as in the approach to QED based on the limit $m_\gamma \to 0$ (see the preceding Chapter 30). More precisely, one does not obtain the same results for the scattering amplitudes, if one takes, naïvely, $\mathscr{H}_{int} = -\mathscr{L}_{int}$. But this is just the point where one should be more cautious. It turns out that here in fact $\mathscr{H}_{int} \neq -\mathscr{L}_{int}$ (similarly as in the case of the interaction of Proca field) and, in the end, the extra $\eta_\mu \eta_\nu$ term in the propagator (31.23), (31.25) gets cancelled in combination with the additional term in $\mathscr{H}_{int}$. We will not discuss the details here (the interested reader can find an explicit treatment of the problem e.g. in the books [2] and [10]); instead, we are going to pursue an alternative route and develop a quantization scheme, which is manifestly covariant (though the price to be paid is the appearance of unphysical degrees of freedom).

So, let us see how one can quantize the Maxwell field and maintain the Lorentz covariance (avoiding thus the cumbersome form of the photon propagator involving (31.25)). To this end, it is certainly desirable to preserve d'Alembert quation (31.5) supplemented, in a proper manner, with the Lorenz condition (31.4). The basic idea is to modify the gauge invariant Lagrangian (31.1) in such a way that Eq. (31.5) is obtained directly. Then, the key invention is the implementation of the Lorenz condition as a constraint imposed on the states of the quantized field. The above-mentioned modification of the Lagrangian consists in the so-called "Fermi trick", which means that Eq. (31.1) is replaced with

$$\mathcal{L} = -\frac{1}{4}F_{\mu\nu}F^{\mu\nu} - \frac{1}{2}(\partial \cdot A)^2 \tag{31.26}$$

(notice that such a form is not gauge invariant any longer; its part proportional to $(\partial \cdot A)^2$ is therefore usually called the **gauge-fixing term**). Let us check that (31.26) yields indeed the d'Alembert equation for $A_\mu$. Utilizing the by now familiar relation (31.3), one gets the Euler–Lagrange equation of motion

$$0 = \partial_\mu \frac{\partial \mathcal{L}}{\partial(\partial_\mu A_\nu)} = \partial_\mu \left[ -F^{\mu\nu} - \frac{1}{2} \cdot 2 \cdot (\partial \cdot A) \frac{\partial(\partial \cdot A)}{\partial(\partial_\mu A_\nu)} \right] = -\Box A^\nu + \partial^\nu(\partial \cdot A) - \partial^\nu(\partial \cdot A)$$

$$= -\Box A^\nu .$$

Note that we have employed here an obvious identity

$$\frac{\partial(\partial \cdot A)}{\partial(\partial_\mu A_\nu)} = g^{\mu\nu} .$$

Thus, Eq. (31.5) is thereby recovered. A caveat is in order here: the four-vector field $A_\mu$ is not the Maxwell field by itself, since the Lorenz condition is yet to be implemented.

Anyway, one may now proceed to quantize such a massless vector field by postulating canonical equal-time commutation relations; all components $A_\mu$ may be treated on an equal footing. So, let us define

$$\pi_\mu \equiv \frac{\partial \mathcal{L}}{\partial(\partial_0 A_\mu)} = -F^{0\mu} - g^{0\mu}\partial \cdot A . \tag{31.27}$$

In particular, one gets

$$\pi_0 = -\partial \cdot A . \tag{31.28}$$

The commutation relations in question read

$$\begin{aligned}
\left[A_\mu(x), \pi_\nu(y)\right]_{\text{E.T.}} &= i\delta_{\mu\nu}\delta^{(3)}(\vec{x} - \vec{y}) , \\
\left[A_\mu(x), A_\nu(y)\right]_{\text{E.T.}} &= 0 , \\
\left[\pi_\mu(x), \pi_\nu(y)\right]_{\text{E.T.}} &= 0 .
\end{aligned} \tag{31.29}$$

Then it is not difficult to show that from Eq. (31.29) one gets

$$\begin{aligned}
\left[A_\mu(\vec{x},t), \dot{A}_\nu(\vec{y},t)\right] &= -ig_{\mu\nu}\delta^{(3)}(\vec{x} - \vec{y}) , \\
\left[\dot{A}_\mu(\vec{x},t), \dot{A}_\nu(\vec{y},t)\right] &= 0 .
\end{aligned} \tag{31.30}$$

The derivation of the identities (31.30) is straightforward and may be left to the reader as an instructive exercise. Now, the general solution of d'Alembert equation can be written as

$$A_\mu(x) = \int d^3k \, N_k \sum_{\lambda=0}^{3} \left[ a(k,\lambda)\epsilon_\mu(k,\lambda)e^{-ikx} + a^\dagger(k,\lambda)\epsilon_\mu^*(k,\lambda)e^{ikx} \right] , \tag{31.31}$$

where the "polarization vectors" are as follows:

$$\begin{aligned}
\epsilon^\mu(k,\lambda) &= \left(0, \vec{\epsilon}(\vec{k},\lambda)\right) , \quad \vec{k} \cdot \vec{\epsilon}(\vec{k},\lambda) = 0 \quad \text{for} \quad \lambda = 1,2 , \\
\epsilon^\mu(k,3) &= \left(0, \frac{\vec{k}}{|\vec{k}|}\right) , \\
\epsilon^\mu(k,0) &= \left(1,0,0,0\right) .
\end{aligned} \tag{31.32}$$

The corresponding orthonormality relations read, obviously,

$$\epsilon(k, \lambda) \cdot \epsilon^*(k, \lambda') = g^{\lambda\lambda'}.$$ (31.33)

Using the conventionally normalized exponentials

$$f_k(x) = N_k e^{-ikx}$$

as in our previous examples of quantized fields (cf. e.g. the case of scalar field), one may recall the orthogonality relations

$$\int d^3x \, f_k^*(x) \, i \overleftrightarrow{\partial_0} f_{k'}(x) = \delta^{(3)}(\vec{k} - \vec{k}'),$$

$$\int d^3x \, f_k(x) \, i \overleftrightarrow{\partial_0} f_{k'}(x) = 0,$$

and express the operators $a(k, \lambda)$, $a^\dagger(k, \lambda)$ in terms of $A_\mu$ and $\dot{A}_\mu$. To make the calculation more transparent, it is convenient to define the combinations

$$a_\mu(k) = \sum_{\lambda=0}^{3} a(k, \lambda) \epsilon_\mu(k, \lambda).$$ (31.34)

One then obtains first

$$a_\mu(k) = i \int d^3x \, f_k^*(x) \overleftrightarrow{\partial_0} A_\mu(x),$$ (31.35)

and thus also

$$a_\mu^\dagger(k) = -i \int d^3x \, f_k(x) \overleftrightarrow{\partial_0} A_\mu(x).$$ (31.36)

Then, employing (31.30), we get

$$\left[a_\mu(k), a_\nu^\dagger(k')\right] = -g_{\mu\nu} \delta^{(3)}(\vec{k} - \vec{k}'),$$
$$\left[a_\mu(k), a_\nu(k')\right] = 0.$$ (31.37)

As a final step, utilizing the orthonormality properties (31.33), one may express $a(k, \lambda)$ as

$$a(k, \lambda) = g_{\lambda\lambda'} \epsilon^{*\mu}(k, \lambda') a_\mu(k),$$ (31.38)

and we thus get, after some simple manipulations,

$$\left[a(k, \lambda), a^\dagger(k', \lambda')\right] = -g_{\lambda\lambda'} \delta^{(3)}(\vec{k} - \vec{k}'),$$
$$\left[a(k, \lambda), a(k', \lambda')\right] = 0.$$ (31.39)

Again, the reader is encouraged to prove that the relations (31.39) indeed follow from (31.37) and (31.38).

The most remarkable feature of the commutation relations (31.39) is that for $\lambda = \lambda' = 0$ one has

$$\left[a(k, 0), a^\dagger(k', 0)\right] = -\delta^{(3)}(\vec{k} - \vec{k}'),$$ (31.40)

i.e. there is an *opposite sign* on the right-hand side in comparison with "normal" commutators of annihilation and creation operators (note that for $\lambda = 1, 2, 3$ the situation is as usual). For convenience, we may resort to a finite box (with volume $V$) instead of the infinite 3-dimensional space, and then one has

$$\left[a(k,0),\, a^\dagger(k',0)\right] = -\delta_{\vec{k},\vec{k}'}\,,$$

in particular,

$$\left[a(k,0),\, a^\dagger(k,0)\right] = -1\,. \tag{31.41}$$

This indicates that the state defined by means of the action of $a^\dagger(k,0)$ on the conventional vacuum should have **negative norm squared**, since Eq. (31.41) yields

$$\langle 0|a(k,0)a^\dagger(k,0)|0\rangle = -1\,. \tag{31.42}$$

The discussion of this intriguing fact is a main theme of the following chapter. It will turn out that a way out of such a conundrum consists in a proper implementation of the Lorenz condition.

# Chapter 32

# Gupta–Bleuler method

---

Before proceeding to the main topic of this chapter, let us calculate the energy of our quantized massless vector field. For this purpose it is useful to notice that the original Lagrangian (31.26) is in fact equivalent to a somewhat simpler form of the "Klein–Gordon type", namely

$$\mathscr{L} = -\frac{1}{2}\partial_\mu A_\nu \partial^\mu A^\nu. \tag{32.1}$$

Indeed, the Lagrangian (31.26) reads

$$\mathscr{L} = -\frac{1}{4}(\partial_\mu A_\nu - \partial_\nu A_\mu)(\partial^\mu A^\nu - \partial^\nu A^\mu) - \frac{1}{2}(\partial \cdot A)^2 = -\frac{1}{2}\partial_\mu A_\nu \partial^\mu A^\nu + \Delta,$$

with

$$\Delta = \frac{1}{2}\partial_\mu A_\nu \partial^\nu A^\mu - \frac{1}{2}(\partial \cdot A)^2, \tag{32.2}$$

and it is easy to verify that

$$\Delta = \frac{1}{2}\partial_\mu(A_\nu \partial^\nu A^\mu - A^\mu \partial^\nu A_\nu). \tag{32.3}$$

Thus, using the notation (32.1), we have

$$\mathscr{L} = \overline{\mathscr{L}} + \partial_\mu \Delta^\mu, \tag{32.4}$$

where

$$\Delta^\mu = \frac{1}{2}(A_\nu \partial^\nu A^\mu - A^\mu \partial^\nu A_\nu).$$

However, we know that two Lagrangian densities differing just by a four-divergence lead to the same physical results; in particular, they give the same energy (and momentum). Now, the $\overline{\mathscr{L}}$ becomes

$$\begin{aligned}\overline{\mathscr{L}} &= -\frac{1}{2}\partial_\mu A_\nu \partial^\mu A^\nu \\ &= -\frac{1}{2}\partial_\mu A_0 \partial^\mu A_0 + \frac{1}{2}\sum_{j=1}^{3}\partial_\mu A_j \partial^\mu A_j,\end{aligned} \tag{32.5}$$

and thus we have an alternating sum of terms of the Klein–Gordon (KG) type (the components $A_\mu$ corresponding to four independent KG fields). So, using our previous knowledge from the theory of quantized KG field, and employing the notation (31.34) in the decomposition (31.31), one gets first

$$H = -\frac{1}{2} \int d^3k \; k_0 \left[ a_0^\dagger(k)a_0(k) + a_0(k)a_0^\dagger(k) \right] + \frac{1}{2} \sum_{j=1}^{3} \int d^3k \; k_0 \left[ a_j^\dagger(k)a_j(k) + a_j(k)a_j^\dagger(k) \right]$$

$$= -\frac{1}{2} \int d^3k \; |\vec{k}| \left[ a_\mu^\dagger(k)a^\mu(k) + a_\mu(k)a^{\mu\dagger}(k) \right] .$$

(32.6)

Next, using (31.34) and the orthogonality relations (31.33), one gets finally

$$H = -\frac{1}{2} \int d^3k \; |\vec{k}| \left[ a^\dagger(k,0)a(k,0) + a(k,0)a^\dagger(k,0) \right]$$

$$+ \frac{1}{2} \sum_{\lambda=1}^{3} \int d^3k \; |\vec{k}| \left[ a^\dagger(k,\lambda)a(k,\lambda) + a(k,\lambda)a^\dagger(k,\lambda) \right] .$$

(32.7)

We will proceed further in the usual way, and try to interpret $a(k,\lambda)$ and $a^\dagger(k,\lambda)$, $\lambda = 0, 1, 2, 3$, as the annihilation and creation operators, respectively. Then we can employ the usual trick of normal ordering; for convenience, we may also resort to the finite volume of 3-dimensional space ("box"), replacing thereby the integrals in the expression (32.7) by infinite sums over discrete values of $\vec{k}$. So, instead of (32.7) we are going to work with the form

$$H = -\sum_{\vec{k}} |\vec{k}| \, a^\dagger(k,0)a(k,0) + \sum_{\lambda=1}^{3} \sum_{\vec{k}} |\vec{k}| \, a^\dagger(k,\lambda)a(k,\lambda) .$$

(32.8)

A remark is in order here. It is instructive to realize that despite the minus sign in the first term in (32.8), the eigenvalues of the operator $H$ are non-negative. This is due to the anomalous commutation relation (31.41)

$$\left[ a(k,0), a^\dagger(k,0) \right] = -1 .$$

(32.9)

Indeed, it is easy to see that if $\left[ a, a^\dagger \right] = -1$, then the operator $-a^\dagger a$ has eigenvalues $0, 1, 2, \ldots$ On the other hand, if one denotes $\hat{N} = -a^\dagger a$ and $|\psi\rangle = a^\dagger|0\rangle$, then $|\psi\rangle$ is an eigenvector of the operator $\hat{N}$ with the eigenvalue 1, $\hat{N}|\psi\rangle = |\psi\rangle$, but the expectation value of $\hat{N}$, $\langle\psi|\hat{N}|\psi\rangle = \langle\psi|\psi\rangle = \langle 0|aa^\dagger|0\rangle = -1$. One should keep in mind such rather counterintuitive facts.

After this somewhat lengthy introduction let us now proceed to the discussion of the Lorenz condition. As a first attempt, one might try to impose the constraint $\partial^\mu A_\mu = 0$ as an operator identity. We have

$$A_\mu(x) = \sum_{\vec{k}} N_k \sum_{\lambda=0}^{3} \left[ a(k,\lambda)\epsilon_\mu(k,\lambda)e^{-ikx} + a^\dagger(k,\lambda)\epsilon_\mu^*(k,\lambda)e^{ikx} \right] ,$$

(32.10)

and in the expression for $\partial^\mu A_\mu$ one thus gets scalar products $k^\mu \epsilon_\mu(k,\lambda)$. Using the definition (31.32), one sees that $k^\mu \epsilon_\mu(k,\lambda) = 0$ for $\lambda = 1, 2$, but

$$k^\mu \epsilon_\mu(k,3) = -\vec{k} \cdot \vec{\epsilon}(k,3) = -|\vec{k}| ,$$

$$k^\mu \epsilon_\mu(k,0) = |\vec{k}| .$$

(32.11)

The operator identity $\partial^\mu A_\mu(x) = 0$ would then obviously mean $a(k,0) - a(k,3) = 0$, i.e. $a(k,0) = a(k,3)$, but this is impossible, since $a(k,0)$ and $a(k,3)$ satisfy different commutation relations (cf. (31.39)).

195

Thus, one is forced to change the strategy. Instead of a constraint for field operators, one may try to impose the Lorenz condition on the states of the quantized massless vector field in question. One may expect, quite naturally, that Lorenz condition should be connected with a selection of physical states (in a vague analogy with the classical theory). So, as a first attempt within such strategy, let us consider the definition of physical states according to

$$\partial^\mu A_\mu(x)|\psi_{\text{phys.}}\rangle = 0 . \tag{32.12}$$

In fact, such a straightforward definition leads to a contradiction: it turns out that then not even the vacuum could satisfy the condition (32.12)). Let us explain this observation in more detail. For the vacuum state one certainly has $\partial^\mu A_\mu^-(x)|0\rangle = 0$, where $A_\mu^-(x)$ denotes the annihilation part of the field operator $A_\mu(x)$. Thus, the condition (32.12) would in fact mean that

$$\partial^\mu A_\mu^+(x)|0\rangle = 0 , \tag{32.13}$$

where $A_\mu^+$ denotes the creation part of $A_\mu$. Among other things, it would then also mean that

$$\langle 0|A_\mu^-(x)\partial^\nu A_\nu^+(y)|0\rangle = 0 . \tag{32.14}$$

On the other hand, one has

$$\langle 0|A_\mu^-(x)A_\nu^+(y)|0\rangle = -g_{\mu\nu}\int \frac{d^3k}{(2\pi)^3 2|\vec{k}|} e^{-ik(x-y)} \tag{32.15}$$

(this is a result of a simple calculation, using in particular the commutation relations (31.37)).[18] Thus, one would get

$$\langle 0|A_\mu^-(x)\partial^\nu A_\nu^+(y)|0\rangle = -ig_{\mu\nu}\partial_{(y)}^\nu D^-(x-y)$$
$$= -i\partial_\mu^{(y)}D^-(x-y) , \tag{32.16}$$

but the right-hand side of Eq. (32.16) is certainly non-zero.

Nevertheless, one may impose the Lorenz condition in a weaker form, namely

$$\partial^\mu A_\mu^-(x)|\psi_{\text{phys.}}\rangle = 0 . \tag{32.17}$$

This is precisely the formulation due to Suraj Gupta and Konrad Bleuler, who suggested this independently in 1950. Needless to say, Eq. (32.17) is automatically valid for the vacuum, as it involves only annihilation operators.

Now, what does the condition (32.17) mean explicitly, in terms of annihilation operators? To see this, let us use again the expansion (32.10) and the relations (32.11). One gets

$$\partial^\mu A_\mu^-(x) = \sum_{\vec{k}}\sum_{\lambda=0}^{3} N_k\, a(k,\lambda)\left(-ik^\mu\epsilon_\mu(k,\lambda)\right)e^{-ikx}$$
$$= -i\sum_{\vec{k}} N_k|\vec{k}|e^{-ikx}\left[a(k,0) - a(k,3)\right] . \tag{32.18}$$

Thus, it is clear that the condition (32.17) for a physical state means

$$\left[a(k,0) - a(k,3)\right]|\psi_{\text{phys.}}\rangle = 0 \tag{32.19}$$

---

[18]Note also that the integral in Eq. (32.15) is usually denoted as $i\mathscr{D}^-(x-y)$ and $\mathscr{D}^-$ is called the Pauli–Jordan commutator function.

for any value of $\vec{k}$.

A remark is in order here. The relation (32.17) also means that

$$\langle\psi_{\text{phys.}}|\partial^\mu A_\mu(x)|\psi_{\text{phys.}}\rangle = 0. \tag{32.20}$$

Why is it so? Obviously, (32.17) implies that $\langle 0|\partial^\mu A_\mu^+ = 0$, and adding these two identities together one obtains (32.20). However, it is not difficult to find out that the conditions (32.17) and (32.20) are not equivalent. An appropriate example of the state vector satisfying (32.20) but not (32.17) is $a^\dagger(k,0)|0\rangle$ (the reader is encouraged to verify this simple statement).

Thus, one may conclude that (32.20) would be an elegant candidate for quantum Lorenz condition, but it turns out that it is weaker than the correct relation (32.17).

The overall picture of the quantization scheme developed so far can be described as follows. We use four types of creation and annihilation operators $a^\dagger(k,\lambda), a(k,\lambda), \lambda = 0, 1, 2, 3$, that act in a rather broad vector space (let's call it $\mathcal{V}$) whose metric, induced by a scalar product (adapted to the $a$, $a^\dagger$ algebra), is *indefinite*, i.e. the would-be "norm squared" $\langle\psi|\psi\rangle$ may be negative for some $|\psi\rangle \in \mathcal{V}$. In particular, defining vacuum state $|0\rangle$ in terms of the annihilation operators in the usual way, then for $|\psi\rangle = a^\dagger(k,0)|0\rangle$ one has $\langle\psi|\psi\rangle < 0$. The operators $a(k,\lambda), a^\dagger(k,\lambda)$ are associated with vectors $\epsilon^\mu(k,\lambda)$ in the expansion (32.10) and this leads naturally to the conventional terminology: the state created by $a^\dagger(k,\lambda)$ is generally called a "photon", which for $\lambda = 1, 2$ is "transverse", for $\lambda = 3$ we use the label "longitudinal", and the case $\lambda = 0$ corresponds to a "scalar" (or "time-like") photon. From our previous discussion of the radiation gauge we know that only transverse photons are physical, so the states with $\lambda = 0, 3$ involved in our present treatment are to be considered unphysical, corresponding to the redundant components of the $A_\mu$ field. The Lorenz condition (32.17), or equivalently (32.19), is intended to select a physical subspace $\mathcal{V}_{\text{phys.}} \subset \mathcal{V}$. As a simple example of properties of the states $|\psi_{\text{phys.}}\rangle$ satisfying Eq. (32.19), let us consider the expectation values of the Hamiltonian (32.8). According to Eq. (32.19) one has

$$a(k,0)|\psi_{\text{phys.}}\rangle = a(k,3)|\psi_{\text{phys.}}\rangle,$$

so

$$\langle\psi_{\text{phys.}}|a^\dagger(k,0) = \langle\psi_{\text{phys.}}|a^\dagger(k,3).$$

Thus one gets

$$\langle\psi_{\text{phys.}}|a^\dagger(k,0)a(k,0)|\psi_{\text{phys.}}\rangle = \langle\psi_{\text{phys.}}|a^\dagger(k,3)a(k,3)|\psi_{\text{phys.}}\rangle. \tag{32.21}$$

From (32.8) it is then clear that the contributions of the scalar and longitudinal photons mutually cancel in the expectation value in question and one gets

$$\langle\psi_{\text{phys.}}|H|\psi_{\text{phys.}}\rangle = \langle\psi_{\text{phys.}}|H_{\text{tr}}|\psi_{\text{phys.}}\rangle, \tag{32.22}$$

where

$$H_{\text{tr}} = \sum_{\vec{k}} \sum_{\lambda=1}^{2} |\vec{k}| \, a^\dagger(k,\lambda)a(k,\lambda)$$

is the transverse part of $H$.

In any case, one would like to know what is the complete structure of the $\mathcal{V}_{\text{phys.}}$. First of all, it is obvious that any state $|\psi_{\text{tr}}\rangle$, consisting of transverse photons only, belongs to $\mathcal{V}_{\text{phys.}}$.

This, of course, is due to the fact that $a(k,0)$ and $a(k,3)$ commute with any $a^\dagger(k,\lambda)$, $\lambda = 1,2$, so $a(k,\lambda)|\psi_{\text{tr}}\rangle = 0$ for $\lambda = 0,3$. Further, let us denote

$$
\begin{aligned}
L^-(k) &= a(k,0) - a(k,3), \\
L^+(k) &= a^\dagger(k,0) - a^\dagger(k,3)
\end{aligned}
\tag{32.23}
$$

(so, $L^-(k)$ and $L^+(k)$ may be provisionally called annihilation and creation part of the "Lorenz operator" in momentum representation). From the commutation relations (31.37) one can see immediately that

$$
\left[L^-(k), L^+(k')\right] = 0
\tag{32.24}
$$

for any $k$, $k'$. From (32.24) it is then clear that a vector $|\psi_{(0)}\rangle$ obtained by a repeated action of the operators $L^+(k)$ on a vector $|\psi_{\text{tr}}\rangle$ (in general, for various values of $k$) is a **zero-norm state**, $\langle \psi_{(0)}|\psi_{(0)}\rangle = 0$. Thus, any $|\psi\rangle$ of the type $|\psi\rangle = |\psi_{\text{tr}}\rangle + |\psi_{(0)}\rangle$ belongs to $\mathscr{V}_{\text{phys.}}$ and for the scalar product of two such vectors one gets

$$
\langle \phi|\psi \rangle = \langle \phi_{\text{tr}}|\psi_{\text{tr}} \rangle .
\tag{32.25}
$$

Moreover, it should be obvious that for any $|\psi_{\text{tr}}\rangle$ one has $\langle \psi_{\text{tr}}|\psi_{\text{tr}}\rangle \geq 0$ (this, of course, is due to the "normal" commutation relations for $a(k,\lambda)$, $a^\dagger(k,\lambda)$ with $\lambda = 1,2$). A key result of the Gupta–Bleuler theory is that the vectors of the type $|\psi_{\text{tr}}\rangle + |\psi_{(0)}\rangle$ in fact constitute the whole $\mathscr{V}_{\text{phys.}}$, i.e. any state in $\mathscr{V}_{\text{phys.}}$ is essentially transverse, with the possible admixture of a zero-norm vector mentioned above. For more technical details concerning this point see e.g. the books [8] and [16].

One may conclude that the Gupta–Bleuler quantization scheme is successful in selecting correctly the physical states corresponding essentially to transverse photons, and in identifying a physical subspace $\mathscr{V}_{\text{phys.}}$ with positive definite metric. $\mathscr{V}_{\text{phys.}}$ is thus a standard Hilbert space — although one has always a whole "tower" of states of the form $|\psi_{\text{tr}}\rangle + |\psi_{(0)}\rangle$, the purely transverse $|\psi_{\text{tr}}\rangle$ may be chosen as a representative of such an equivalence class in practical perturbative calculations.

One must admit that the Gupta–Bleuler theory is rather sophisticated and quite complicated. However, for practical calculations it is sufficient to know the polarization sum (31.13) for the physical photons (of course, this is the same for any approach to the quantization of the Maxwell field) and, last but not least, the propagator.

In fact, evaluating the propagator of the field $A_\mu$ within our covariant quantization scheme is quite simple — it is almost identical to the case of the (massless) Klein–Gordon scalar field. Indeed, $A_\mu(x)$ is written as

$$
A_\mu(x) = \int d^3k \, N_k \left[ a_\mu(k)e^{-ikx} + a_\mu^\dagger(k)e^{ikx} \right],
$$

and from the commutation relations (31.37) one gets

$$
\langle 0|a_\mu(k)a_\nu^\dagger(l)|0\rangle = -g_{\mu\nu}\delta^{(3)}(\vec{k} - \vec{l}) .
\tag{32.26}
$$

Thus, one may proceed in the same way as in the case of the scalar field, the only exception being the extra factor $-g_{\mu\nu}$ in (32.26). The result can therefore be written down immediately:

$$
\langle 0|T\big(A_\mu(x)A_\nu(y)\big)|0\rangle = i \int \frac{d^4q}{(2\pi)^4} D_{\mu\nu}(q)e^{iq(x-y)} ,
\tag{32.27}
$$

with

$$D_{\mu\nu}(q) = \frac{-g_{\mu\nu}}{q^2 + i\epsilon}.$$ (32.28)

It is reassuring that this result coincides with the form following from the massless limit of the covariant part of the propagator of Proca field within scattering amplitudes of physical processes like $e^+e^- \rightarrow \mu^+\mu^-$, $ep \rightarrow ep$, etc. In the Dyson perturbation series one can also use $\mathcal{H}_{int} = -\mathcal{L}_{int}$ in a straightforward way.

# Chapter 33

# Compton scattering: Klein–Nishina formula

As a prelude to the main topic of this chapter, we are going to derive first a useful formula for the cross section of a general binary process in variables of the laboratory system (i.e. in the rest system of a target particle).

So, let us consider a process $1 + 2 \rightarrow 3 + 4$, denoting the corresponding four-momenta as $p_1$, $p_2$, $p_3$, $p_4$ and the masses as $m_1$, ..., $m_4$. We choose the laboratory system to be the rest frame of the particle 2, i.e.

$$p_2 = (m_2, \vec{0}) \,. \tag{33.1}$$

According to the general formula (23.38) for differential cross section we have

$$d\sigma = \frac{1}{|\vec{v}_1|} \frac{1}{2E_1} \frac{1}{2E_2} |\mathcal{M}|^2 \frac{d^3 p_3}{(2\pi)^3 \, 2E_3} \frac{d^3 p_4}{(2\pi)^3 \, 2E_4} (2\pi)^4 \delta^{(4)}(p_3 + p_4 - p_1 - p_2) \,. \tag{33.2}$$

Our goal is to derive the angular distribution of the particle 3. The integration over $d^3 p_4$ in (33.2) is trivial and one thus gets first (using also the familiar relation $|\vec{v}_1| = |\vec{p}_1|/E_1$)

$$d\sigma = \frac{E_1}{|\vec{p}_1|} \frac{1}{2E_1} \frac{1}{2E_2} |\mathcal{M}|^2 \frac{d^3 p_3}{(2\pi)^3 \, 2E_3} \frac{1}{(2\pi)^3 \, 2E_4}$$
$$\times (2\pi)^4 \delta\left( \sqrt{|\vec{p}_3|^2 + m_3^2} + \sqrt{|\vec{p}_4|^2 + m_4^2} - E_1 - E_2 \right), \tag{33.3}$$

where $\vec{p}_4 = \vec{p}_1 - \vec{p}_3$ (see Fig. 33.1). Thus, using the above definition of the scattering angle and

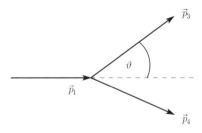

**Fig. 33.1:** Kinematics of the $1 + 2 \rightarrow 3 + 4$ process in the laboratory system.

setting also $E_2 = m_2$, one has

$$
d\sigma = \frac{1}{2|\vec{p}_1|} \frac{1}{2m_2} |\mathcal{M}|^2 \frac{d^3 p_3}{(2\pi)^3 2E_3} \frac{1}{(2\pi)^3 2E_4}
$$
$$
\times (2\pi)^4 \delta\left( \sqrt{|\vec{p}_3|^2 + m_3^2} + \sqrt{|\vec{p}_3|^2 - 2|\vec{p}_1||\vec{p}_3| \cos\vartheta + |\vec{p}_1|^2 + m_4^2} - E_1 - m_2 \right). \tag{33.4}
$$

For brevity, let us now denote $x \equiv |\vec{p}_3|$. Eq. (33.4) may be recast as

$$
d\sigma = \frac{1}{2|\vec{p}_1|} \frac{1}{2m_2} |\mathcal{M}|^2 \frac{x^2 dx \, d\Omega}{(2\pi)^3 2E_3} \frac{1}{(2\pi)^3 2E_4} (2\pi)^4 \delta\left[ f(x) \right], \tag{33.5}
$$

where

$$
f(x) = \sqrt{x^2 + m_3^2} + \sqrt{x^2 - 2x|\vec{p}_1| \cos\vartheta + |\vec{p}_1|^2 + m_4^2} - E_1 - m_2. \tag{33.6}
$$

Then one may utilize the standard formula

$$
\delta\left[ f(x) \right] = \frac{1}{|f'(x_0)|} \delta(x - x_0), \tag{33.7}
$$

where $x_0$ is the zero of $f(x)$, i.e. $f(x_0) = 0$. Differentiating the function (33.6), one gets

$$
f'(x) = \frac{x}{\sqrt{x^2 + m_3^2}} + \frac{x - |\vec{p}_1| \cos\vartheta}{\sqrt{x^2 - 2x|\vec{p}_1| \cos\vartheta + |\vec{p}_1|^2 + m_4^2}}, \tag{33.8}
$$

i.e.

$$
f'(x_0) = \frac{x_0}{E_3} + \frac{x_0 - |\vec{p}_1| \cos\vartheta}{E_4} = \frac{|\vec{p}_3|(E_1 + m_2) - E_3|\vec{p}_1| \cos\vartheta}{E_3 E_4}, \tag{33.9}
$$

where we have taken into account the energy conservation $E_3 + E_4 = E_1 + m_2$. The integration over the variable $x$ in the expression (33.5) thus yields

$$
d\sigma = \frac{1}{2|\vec{p}_1|} \frac{1}{2m_2} |\mathcal{M}|^2 \frac{|\vec{p}_3|^2 d\Omega}{(2\pi)^3 2E_3} \frac{1}{(2\pi)^3 2E_4} \frac{E_3 E_4 (2\pi)^4}{|\vec{p}_3|(E_1 + m_2) - E_3|\vec{p}_1| \cos\vartheta},
$$

and the resulting formula for the angular distribution in question becomes

$$
\frac{d\sigma}{d\Omega_{\text{lab}}} = \frac{1}{64\pi^2} \frac{1}{|\vec{p}_1| m_2} |\mathcal{M}|^2 \frac{|\vec{p}_3|}{E_1 + m_2 - \dfrac{|\vec{p}_1|}{|\vec{p}_3|} E_3 \cos\vartheta}. \tag{33.10}
$$

Let us add that $|\vec{p}_3|$ and $E_3$ are given as functions of the scattering angle $\vartheta$ and masses through the condition of energy conservation, i.e.

$$
\sqrt{x_0^2 + m_3^2} + \sqrt{x_0^2 - 2x_0|\vec{p}_1| \cos\vartheta + |\vec{p}_1|^2 + m_4^2} = E_1 + m_2, \tag{33.11}
$$

where we have kept the notation $x_0 = |\vec{p}_3|$. In a general case of arbitrary masses $m_1, \ldots, m_4$ the result for $x_0$ is rather complicated, but it is simplified substantially e.g. for the elastic scattering of a massless particle (which is precisely the envisaged case of the Compton scattering).

So, let us now consider the case $m_1 = m_3 = 0$, $m_2 = m_4 = m \neq 0$ and rename the four-momenta as $p_1 \equiv k$, $p_2 \equiv p$, $p_3 \equiv k'$, $p_4 \equiv p'$. The energy conservation (33.11) then means

$$
|\vec{k}'| + \sqrt{|\vec{k}'|^2 - 2|\vec{k}||\vec{k}'| \cos\vartheta + |\vec{k}|^2 + m^2} = |\vec{k}| + m. \tag{33.12}
$$

For brevity, let us denote provisionally $\omega = |\vec{k}|$, $\omega' = |\vec{k}'|$. The solution of Eq. (33.12) can be then written as

$$\omega' = \frac{\omega}{1 + \dfrac{\omega}{m}(1 - \cos\vartheta)}, \tag{33.13}$$

which is the famous Compton relation for the change of energy (or frequency) of the photon in the scattering process.[19] From (33.10) one then gets, after some simple manipulations,

$$\frac{d\sigma}{d\Omega_{\text{lab}}} = \frac{1}{64\pi^2}\frac{1}{m^2}|\mathscr{M}|^2\frac{|\vec{k}'|^2}{|\vec{k}|^2}, \tag{33.14}$$

where $|\vec{k}'| = \omega'$ is given by (33.13).

Let us now consider the Compton scattering, i.e. the scattering of the photon on a charged particle — for definiteness we will have in mind the electron. So, the process in question is

$$\gamma(k) + e^-(p) \rightarrow \gamma(k') + e^-(p'), \tag{33.15}$$

where we have also specified the corresponding four-momenta. In what follows, we are going to work in the laboratory frame where $p = (m, \vec{0})$. The interaction Lagrangian has the familiar form $\mathscr{L}_{\text{int}} = e\bar{\psi}\gamma^\mu\psi A_\mu$ and the lowest order Feynman diagrams describing the scattering amplitude are shown in Fig. 33.2 (the reader is recommended to find out how these diagrams emerge from the 2nd order of Dyson expansion of the $S$-matrix).

**Fig. 33.2:** 2nd order Feynman diagrams for Compton scattering.

Utilizing standard Feynman rules, the matrix element $\mathscr{M} = \mathscr{M}_a + \mathscr{M}_b$ is given by

$$i\mathscr{M} = i^3 e^2 \bar{u}(p')\gamma_\mu \frac{1}{\slashed{p} + \slashed{k} - m}\gamma_\nu u(p)\epsilon'^\mu(k')\epsilon^\nu(k)$$

$$+ i^3 e^2 \bar{u}(p')\gamma_\mu \frac{1}{\slashed{p} - \slashed{k}' - m}\gamma_\nu u(p)\epsilon^\mu(k)\epsilon'^\nu(k'). \tag{33.16}$$

Note that in general one has to use $\epsilon(k)$ for the incoming photon and $\epsilon^*(k')$ for the outgoing one. For simplicity, we consider here real values of $\epsilon$'s (having thus in mind linear polarizations). In what follows, we are going to use the shorthand notation $\varepsilon$ and $\varepsilon'$ for the relevant polarization

---

[19]The relation (33.13) can be easily recast in terms of the corresponding wavelengths; in the ordinary system of units it reads $\lambda' - \lambda = (h/mc)(1 - \cos\vartheta)$, where $h = 2\pi\hbar$. In 1922, Arthur Holly Compton (1892-1962) discovered experimentally such a shift of the X-rays wavelength due to the scattering by free electrons and this is why the term Compton wavelength refers traditionally to the quantity $h/mc$. This observation was of fundamental importance, since it provided a truly convincing evidence for photons as the quanta of the electromagnetic field. Compton received the Nobel Prize in 1927, according to the original (rather terse) citation of the Nobel committee, "for his discovery of the effect named after him".

vectors. Recall that the physical (transverse) polarizations have the form $\epsilon = (0, \vec{\epsilon})$ and it holds $k \cdot \epsilon(k) = 0$, $k' \cdot \epsilon'(k') = 0$. So, we have

$$\mathcal{M} = -e^2\left[\bar{u}(p')\not{\epsilon}'\,\frac{\not{p} + \not{k} + m}{(p+k)^2 - m^2}\,\not{\epsilon}u(p) + \bar{u}(p')\not{\epsilon}\,\frac{\not{p} - \not{k}' + m}{(p-k')^2 - m^2}\,\not{\epsilon}'u(p)\right],$$

and using $p^2 = m^2$, $k^2 = 0$, $k'^2 = 0$, it becomes

$$\mathcal{M} = -e^2\bar{u}(p')\left[\frac{\not{\epsilon}'(\not{p} + \not{k} + m)\not{\epsilon}}{2p\cdot k} + \frac{\not{\epsilon}(\not{p} - \not{k}' + m)\not{\epsilon}'}{-2p\cdot k'}\right]u(p). \tag{33.17}$$

The last expression can be further simplified. Indeed, taking into account that $p = (m, \vec{0})$, one has $p \cdot \epsilon = 0$, $p \cdot \epsilon' = 0$, and this in turn means that

$$\not{p}\not{\epsilon} = -\not{\epsilon}\not{p}, \quad \not{p}\not{\epsilon}' = -\not{\epsilon}'\not{p}. \tag{33.18}$$

Then it becomes clear that $(\not{p} + m)\not{\epsilon}u(p) = \not{\epsilon}(-\not{p} + m)u(p) = 0$ due to the Dirac equation for $u(p)$, and in the same way one gets $(\not{p} + m)\not{\epsilon}'u(p) = 0$. Moreover, because of $k \cdot \epsilon = 0$ and $k' \cdot \epsilon' = 0$ one has

$$\not{k}\not{\epsilon} = -\not{\epsilon}\not{k}, \quad \not{k}'\not{\epsilon}' = -\not{\epsilon}'\not{k}'. \tag{33.19}$$

Employing all this in (33.17), one obtains

$$\mathcal{M} = e^2\bar{u}(p')\left(\frac{\not{\epsilon}'\not{\epsilon}\not{k}}{2p\cdot k} + \frac{\not{\epsilon}\not{\epsilon}'\not{k}'}{2p\cdot k'}\right)u(p). \tag{33.20}$$

Next, we will calculate the matrix element squared $|\mathcal{M}|^2$ for unpolarized electrons; it amounts to summing over the spin states of the outgoing electron and averaging over the spins of the initial (target) electron. Using the standard trace technique, one gets first

$$\overline{|\mathcal{M}|^2} = \frac{1}{2}e^4\,\text{Tr}\left[(\not{p}' + m)\left(\frac{\not{\epsilon}'\not{\epsilon}\not{k}}{2p\cdot k} + \frac{\not{\epsilon}\not{\epsilon}'\not{k}'}{2p\cdot k'}\right)(\not{p} + m)\left(\frac{\not{k}\not{\epsilon}\not{\epsilon}'}{2p\cdot k} + \frac{\not{k}'\not{\epsilon}'\not{\epsilon}}{2p\cdot k'}\right)\right]. \tag{33.21}$$

In this first step, we keep photon polarizations fixed, the case of unpolarized photons will be discussed at the end of our calculation. So, in (33.21) one can see four types of traces, namely

$$\begin{aligned}
T_1 &= \text{Tr}\left[(\not{p}' + m)\not{\epsilon}'\not{\epsilon}\not{k}(\not{p} + m)\not{k}\not{\epsilon}\not{\epsilon}'\right], \\
T_2 &= \text{Tr}\left[(\not{p}' + m)\not{\epsilon}\not{\epsilon}'\not{k}'(\not{p} + m)\not{k}'\not{\epsilon}'\not{\epsilon}\right], \\
T_3 &= \text{Tr}\left[(\not{p}' + m)\not{\epsilon}'\not{\epsilon}\not{k}(\not{p} + m)\not{k}'\not{\epsilon}'\not{\epsilon}\right], \\
T_4 &= \text{Tr}\left[(\not{p}' + m)\not{\epsilon}\not{\epsilon}'\not{k}'(\not{p} + m)\not{k}\not{\epsilon}\not{\epsilon}'\right].
\end{aligned} \tag{33.22}$$

These expressions may look rather complicated at first sight, as there are products of up to eight $\gamma$-matrices under the trace symbol. Fortunately, the situation is in fact far better than it might seem. Let us illustrate the feasibility of the algebraic calculations on the example of the trace $T_1$. One gets first

$$T_1 = \text{Tr}\left(\not{p}'\not{\epsilon}'\not{\epsilon}\not{k}\not{p}\not{k}\not{\epsilon}\not{\epsilon}'\right) + m^2\,\text{Tr}\left(\not{\epsilon}'\not{\epsilon}\not{k}\not{k}\not{\epsilon}\not{\epsilon}'\right).$$

The second term is zero because of $\not{k}\not{k} = k^2 = 0$. In the first term, we can write

$$\not{k}\not{p}\not{k} = (2k\cdot p - \not{p}\not{k})\not{k} = 2k\cdot p\,\not{k}.$$

Thus,

$$
\begin{aligned}
T_1 &= 2k \cdot p \ \mathrm{Tr}\big(p'\epsilon'\epsilon k k \epsilon\epsilon'\big), \\
&= -2k \cdot p \ \mathrm{Tr}\big(p'\epsilon' k \epsilon\epsilon\epsilon'\big), \\
&= 2k \cdot p \ \mathrm{Tr}\big(p'\epsilon' k \epsilon'\big).
\end{aligned}
\tag{33.23}
$$

Note that in arriving at (33.23) we have used (33.19) and $\epsilon\epsilon = \epsilon^2 = -1$. The trace of the product of four $\gamma$-matrices is worked out easily:

$$
\begin{aligned}
\mathrm{Tr}\big(p'\epsilon' k \epsilon'\big) &= 4\big[2(p' \cdot \epsilon')(k \cdot \epsilon') - (k \cdot p')(\epsilon' \cdot \epsilon')\big] \\
&= 4\big[2(k + p - k') \cdot \epsilon'(k \cdot \epsilon') + k \cdot p'\big] \\
&= 4\big[2(k \cdot \epsilon')^2 + k' \cdot p\big].
\end{aligned}
$$

Here we have used the four-momentum conservation $p' = k + p - k'$, and also the relations $p \cdot \epsilon' = 0$, $k' \cdot \epsilon' = 0$, as well as $k \cdot p' = k' \cdot p$ (this follows easily from $(k - p')^2 = (k' - p)^2$). Putting all this together, one has finally

$$
T_1 = 8k \cdot p \ \big[2(k \cdot \epsilon')^2 + k' \cdot p\big].
\tag{33.24}
$$

The remaining traces in (33.22) can be evaluated in a similar way. In fact, it is a good topic for a homework or a boring exercise, so I will just summarize the relevant results:

$$
\begin{aligned}
T_2 &= 8(k' \cdot p)\big[-2(k' \cdot \epsilon)^2 + k \cdot p\big], \\
T_3 &= 8(k \cdot p)(k' \cdot p)\big[2(\epsilon \cdot \epsilon')^2 - 1\big] - 8(k \cdot \epsilon')^2(k' \cdot p) + 8(k' \cdot \epsilon)^2(k \cdot p), \\
T_4 &= T_3.
\end{aligned}
\tag{33.25}
$$

Now, the trace in (33.21) is

$$
\mathrm{Tr}\big[\dots\big] = \frac{1}{(2p \cdot k)^2} T_1 + \frac{1}{(2p \cdot k')^2} T_2 + 2\frac{1}{(2p \cdot k)(2p \cdot k')} T_3.
\tag{33.26}
$$

Using the results (33.24) and (33.25), the expression (33.26) becomes, after some elementary manipulations,

$$
\mathrm{Tr}\big[\dots\big] = 2\left[\frac{k' \cdot p}{k \cdot p} + \frac{k \cdot p}{k' \cdot p} + 4(\epsilon \cdot \epsilon')^2 - 2\right],
$$

and $\overline{|\mathscr{M}|^2}$ given by (33.21) is then

$$
\overline{|\mathscr{M}|^2} = e^4\left[\frac{k' \cdot p}{k \cdot p} + \frac{k \cdot p}{k' \cdot p} + 4(\epsilon \cdot \epsilon')^2 - 2\right].
\tag{33.27}
$$

Thus, the calculation has been somewhat tedious, but the result is quite simple and rewarding (note that the first two terms in the square bracket in (33.27) are in fact $\omega'/\omega$ and $\omega/\omega'$, respectively, if we return to the notation used in (33.13)). Using now the formula (33.14) for the cross section, we get finally

$$
\frac{d\sigma}{d\Omega} = \frac{\alpha^2}{4m^2}\left(\frac{\omega'}{\omega}\right)^2\left[\frac{\omega'}{\omega} + \frac{\omega}{\omega'} + 4(\epsilon \cdot \epsilon')^2 - 2\right],
\tag{33.28}
$$

where we have introduced the fine-structure constant $\alpha = e^2/(4\pi)$. The result (33.28) is traditionally called the **Klein–Nishina formula**, in honour of Oskar Klein and Yoshio Nishina,

204

who derived the formula for Compton scattering of unpolarized photons in 1930 (it was done independently also by Igor Tamm). The formula (33.28) for polarized photons was obtained first by Ugo Fano in 1949.

Let us now proceed to the case of unpolarized photons. It means that we have to sum the expression (33.28) over polarizations of the initial and final photon and multiply by 1/2 (this corresponds to averaging over the polarizations of the incident photon). For this purpose we may write the scalar product of polarizations on the right-hand side of (33.28) explicitly as

$$\epsilon \cdot \epsilon' = \epsilon(k,\lambda) \cdot \epsilon(k',\lambda') = -\vec{\epsilon}(k,\lambda) \cdot \vec{\epsilon}(k',\lambda') . \tag{33.29}$$

Then

$$\sum_{\lambda,\lambda'=1}^{2} \left( \vec{\epsilon}(k,\lambda) \cdot \vec{\epsilon}(k',\lambda') \right)^2 = \sum_{\lambda,\lambda'=1}^{2} \epsilon_i(k,\lambda)\epsilon_i(k',\lambda')\epsilon_j(k,\lambda)\epsilon_j(k',\lambda')$$

$$= \sum_{\lambda=1}^{2} \epsilon_i(k,\lambda)\epsilon_j(k,\lambda) \sum_{\lambda'=1}^{2} \epsilon_i(k',\lambda')\epsilon_j(k',\lambda') \tag{33.30}$$

$$= \left( \delta_{ij} - \frac{k_i k_j}{|\vec{k}|^2} \right)\left( \delta_{ij} - \frac{k'_i k'_j}{|\vec{k}'|^2} \right) .$$

Note that in the last line we have utilized the formula (31.24) for the photon polarization sum. Now, taking into account that $\delta_{ij}\delta_{ij} = 3$ and $k_i k'_i = \vec{k} \cdot \vec{k}' = |\vec{k}||\vec{k}'| \cos\vartheta$, the expression (33.30) becomes

$$3 - 1 - 1 + \cos^2\vartheta = 1 + \cos^2\vartheta . \tag{33.31}$$

For the unpolarized cross section one then gets

$$\frac{d\bar{\sigma}}{d\Omega} = \frac{\alpha^2}{2m^2} \left( \frac{\omega'}{\omega} \right)^2 \left( \frac{\omega'}{\omega} + \frac{\omega}{\omega'} - \sin^2\vartheta \right) . \tag{33.32}$$

For passing from (33.28) to (33.32) don't forget that upon summing over photon polarizations, the polarization-independent terms in (33.28) are simply multiplied by four and for the term involving $(\epsilon \cdot \epsilon')^2$ we employ the results (33.30), (33.31). Taking into account the relation (33.13), it is clear that the angular distribution (33.32) is rather complicated and its form also depends strongly on the photon energy. An instructive picture can be found in the literature, see e.g. the book [15]. One prominent feature of the angular dependence in (33.32) is that at high energy, there is a pronounced forward peak. Let us stress that such a picture is valid in the laboratory frame; in the c.m. system the angular distribution is quite different. Anyway, an interesting point is what becomes of the expression (33.32) in the low-energy limit, i.e. for $\omega \ll m$. In such a case one may expect that the classical result could be reproduced, since for $\omega \ll m$ one gets from (33.13) that $\omega' \approx \omega$. Note that in classical electrodynamics $\omega' = \omega$, i.e. the frequency of the scattered radiation is the same as the initial one: it is an inevitable consequence of the scattering mechanism in classical theory, which, of course, is quite different from the quantum case.

So, setting simply $\omega' = \omega$ in the formula (33.32), one gets

$$\frac{d\bar{\sigma}}{d\Omega}\bigg|_{\omega \ll m} = \frac{\alpha^2}{2m^2} \left( 1 + \cos^2\vartheta \right) , \tag{33.33}$$

and this is the famous classical formula for **Thomson scattering** (named in honour of J. J. Thomson, by the way the Nobel Prize winner for 1906, for the discovery of the electron). The expression

(33.33) can be easily integrated over the scattering angle and one gets

$$\bar{\sigma}\big|_{\omega \ll m} \simeq \frac{8\pi}{3} \frac{\alpha^2}{m^2} \,. \tag{33.34}$$

For reader's convenience, let us recast the result (33.34) in ordinary units, in which $\alpha = e^2/\hbar c$ and $1/m$ becomes $\hbar/mc$. Then $\alpha/m$ turns into $e^2/mc^2$, which is the classical electron radius $r_0$. Putting in numbers, in particular $e = 4.8 \times 10^{-10}$ esu, one has $r_0 \simeq 2.8 \times 10^{-13}$ cm and thus

$$\bar{\sigma}\big|_{\omega \ll m} \approx \frac{8\pi}{3} r_0^2 \approx 6.67 \times 10^{-25} \text{ cm}^2 \,. \tag{33.35}$$

# Chapter 34

# *S*-matrix and Wick's theorems: an overview

In our previous calculations of *S*-matrix elements for various decay and scattering processes we have seen that any such matrix element can be recast as the vacuum expectation value (v. e. v.) of a product of annihilation and creation operators and the non-zero contributions to such a v. e. v. correspond to complete "pairing" of annihilation and creation operators that match each other. The procedure of pairing is based on the observation that a non-zero contribution to the considered *S*-matrix element originates in a commutator (or anticommutator) of a pair of operators of the same sort. In doing this, one in fact uses, though only implicitly, a particular consequence of famous **Wick's theorems**, which are the main topic of this chapter.[20]

So, now it is the right time to discuss products of creation and annihilation operators in a systematic way. To begin with, we are going to formulate some basic definitions. First, we will consider bosonic operators in the Fock space; more precisely, the operators in question are in general linear combinations of annihilation and creation operators (satisfying standard commutation relations) and may depend on spacetime coordinates (a typical example: operator of a quantized bosonic field, e.g. scalar or another one).

Let $A$ and $B$ be such operators. We define the **contraction** (pairing) of $A$ and $B$, denoted as $\overset{\smile}{AB}$, according to

$$AB = \; :AB: + \overset{\smile}{AB} \,, \tag{34.1}$$

where $: AB :$ is the **normal product**, in which creation operators stand on the left of the annihilation operators. Since we know that the basic commutators are "c-numbers" (i.e. multiples of unit operator — Kronecker deltas, or delta functions), it is clear that the contraction $\overset{\smile}{AB}$ is a c-number. Then, it is also obvious that $\overset{\smile}{AB}$ is the v. e. v. of $AB$,

$$\overset{\smile}{AB} = \langle 0|AB|0\rangle \,. \tag{34.2}$$

An important point to be mentioned here is that within the normal product the operators commute. Indeed, any operator we are dealing with can be written as a sum of its creation and annihilation parts, i.e. $A = A^- + A^+$, $B = B^- + B^+$. Then,

$$:AB: \; = \; :(A^- + A^+)(B^- + B^+): \; = \; A^-B^- + B^+A^- + A^+B^- + A^+B^+ \,,$$

while

$$:BA: \; = \; :(B^- + B^+)(A^- + A^+): \; = \; B^-A^- + A^+B^- + B^+A^- + B^+A^+ \,.$$

---

[20]Gian-Carlo Wick (1909–1992) was an Italian theoretical physicist. He was an assistant of Enrico Fermi in Rome, later worked in United States, in particular at the Columbia University, N. Y.

However, $A^-B^- = B^-A^-$ and $A^+B^+ = B^+A^+$ (commutators between annihilation operators are trivial and the same is true for the creation operators). Thus, it is clear that one has

$$: AB := : BA :. \tag{34.3}$$

In a similar way, one may define **chronological contraction** (pairing) through

$$T\big(A(x)B(y)\big) = : A(x)B(y) : + \underline{A(x)B(y)}. \tag{34.4}$$

It is not difficult to realize that $\underline{A(x)B(y)}$ is also a c-number.[21] Indeed, one has

$$
\begin{aligned}
T\big(A(x)B(y)\big) &= \vartheta(x_0 - y_0)A(x)B(y) + \vartheta(y_0 - x_0)B(y)A(x) \\
&= \vartheta(x_0 - y_0)\big(: A(x)B(y) : + \overbrace{A(x)B(y)}\big) + \vartheta(y_0 - x_0)\big(: B(y)A(x) : + \overbrace{B(y)A(x)}\big).
\end{aligned}
$$

Then, using (34.3) and taking into account that

$$\vartheta(x_0 - y_0) + \vartheta(y_0 - x_0) = 1,$$

one gets

$$T\big(A(x)B(y)\big) = : A(x)B(y) : + \vartheta(x_0 - y_0)\,\overbrace{A(x)B(y)} + \vartheta(y_0 - x_0)\,\overbrace{B(y)A(x)}.$$

So,

$$\underline{A(x)B(y)} = \vartheta(x_0 - y_0)\,\overbrace{A(x)B(y)} + \vartheta(y_0 - x_0)\,\overbrace{B(y)A(x)}, \tag{34.5}$$

and that's it.

Remark on the notation: We have distinguished the symbols for "ordinary" and "chronological" contractions as $\overbrace{\phantom{xx}}$ and $\underline{\phantom{xx}}$, respectively. In fact, there is no firmly established notation used in current literature, so the reader can take the present convention as provisional and rather *ad hoc*.

Now it is also clear that

$$\underline{A(x)B(y)} = \langle 0|T\big(A(x)B(y)\big)|0\rangle. \tag{34.6}$$

Thus, we see that (34.4) in fact coincides with our earlier definition of chronological pairing that led us to the concept of the propagator of a quantized field.

So far we have considered boson fields. For fermion (Dirac) fields one may formulate analogous definitions, except that commutators are replaced by anticommutators. This in turn means that in the definition of normal product one has to incorporate a sign change for transposition of creation and annihilation operators and the same rule concerns the definition of the T-product. Thus, in particular, fermion operators anticommute within the normal product.

Before proceeding to the announced Wick's theorems, let us work out, for the sake of motivation, an instructive example. We will use bosonic annihilation and creation operators $a_k$, $a_k^\dagger$, with $k$ being the standard label for the momentum; for brevity, we will denote $a_1 \equiv a_{k_1}$, $a_2 \equiv a_{k_2}$, etc. Now, consider the product $a_1 a_2 a_3^\dagger a_4^\dagger$ and rearrange the operator factors so as to

---

[21] The symbol for the chronological contraction might be called the "cramp" or "clamp", alluding to the form of a simple carpenter's tool (the reader fluent in Czech might appreciate the colloquial expression "kramle"). This linguistic extempore could be understood as the author's self-made contribution to QFT terminology.

have creation operators on the left and annihilation operators on the right (i.e. to accomplish "normal ordering"). We get, step by step:

$$
\begin{aligned}
a_1 a_2 a_3^\dagger a_4^\dagger &= a_1 (a_3^\dagger a_2 + \delta_{23}) a_4^\dagger \\
&= a_1 a_3^\dagger (a_4^\dagger a_2 + \delta_{24}) + (a_4^\dagger a_1 + \delta_{14}) \delta_{23} = \dots \\
&= a_3^\dagger a_4^\dagger a_1 a_2 + \delta_{14} a_3^\dagger a_2 + \delta_{13} a_4^\dagger a_2 + \delta_{24} a_3^\dagger a_1 + \delta_{23} a_4^\dagger a_1 + \delta_{13}\delta_{24} + \delta_{23}\delta_{14} .
\end{aligned}
\tag{34.7}
$$

From (34.7) we see that the original operator product can be rearranged as a sum of normal products accompanied by the appropriate contractions, including the term without contraction and completely contracted terms. The result (34.7) is leading us to the following general definition of a **normal product with contractions**. Symbolically,

$$
\underbrace{AB \dots R \dots XYZ} = \pm AZ\, BX : \dots R \dots Y :,
\tag{34.8}
$$

where the sign $\pm$ is included depending on whether the permutation of fermionic factors corresponding to the passage from the left-hand side to right-hand side is even or odd. The definition (34.8) is valid also for a "normal product with chronological contractions".

Now we are in a position to formulate the basic Wick's theorems (they are valid for the sort of operators specified at the beginning of this chapter).

**T1 (Wick's theorem for ordinary products)**: *A product of operators is equal to the sum of their normal products with all possible contractions, including the normal product without contractions.*

For T-products, the statement is almost identical:

**T2 (Wick's theorem for T-products)**: *T-product of operators is equal to the sum of their normal products with all possible chronological contractions, including the normal product without contractions.*

Hopefully, the validity of the theorem T1 is plausible, on the basis of our explicit motivation example. In general, the theorems are proved by induction (which is not a very inspiring procedure, as we know).

Let us also remark that in our previous straightforward calculations of the $S$-matrix elements we have obviously employed particular examples of the theorem T1, namely its application to the vacuum expectation value of operator products in question (this is tantamount to considering just the fully contracted terms).

The great value of Wick's theorems consists in the systematic reordering of operator products in terms of normal products, since the matrix elements of the latter have some important specific properties. Let us illustrate it on a simple example. Consider the operator

$$
O = a_{k_1}^\dagger a_{k_2}^\dagger a_{k_3} a_{k_4} .
\tag{34.9}
$$

It turns out that the only non-trivial matrix elements of (34.9) are of the type $\langle k_1' k_2' | O | k_3' k_4' \rangle$, i.e. (34.9) can only have non-zero matrix elements between two-particle states. In more detail: for instance, there is no matrix element of the operator (34.9) between one-particle states. Indeed,

$$
\begin{aligned}
\langle p | a_{k_1}^\dagger a_{k_2}^\dagger a_{k_3} a_{k_4} | k \rangle &= \langle 0 | a_p a_{k_1}^\dagger a_{k_2}^\dagger a_{k_3} a_{k_4} a_k^\dagger | 0 \rangle \\
&= \langle 0 | (\delta_{k_1 p} + a_{k_1}^\dagger a_p) a_{k_2}^\dagger a_{k_3} (\delta_{k_4 k} + a_k^\dagger a_{k_4}) | 0 \rangle \\
&= \delta_{k_1 p} \delta_{k_4 k} \langle 0 | a_{k_2}^\dagger a_{k_3} | 0 \rangle \\
&= 0 .
\end{aligned}
$$

On the other hand, if one takes

$$\langle pq|O|kl\rangle = \langle 0|a_p a_q a_{k_1}^\dagger a_{k_2}^\dagger a_{k_3} a_{k_4} a_k^\dagger a_l^\dagger|0\rangle , \qquad (34.10)$$

then there is a non-zero contribution: according to the Wick's theorem T1, there is a completely contracted term in the expression (34.10), namely

$$\langle 0|a_p a_q a_{k_1}^\dagger a_{k_2}^\dagger a_{k_3} a_{k_4} a_k^\dagger a_l^\dagger|0\rangle .$$

So, what is the substantial difference in comparison with the preceding case? For one-particle states we have only a fully contracted term like e.g.

$$\langle 0|a_p a_{k_1}^\dagger a_{k_2}^\dagger a_{k_3} a_{k_4} a_k^\dagger|0\rangle ,$$

but $a_{k_2}^\dagger a_{k_3} = 0$, since $\langle 0|a_{k_2}^\dagger a_{k_3}|0\rangle = 0$. In other words, we cannot avoid here trivial contractions!

On the other hand, for an operator of the sort (34.9) that is not normally ordered, one does obtain non-zero matrix elements even for one-particle states. Indeed, as an example, let us consider the operator

$$\widetilde{O} = a_{k_1} a_{k_2}^\dagger a_{k_3} a_{k_4}^\dagger . \qquad (34.11)$$

For its matrix element between one-particle states one gets e.g. the contribution

$$\langle 0|a_p a_{k_1} a_{k_2}^\dagger a_{k_3} a_{k_4}^\dagger a_k^\dagger|0\rangle$$

and also some other possible fully contracted terms.

Thus, it is clear that Wick's theorems provide an efficient tool for a systematic investigation of matrix elements of operator products in the Fock space and thereby may serve as a basis for the derivation and construction of Feynman diagrams.

Chapter 35

# *S*-matrix and Wick's theorems: some applications

Before considering applications for the *S*-matrix calculations, we should clarify some remaining problems. We know that in the Dyson expansion of the *S*-matrix, one encounters T-products of Lagrangian densities, i.e. whole "blocks" of field operators like $\bar{\psi}\psi\varphi$, or $\bar{\psi}\gamma^\mu\psi A_\mu$, etc. The fields constituting such monomials are taken at the same spacetime point, so their time ordering is in a sense arbitrary. Let us now concentrate on quantum electrodynamics (QED), which is the most important QFT model discussed throughout these lecture notes. It turns out that a good idea is to improve the definition of the QED interaction Lagrangian so as to replace the current $\bar{\psi}\gamma^\mu\psi$ by its normal-ordered form, i.e.

$$\bar{\psi}(x)\gamma^\mu\psi(x) \longrightarrow \; :\bar{\psi}(x)\gamma^\mu\psi(x): \; . \tag{35.1}$$

The redefined interaction Lagrangian thus becomes

$$\mathscr{L}_{\text{int}} = e :\bar{\psi}(x)\gamma^\mu\psi(x): A_\mu(x), \tag{35.2}$$

which in fact is the same as

$$\mathscr{L}_{\text{int}} = e :\bar{\psi}(x)\gamma^\mu\psi(x)A_\mu(x): , \tag{35.3}$$

since the operator of the photon field $A_\mu$ commutes with $\psi$ and $\bar{\psi}$. We will be able to appreciate the improvement (35.3) in the Dyson expansion somewhat later, but there are at least two assets connected with the redefinition of the current shown in (35.2) that one can notice right now.

First, the charge $Q$ corresponding to the current $:\bar{\psi}\gamma^\mu\psi:$ is

$$Q = \int \mathrm{d}^3x :\bar{\psi}(x)\gamma_0\psi(x): = \int \mathrm{d}^3x :\psi^\dagger(x)\psi(x): , \tag{35.4}$$

and the normal product in (35.4) thus guarantees that $Q$ annihilates the vacuum, which is certainly desirable (in fact, we have used such a prescription in our earlier discussion of the quantization of free Dirac field). Second, one may consider the transformation of the current under charge conjugation $C$. We know that for the Dirac field this transformation means

$$\psi \longrightarrow \psi_C = C\bar{\psi}^{\mathsf{T}}, \tag{35.5}$$

where T denotes transposition and $C$ is the matrix $i\gamma_2\gamma_0$ in the standard representation of $\gamma$-matrices. For the current $\bar{\psi}\gamma_\mu\psi$ made of classical fields one gets, after some simple manipulations (and using the known properties of the matrix $C$, like $C^\dagger = C^{-1} = -C$, $C^{-1}\gamma_\mu C = -\gamma_\mu^{\mathsf{T}}$):

$$\bar{\psi}_C\gamma_\mu\psi_C = \psi^{\mathsf{T}}C\gamma_\mu C\bar{\psi}^{\mathsf{T}} = \psi^{\mathsf{T}}\gamma_\mu^{\mathsf{T}}\bar{\psi}^{\mathsf{T}} . \tag{35.6}$$

For classical Dirac fields, the expression (35.6) obviously becomes $\bar{\psi}\gamma_\mu\psi$, since in such a case there is no obstacle to commuting $\psi$ and $\bar{\psi}$. However, for quantized fields we have to take into account the relevant anticommutation relations; these lead to a sign change when passing from $\psi^T \gamma_\mu^T \bar{\psi}^T$ to $\bar{\psi}\gamma_\mu\psi$ and also produce an additional awkward c-number term. Using the normal-ordered form (35.1) one can simply anticommute the Dirac fields, and the result is then

$$: \bar{\psi}_C \gamma_\mu \psi_C := -: \bar{\psi}\gamma_\mu\psi : . \tag{35.7}$$

Intuitively, this is a desirable relation, since from the physical point of view one would certainly prefer such a change of sign for the electromagnetic current under charge conjugation.

Now the main problem is what is the variant of a Wick's theorem relevant for the "mixed products" appearing in the S-matrix expansion, namely

$$T\big(: \bar{\psi}(x_1)\gamma^\alpha\psi(x_1)A_\alpha(x_1): \ldots : \bar{\psi}(x_n)\gamma^\omega\psi(x_n)A_\omega(x_n):\big)$$

(explicitly, by a "mixed product" we mean a T-product containing already some normal products inside). The answer to this question is provided by the Wick's theorem for mixed products, which states:

**T3**: *A mixed (Dyson) T-product can be decomposed into a sum of normal products with chronological contractions, omitting contractions between operators that are already normally ordered.*

A proof of the theorem can be found e.g. in [16].

Now we may proceed to some applications, employing the formalism based on the Wick's theorems. We will consider the S-matrix in a given perturbative order (i.e. a definite term of the Dyson expansion) and represent it as a Wick series of normal products with contractions, involving field operators that constitute the relevant interaction Lagrangian. Let us start with the second order of the S-matrix for QED (as for the coupling constant, we will set temporarily $e = 1$). One has

$$S^{(2)} = \frac{i^2}{2!} \iint d^4x\, d^4y\; T\big(: \bar{\psi}(x)\gamma^\mu\psi(x)A_\mu(x) : : \bar{\psi}(y)\gamma^\nu\psi(y)A_\nu(y):\big).$$

Using the Wick's theorem for mixed products (T3), one can examine consecutively normal products with no contraction, one contraction, etc. Then one may utilize our earlier finding that in order to get a non-zero matrix element, the number of particles involved in the states in question should coincide with the number of the corresponding operators in the normal product.

As for the term without any contraction, it is not difficult to realize that it cannot describe any real physical process (the reader is recommended to verify this negative statement). Concerning terms with one contraction, there are essentially two possibilities, namely

$$: \bar{\psi}(x)\gamma^\mu\psi(x)A_\mu(x)\bar{\psi}(y)\gamma^\nu\psi(y)A_\nu(y):$$

$$= \langle 0|T\big(A_\mu(x)A_\nu(y)\big)|0\rangle : \bar{\psi}(x)\gamma^\mu\psi(x)\bar{\psi}(y)\gamma^\nu\psi(y): \tag{35.8}$$

and

$$: \bar{\psi}(x)\gamma^\mu\psi(x)A_\mu(x)\bar{\psi}(y)\gamma^\nu\psi(y)A_\nu(y):$$

$$= \langle 0|T\big(\psi(x)\bar{\psi}(y)\big)|0\rangle : \bar{\psi}(x)\gamma^\mu A_\mu(x)\gamma^\nu\psi(y)A_\nu(y): . \tag{35.9}$$

212

From our previous observations it should be clear that the expression (35.8) corresponds to processes involving four fermions (electrons and/or positrons) only, so one may anticipate Feynman diagrams with four external fermion lines and one internal photon line (photon propagator). Similarly, the form (35.9) describes processes involving two photons and two fermions (i.e. either the Compton scattering, or the two-photon annihilation of the electron–positron pair).

Next, there are two terms with two contractions; let us show explicitly one of them:

$$: \overline{\psi}_j(x)\gamma^\mu_{jk}\psi_k(x)A_\mu(x)\overline{\psi}_r(y)\gamma^\nu_{rs}\psi_s(y)A_\nu(y):$$

$$= (-1)\langle 0|T\big(\psi_s(y)\overline{\psi}_j(x)\big)|0\rangle\langle 0|T\big(\psi_k(x)\overline{\psi}_r(y)\big)|0\rangle\gamma^\mu_{jk}\gamma^\nu_{rs} : A_\mu(x)A_\nu(y): . \quad (35.10)$$

In arriving at (35.10), we have used our previous definition of the normal product with chronological contractions and the minus sign has been included so as to take into account the signature of the relevant permutation of fermion fields. The "coefficient function" multiplying the normal product $: A_\mu(x)A_\nu(y):$ in (35.10) is obviously equal to

$$(-1)\,\mathrm{Tr}\big[i\mathscr{S}_F(y - x)\gamma^\mu i\mathscr{S}_F(x - y)\gamma^\nu\big], \quad (35.11)$$

where we have used our earlier notation for the propagator of Dirac field. From (35.10) one can obviously get only a trivial process of photon–photon transition. Nevertheless, the structures (35.10) and (35.11) suggest the graphical representation corresponding to a Feynman diagram (here in the coordinate space) shown in Fig. 35.1. While the photon–photon transition is

**Fig. 35.1:** QED fermion bubble: the simplest correction to the photon propagator.

physically trivial, we will see soon that the closed loop in Fig. 35.1 made of internal fermion lines plays an important role as a subdiagram in higher-order contributions to the $S$-matrix. Let us also mention that apart from the configuration of contractions (35.10) there is a term involving one contraction of Dirac fields along with the contraction of the electromagnetic fields $A_\mu$ and $A_\nu$. In analogy with the picture shown in Fig. 35.1 one may guess easily that the corresponding diagram is the one shown in Fig. 35.2. Again, Fig. 35.2 would describe a

**Fig. 35.2:** One-loop correction to the fermion propagator.

physically trivial process of electron–electron transition, but, similarly to Fig. 35.1, it will also become an important subdiagram in higher-order Feynman graphs.

Finally, there is the fully contracted term that would correspond to a physically irrelevant term of vacuum–vacuum transition. It can be visualized by means of a diagram that reminds one of a cracked egg, see Fig. 35.3 (the reader may find some relevant comments on the role played by those vacuum graphs e.g. in the book [10]).

**Fig. 35.3:** "Cracked-egg" vacuum diagram.

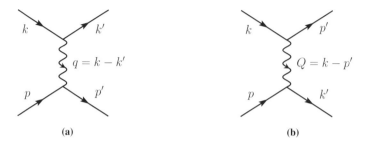

**Fig. 35.4:** Feynman diagrams for Møller scattering in the second order of QED.

Let us now come back briefly to the normal product with the contraction (35.8). Its operator content indicates that it can describe elastic scattering processes $e^- e^- \to e^- e^-$ (**Møller scattering**) and $e^- e^+ \to e^- e^+$ (**Bhabha scattering**).[22] As an illustration, let us consider the first case. The initial and final states can be written as

$$
\begin{aligned}
|i\rangle &= b^\dagger(k)b^\dagger(p)|0\rangle, \\
|f\rangle &= b^\dagger(k')b^\dagger(p')|0\rangle.
\end{aligned}
\tag{35.12}
$$

Now, in order to work out the matrix element

$$
\langle f | : \bar\psi(x)\gamma^\mu\psi(x)\bar\psi(y)\gamma^\nu\psi(y) : |i\rangle,
\tag{35.13}
$$

one has to consider the possible pairings in the usual way. Then it is not difficult to find out that a non-zero contribution can only originate from

$$
\langle 0|b(p')b(k') : \bar\psi^+(x)\gamma^\mu\psi^-(x)\bar\psi^+(y)\gamma^\nu\psi^-(y) : b^\dagger(k)b^\dagger(p)|0\rangle,
\tag{35.14}
$$

where $+$ and $-$ denote the creation and annihilation parts, respectively. It is clear that one has to calculate the fully contracted terms; it is easy to see that there are four possibilities: $b(k')$ and $b(p')$ can be contracted with $\bar\psi^+(x)$ or $\bar\psi^+(y)$, while $b^\dagger(k)$ and $b^\dagger(p)$ should be contracted with $\psi^-(x)$ or $\psi^-(y)$. The procedure is straightforward and working out its details may serve as a useful exercise for any serious reader. The net result is that the factor $1/2!$ from the Dyson expansion is cancelled, and one is left with two types of contributions, which can be depicted as Feynman diagrams in Fig. 35.4. An important point is that there is a **relative minus sign** for contributions (a) and (b).

In a similar way, one can treat the process of Bhabha scattering

$$
e^-(k) + e^+(p) \to e^-(k') + e^+(p').
\tag{35.15}
$$

In this case, one ends up with two diagrams, namely those shown in Fig. 35.5. The corresponding

---

[22]Christian Møller (1904–1980) was a Danish physicist, he spent his scientific career at the University of Copenhagen (he also studied there, with Niels Bohr). Homi Bhabha (1909–1966) was a prominent Indian physicist, father of the Indian nuclear program. He died at Mont Blanc in an airplane accident.

214

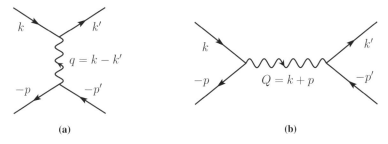

**Fig. 35.5:** Feynman diagrams for Bhabha scattering in the second order of QED.

matrix elements can be written down by using the familiar Feynman rules that we already know from our previous experience with other processes. Again, as in the case of Møller scattering, one has to include an additional relative minus sign between the contributions of the diagrams (a) and (b).

With the matrix elements for these scattering processes at hand, one can also evaluate the corresponding cross sections. Such a calculation is straightforward, though somewhat tedious. As usual, one can employ the standard trace technique, as well as the elementary kinematic relations. Let us display here the results for both processes in the high-energy limit, i.e. for $E_{\text{c.m.}} = s^{1/2} \gg m$. One has, in the variables of the c.m. system,

$$\left(\frac{d\sigma}{d\Omega}\right)_{\text{M}} = \frac{\alpha^2}{2s} \left[ \frac{1 + \cos^4 \frac{\vartheta}{2}}{\sin^4 \frac{\vartheta}{2}} + \frac{2}{\sin^2 \frac{\vartheta}{2} \cos^2 \frac{\vartheta}{2}} + \frac{1 + \sin^4 \frac{\vartheta}{2}}{\cos^4 \frac{\vartheta}{2}} \right] \tag{35.16}$$

for the Møller scattering and

$$\left(\frac{d\sigma}{d\Omega}\right)_{\text{B}} = \frac{\alpha^2}{2s} \left[ \frac{1 + \cos^4 \frac{\vartheta}{2}}{\sin^4 \frac{\vartheta}{2}} - \frac{2 \cos^4 \frac{\vartheta}{2}}{\sin^2 \frac{\vartheta}{2}} + \frac{1 + \cos^2 \vartheta}{2} \right] \tag{35.17}$$

for the Bhabha scattering (cf. e.g. the book [1]).

The derivation of the formulae (35.16) and (35.17) should be a challenge for any diligent reader. In any case, it is a good topic for a tutorial.

# Chapter 36

# S-matrix in fourth order: QED example

In previous chapters, we have discussed several examples of processes described by 2nd order contributions to the Dyson expansion of the S-matrix. These are graphically represented by **tree diagrams** that are made of external and internal lines connecting the interaction vertices and do not contain closed loops of internal lines (precisely this is the defining feature of the tree-level diagrams). In the preceding chapter, we also come across two simple examples of closed-loop diagrams, in the S-matrix contributions corresponding to processes that are in a sense physically trivial.

In the present chapter, we are going to examine one-loop diagrams that arise as contributions to the 4th order S-matrix elements within QED, for a particular physical process, namely electron–muon elastic scattering (chosen here as one instructive example from the variety of possibilities provided by QED). An astute reader might wonder why we have skipped the 3rd order; we will comment briefly on this point later on.

For our present purpose, we will work with the interaction Lagrangian

$$\mathscr{L}_{\text{int}} = e \left( :\bar{\psi}_1 \gamma^\mu \psi_1 : + :\bar{\psi}_2 \gamma^\mu \psi_2 : \right) A_\mu , \tag{36.1}$$

where $\psi_1$ and $\psi_2$ correspond to electron and muon fields, respectively. Obviously, in the 2nd order the process

$$e^-(k) + \mu^-(p) \rightarrow e^-(k') + \mu^-(p') \tag{36.2}$$

is described by a single tree diagram, namely the one shown in Fig. 36.1. The 4th order term in

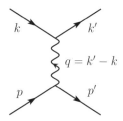

**Fig. 36.1:** The lowest-order contribution to the process (36.2) represented by a tree-level diagram.

the S-matrix Dyson expansion reads

$$S^{(4)} = \frac{i^4}{4!} e^4 \int d^4x_1 \dots d^4x_4 \, T(J_\mu(x_1) A^\mu(x_1) \dots J_\sigma(x_4) A^\sigma(x_4)) , \tag{36.3}$$

where we have introduced the usual notation for the current

$$J_\alpha = :\bar{\psi}_1 \gamma_\alpha \psi_1 : + :\bar{\psi}_2 \gamma_\alpha \psi_2 := J_\alpha^{(1)} + J_\alpha^{(2)} . \tag{36.4}$$

216

In order to get from the expression (36.3) a contribution to the process (36.2), one has to take the muon part from at least one current in (36.3) (but not from all of them). There are three possibilities, namely

(i) three $J^{(1)}$ and one $J^{(2)}$,

(ii) two $J^{(1)}$ and two $J^{(2)}$, $\qquad\qquad\qquad$ (36.5)

(iii) one $J^{(1)}$ and three $J^{(2)}$.

Let us start with the variant (i). There are four possibilities of how to get a term of such a type (taking $J_\mu^{(2)}$ at $x_1, x_2, x_3$, or $x_4$). Obviously, the corresponding contributions are the same. Thus, one may write

$$S_{(i)}^{(4)} = 4\,\frac{i^4}{4!}\,e^4 \int d^4x_1 \ldots d^4x_4\, T\big(J_\mu^{(2)}(x_1)A^\mu(x_1)J_\nu^{(1)}(x_2)A^\nu(x_2)J_\rho^{(1)}(x_3)A^\rho(x_3)J_\sigma^{(1)}(x_4)A^\sigma(x_4)\big).$$
(36.6)

Now, if one wants to describe the four-fermion process like (36.2), photon fields must be contracted completely. Furthermore, two pairs of Dirac fields must be contracted as well, since only two electron and two muon fields should survive as constituents of the relevant normal product (needless to say, we are relying here on the Wick's theorem T3 for the mixed operator products). In this way, the basic strategy for constructing the individual Feynman diagrams is given. As a first step, let us consider the configuration of contractions that would lead to a closed purely fermion loop (this amounts to taking Dirac field contractions at a single pair of spacetime points). So, the anticipated structure of the emerging Feynman graph we have in mind is e.g. the one shown in Fig. 36.2.

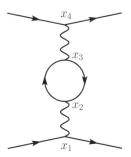

**Fig. 36.2:** The contribution of the closed fermion loop to the process (36.2) in QED at the fourth order.

For convenience, let us denote the electron field simply as (lower case) $\psi$ and the muon field as (capital) $\Psi$. In fact, there are 3 possibilities of how to contract the $A^\mu(x_1)$ (coupled to $J_\mu^{(2)}(x_1)$), but for a chosen variant, e.g. $A^\mu(x_1)A^\nu(x_2)$, the second contraction $AA$ is determined uniquely. With those bosonic contractions fixed, there are still two possibilities of how to contract electron fields to arrive at the closed loop as in Fig. 36.2 (making it either at $x_2, x_3$ or $x_2, x_4$). Taking into account all those observations, one may conclude that it is sufficient to consider the diagram in Fig. 36.2 and cancel the factor $1/4!$ from Dyson expansion by our combinatorial factor $4 \times 3 \times 2$. Notice also that one might recover our careful combinatorics as the 4! permutations of points $x_1, \ldots, x_4$ in Fig. 36.2. Thus, the diagram in Fig. 36.2 represents

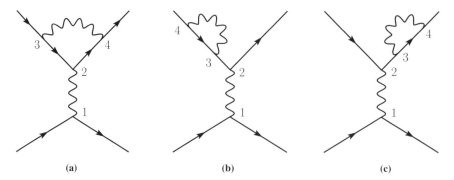

**Fig. 36.3:** Other 4th-order contributions to the process (36.2).

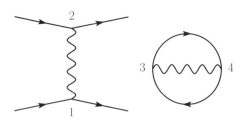

**Fig. 36.4:** A disconnected fourth-order graph contributing to the process (36.2).

the normal product with contractions

$$:\overline{\Psi}(x_1)\gamma_\mu\Psi(x_1)A^\mu(x_1)\overline{\psi}(x_2)\gamma_\nu\psi(x_2)A^\nu(x_2)\overline{\psi}(x_3)\gamma_\rho\psi(x_3)A^\rho(x_3)\overline{\psi}(x_4)\gamma_\sigma\psi(x_4)A^\sigma(x_4):\,.$$
(36.7)

The explicit expression corresponding to (36.7) is, according to our previous knowledge,

$$i\mathscr{D}_{\mathrm{F}}^{\mu\nu}(x_1-x_2)\,(-1)\,\mathrm{Tr}\big[i\mathscr{S}_{\mathrm{F}}(x_3-x_2)\gamma_\nu\,i\mathscr{S}_{\mathrm{F}}(x_2-x_3)\gamma_\rho\big]i\mathscr{D}_{\mathrm{F}}^{\rho\sigma}(x_3-x_4)$$
$$\times:\overline{\Psi}(x_1)\gamma_\mu\Psi(x_1)\overline{\psi}(x_4)\gamma_\sigma\psi(x_4):\,.\quad(36.8)$$

Now one might proceed further by considering the relevant matrix element and using Fourier representations of the propagators in the coefficient function in (36.8). We will do it later (in the next chapter); now, let us examine other diagrams corresponding to the remaining possible configurations of chronological contractions according to the Wick's theorem. We are going to avoid writing explicitly further long expressions like (36.7); hopefully, one such instructive example is enough. Instead, let us show the relevant graphical representations — the rest of this chapter should thus be understood as an eulogy on Feynman diagrams.

Employing the shorthand notation $1 \equiv x_1$, $2 \equiv x_2$, etc., the diagrams we have in mind are guessed quite easily (see Fig. 36.3). This is essentially all, as far as the topologically connected graphs are concerned. Apart from these, one can also obtain a disconnected diagram as shown in Fig. 36.4, but this is irrelevant for the evaluation of the scattering amplitude in question.

Next, there are some other connected diagrams that originate in the options (ii) and (iii) in (36.5). As for the variant (ii), one gets a new type of Feynman graph, namely the "box" (see Fig. 36.5), where the lines in the lower part correspond to muons and the upper part represents

218

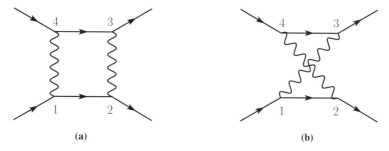

**Fig. 36.5:** Box-type contribution to the process (36.2).

**Fig. 36.6:** Another disconnected contribution to the process (36.2).

electrons. Here one could also consider the disconnected graph shown in Fig. 36.6, but this would obviously lead to a kinematically trivial matrix element (in fact, no scattering at all).

Finally, the option (iii) in (36.5) is essentially the same like (i), with graphs from Fig. 36.4 turned "upside down" and with the electron loop in Fig. 36.2 being replaced by muon loop (hopefully, such an observation is more or less obvious).

Let us now return to our remark at the beginning of this chapter, namely why did we skip the third order of the Dyson expansion? Well, in the 3rd order in QED, one certainly encounters some new tree-level diagrams, e.g. for $e^- + e^+ \to \gamma\gamma\gamma$ (see Fig. 36.7), or $e + \mu \to e + \mu + \gamma$ (the so-called **bremsstrahlung**, see Fig. 36.8).

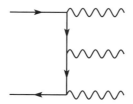

**Fig. 36.7:** The process $e^- + e^+ \to \gamma\gamma\gamma$ at the tree level.

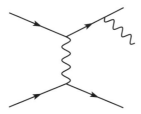

**Fig. 36.8:** The process $e + \mu \to e + \mu + \gamma$ at the tree level.

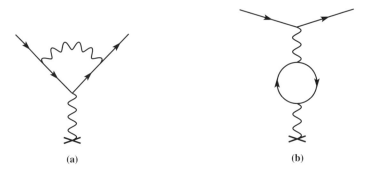

(a)                           (b)

**Fig. 36.9:** Processes involving classical external field in the 3rd order in QED.

In fact, one could also get some one-loop diagrams, if only an interaction with a classical external (e.g. Coulomb) field is included along with the quantized photon field (remember our earlier derivation of the famous Mott formula). Then one could consider diagrams like e.g. those shown in Fig. 36.9.

For the main part of the present chapter, our primary option was an analysis of the 4th order of the Dyson expansion because of the methodology, since this gives a fuller view of the relevant one-loop diagrams (including the box in Fig. 36.5). Nevertheless, we will discuss the graphs in Fig. 36.9 later on as well, since they enable one to elucidate some important results in QED applications. The bremsstrahlung graphs of the type shown in Fig 36.8 also play an extremely important role in QED, in the analysis of the problem of the so-called **infrared divergences** due to the zero mass of the photon (for more details, the reader is referred e.g. to the books [1], [6], or [7]).

A final remark is perhaps in order here. One should keep in mind that the diagrams like those in Figs. 36.2 and 36.3, etc. represent contributions to the **S-matrix operator**. So, the external lines represent operators $\psi$, $\bar{\psi}$, $\Psi$, $\bar{\Psi}$ and internal lines are propagators in the coordinate space. We will proceed to the usual Feynman diagrams for scattering matrix elements (amplitudes) in the next chapter.

# Chapter 37

# One-loop QED diagrams
# in momentum space

In the preceding chapter, we have found several types of one-loop diagrams characteristic for the QED $S$-matrix. As we have emphasized there, internal lines in those diagrams depict spacetime propagators and the external lines represent some field operators (or a classical field like in Fig. 36.9). Needless to say, the integration over the relevant set of spacetime coordinates is tacitly assumed for the evaluation of the whole diagram contribution (i.e. over $x_1, \ldots, x_4$ in our examples).

Thus, as the next step, we would like to calculate the one-loop contributions to the matrix element (scattering amplitude) for our sample process $e + \mu \to e + \mu$, in an analogy with what we have done before at the tree level. For the purpose of an illustration we will perform the calculation in detail for the diagram in Fig. 36.2 involving the purely fermion loop. Hopefully, such a sample calculation should be sufficient for understanding the basic technicalities; in fact, an evaluation of the other diagrams, like those in Fig. 36.3, etc. proceeds essentially along the same lines and does not bring any new ingredients.

So, let us start with the expression (36.8). It must be integrated over all possible values of $x_1, \ldots, x_4$, and the full contribution of the diagram in Fig. 36.2 to the $S$-matrix thus reads

$$S_a^{(4)} = e^4 \int \mathrm{d}^4x_1 \ldots \mathrm{d}^4x_4 \, \mathscr{D}_{\mathrm{F}}^{\alpha\mu}(x_1 - x_2) \, (-1) \, \mathrm{Tr}\big[\gamma_\mu \mathscr{S}_{\mathrm{F}}(x_2 - x_3)\gamma_\nu \mathscr{S}_{\mathrm{F}}(x_3 - x_2)\big] \mathscr{D}_{\mathrm{F}}^{\nu\beta}(x_3 - x_4)$$

$$\times : \overline{\Psi}(x_1)\gamma_\alpha \Psi(x_1)\overline{\psi}(x_4)\gamma_\beta\psi(x_4) : . \quad (37.1)$$

Note that when passing from (36.8) to (37.1), we have employed the trace cyclicity and changed, for later convenience, the labelling of the summation indices; we have also used the obvious fact that $i^4 \cdot i^4 = 1$. Now one may employ the Fourier representations of the spacetime propagators, given by

$$\mathscr{D}_{\mathrm{F}}^{\alpha\mu}(x_1 - x_2) = \int \frac{\mathrm{d}^4p}{(2\pi)^4} \, D_{\mathrm{F}}^{\alpha\mu}(p) \, e^{ip(x_1-x_2)} ,$$

$$\mathscr{D}_{\mathrm{F}}^{\beta\nu}(x_3 - x_4) = \int \frac{\mathrm{d}^4q}{(2\pi)^4} \, D_{\mathrm{F}}^{\beta\nu}(q) \, e^{iq(x_3-x_4)} ,$$

$$\mathscr{S}_{\mathrm{F}}(x_2 - x_3) = \int \frac{\mathrm{d}^4k}{(2\pi)^4} \, S_{\mathrm{F}}(k) \, e^{-ik(x_2-x_3)} ,$$

$$\mathscr{S}_{\mathrm{F}}(x_3 - x_2) = \int \frac{\mathrm{d}^4l}{(2\pi)^4} \, S_{\mathrm{F}}(l) \, e^{-il(x_3-x_2)}$$

(37.2)

(where $D_F^{\rho\sigma}(q)$ and $S_F(p)$ are the functions found in chapters 27 and 32). Further, for the evaluation of the matrix element of the operator in (37.1) we will also change, for convenience, the labelling of the electron and muon four-momenta, so that the initial and final state will be

$$|i\rangle = b_1^\dagger(p_1)b_2^\dagger(p_2)|0\rangle\,,$$
$$|f\rangle = b_1^\dagger(p_1')b_2^\dagger(p_2')|0\rangle\,,$$

(37.3)

where 1 and 2 denote the electron and muon, respectively. Now, one may take into account the contractions of the operators in (37.3) and field operators appearing in the expression (37.1):

$$b_1(p_1')\overline{\psi}(x_4) = N_{p_1'}\overline{u}(p_1')\,e^{ip_1'x_4}\,,$$

$$\psi(x_4)b_1^\dagger(p_1) = N_{p_1}u(p_1)\,e^{-ip_1x_4}\,,$$

$$b_2(p_2')\overline{\Psi}(x_1) = N_{p_2'}\overline{u}(p_2')\,e^{ip_2'x_1}\,,$$

$$\Psi(x_1)b_2^\dagger(p_2) = N_{p_2}u(p_2)\,e^{-ip_2x_1}\,.$$

(37.4)

The matrix element of the operator $:\overline{\Psi}(x_1)\ldots\psi(x_4):$ in (37.1) between the states (37.3) thus becomes

$$N_{p_1}N_{p_2}N_{p_1'}N_{p_2'}\big[\overline{u}(p_1')\gamma_\beta u(p_1)\big]\big[\overline{u}(p_2')\gamma_\alpha u(p_2)\big]e^{i(p_2'-p_2)x_1}e^{i(p_1'-p_1)x_4}\,.$$

(37.5)

Using the representation (37.2) in (37.1), one has

$$S_a^{(4)} = \int d^4x_1\ldots d^4x_4 \int \frac{d^4p}{(2\pi)^4} \int \frac{d^4k}{(2\pi)^4} \int \frac{d^4l}{(2\pi)^4} \int \frac{d^4q}{(2\pi)^4}$$
$$\times\, e^{ip(x_1-x_2)}e^{-ik(x_2-x_3)}e^{-il(x_3-x_2)}e^{iq(x_3-x_4)}$$
$$\times\, D_F^{\alpha\mu}(p)(-1)\,\mathrm{Tr}\big[\gamma_\mu S_F(k)\gamma_\nu S_F(l)\big]D_F^{\nu\beta}(q)$$
$$\times :\overline{\Psi}(x_1)\gamma_\alpha\Psi(x_1)\overline{\psi}(x_4)\gamma_\beta\psi(x_4):\,.$$

(37.6)

Such an expression may look intimidating, but most of the integrations are actually trivial. When we form the matrix element $\langle f|S_a^{(4)}|i\rangle$, the exponential factors can be accumulated as

$$e^{i(p_2'-p_2)x_1}e^{i(p_1'-p_1)x_4}e^{ip(x_1-x_2)}e^{-ik(x_2-x_3)}e^{-il(x_3-x_2)}e^{iq(x_3-x_4)}\,,$$

(37.7)

and the integration with $d^4x_1\ldots d^4x_4$ leads to the product of delta functions:

$$(2\pi)^4\delta^{(4)}(p_2'+p-p_2)(2\pi)^4\delta^{(4)}(-p-k+l)(2\pi)^4\delta^{(4)}(k-l+q)(2\pi)^4\delta^{(4)}(-q+p_1'-p_1)\,.$$

(37.8)

Now, the integration over the variable $k$ reduces the third $\delta$-function to $\delta^{(4)}(q-p)$ (it means that photon propagators in (37.6) carry the same four-momentum $p = q$). The subsequent integrations over $p$ and $q$ then lead finally to $\delta^{(4)}(p_1'+p_2'-p_1-p_2)$, which embodies the anticipated overall energy and momentum conservation. Note also that the second and third $\delta$-functions in (37.8) mean that $l = k+q$ and $l = k+p$ (and we also know that $p = q$). The variable $l$ remains unconstrained and the integral over the infinite domain of $l$ stays with us in the contribution of the considered matrix element. Precisely this is a substantially new ingredient concerning closed-loop diagrams, in comparison with tree-level graphs (where all four-momenta flowing through the diagram are determined by their values for incoming and outgoing particles — simply because of the conservation laws acting in the interaction vertices).

So, putting all things together, in the matrix element in question one sees the usual factorization of the normalization factors

$$N_{p_1}N_{p_2}N_{p'_1}N_{p'_2} = \prod_{i,f} N_{i,f} \tag{37.9}$$

and the delta function for overall four-momentum conservation

$$(2\pi)^4 \delta^{(4)}(p'_1 + p'_2 - p_1 - p_2) = (2\pi)^4 \delta^{(4)}(P_f - P_i). \tag{37.10}$$

Using the conventional form (37.10), one may then attach a factor of $1/(2\pi)^4$ to $d^4l$ in the integral over the arbitrary loop momentum $l$.

The result of our calculation may thus be written as

$$\langle f|S_a^{(4)}|i\rangle = \prod_{i,f} N_{i,f} \, i\mathcal{M}_{fi} \, (2\pi)^4 \delta^{(4)}(P_f - P_i), \tag{37.11}$$

with

$$i\mathcal{M}_{fi} = e^4 \left[ \bar{u}(p'_2)\gamma_\alpha u(p_2) \right] D_F^{\alpha\mu}(q)\left( -i\Pi_{\mu\nu}(q) \right) D_F^{\nu\beta}(q) \left[ \bar{u}(p'_1)\gamma_\beta u(p_1) \right], \tag{37.12}$$

where we have denoted, for convenience,

$$-i\Pi_{\mu\nu}(q) = (-1)\int \frac{d^4l}{(2\pi)^4} \, \mathrm{Tr}\left( \gamma_\mu \frac{1}{\slashed{l} - \slashed{q} - m} \gamma_\nu \frac{1}{\slashed{l} - m} \right). \tag{37.13}$$

The expression (37.12) can be represented by the Feynman graph shown in Fig. 37.1. Note that

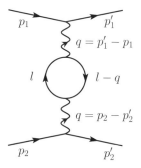

**Fig. 37.1:** Fourth-order contribution to the electron-muon scattering matrix element involving closed fermion loop as a correction to the photon propagator.

in (37.13) we have already used the explicit form of the propagator of Dirac field in momentum space; for the photon propagators one can use e.g. the simple expression

$$D_F^{\mu\nu}(q) = \frac{-g^{\mu\nu}}{q^2}. \tag{37.14}$$

Let us also mention that when "reading" the diagram in Fig. 37.1, one maintains, apart from other familiar rules, also the factors of $i$ for any vertex and any internal line (propagator).

As for the other one-loop diagrams displayed in the preceding chapter (see Figs. 36.3(a),(b),(c)), one may proceed in much the same way. Instead of repeating the rather boring steps in the evaluation of the relevant $S$-matrix elements, let us just show some final results for the momentum-space Feynman diagrams.

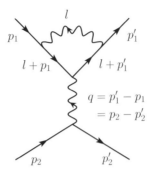

**Fig. 37.2:** Fourth-order contribution to the electron-muon scattering matrix element involving the one-loop correction to the interaction vertex.

From Fig. 36.3(a) one gets the Feynman graph shown in Fig. 37.2 and its contribution (the corresponding matrix element $\mathcal{M}$) is given by

$$i\mathcal{M} = e^4 i^2 \left[ \bar{u}(p_1')i\Gamma^\mu(p_1', p_1)u(p_1) \right] \frac{-g_{\mu\nu}}{q^2} \left[ \bar{u}(p_2')\gamma^\nu u(p_2) \right], \tag{37.15}$$

where we have denoted, for later convenience,

$$i\Gamma^\mu(p_1', p_1) = -\int \frac{d^4l}{(2\pi)^4} \gamma_\alpha \frac{1}{\slashed{l} + \slashed{p}_1' - m} \gamma^\mu \frac{1}{\slashed{l} + \slashed{p}_1 - m} \gamma_\beta \frac{-g^{\alpha\beta}}{l^2}. \tag{37.16}$$

Here the prefactor $-1$ is $i^6$, since we observe the "book-keeping rule" of $i$'s for vertices and propagators.

Finally, concerning the diagrams (c) and (d) in Fig. 36.3, let us write down explicitly just the expression for the one-loop subdiagram shown in Fig. 37.3. Employing a conventional (quite

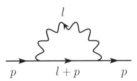

**Fig. 37.3:** One-loop graph that appears as a subdiagram in various higher-order Feynman graphs. It is called commonly the "electron-self energy graph".

standard) symbol $i\Sigma(p)$ for the graph in Fig. 37.3, we have (suppressing the coupling factor $e^2$)

$$i\Sigma(p) = \int \frac{d^4l}{(2\pi)^4} \gamma_\alpha \frac{1}{\slashed{l} + \slashed{p} - m} \gamma_\beta \frac{-g^{\alpha\beta}}{l^2}. \tag{37.17}$$

An important lesson to be learnt from the preceding discussion is that while the original mathematical expressions for the higher-order $S$-matrix contributions are rather long and perhaps somewhat awkward, the resulting picture in terms of Feynman diagrams is quite elegant and easy to recover almost by heart. There is a set of relatively simple rules that work both at the tree level and for closed-loop graphs; a substantially new extra rule for loop diagrams is that one has to integrate over the arbitrary values of a four-momentum circulating around the loop. Thus, there is a good reason to consider the main part of the forthcoming chapters as a "canticle for Feynman diagrams".

In this chapter we have already had a lot of good news, but there are bad news as well — that's life! The bad news about the nice expressions (37.13), (37.16) and (37.17) is that the integrals *diverge*[23]. Indeed, e.g. in (37.13) the integrand behaves like $l^{-2}$ at infinity and the integration volume element $d^4 l$ amounts to $l^3 dl$ (if one employs e.g. hyperspherical coordinates in the four-dimensional space). Thus, for $l \to \infty$, the integral in question looks like $\int l dl$ for the neighbourhood of infinity; if one introduces a "cut-off" $\Lambda$ (a provisional upper limit in this integral), one gets

$$\int^{\Lambda} l dl \simeq \Lambda^2 , \tag{37.18}$$

for $\Lambda \to \infty$. Thus, one may say that the integral in the formula (37.13) is quadratically divergent in the asymptotic region $l \to \infty$. In QFT, a divergence of such a type is generally called ultraviolet (UV) divergence (since it is due to the large values of loop energies and momenta). In a similar way, one may conclude that the integral in (37.16) is logarithmically divergent and the expression (37.17) exhibits a linear divergence.

On the other hand, Feynman diagrams like the one in Fig. 37.1, etc., being of the order $\mathcal{O}(e^4)$ (i.e. $\mathcal{O}(\alpha^2)$), should represent higher-order corrections to the contribution of the lowest-order (tree-level) graph(s) (that are of the order $\mathcal{O}(e^2)$). Thus, one would like to tame the annoying UV divergences and give some reasonable meaning to the UV finite parts of the loop diagrams. This is precisely the program of the **regularization** and **renormalization** to be pursued in the subsequent chapters.

---

[23] So, if you know the popular animated sitcom South Park, you might be tempted to exclaim: "Oh my God, they killed Kenny!"

# Chapter 38

# Regularization of UV divergences

In several forthcoming chapters, we will discuss the problem of the regularization of UV divergences, arising in QED one-loop diagrams. We are going to start with the closed purely fermion loop (see Fig. 38.1) that emerged as a subgraph of the 4th order S-matrix diagram shown in Fig. 37.1. Let us remark already now that it is usually called "vacuum polarization graph" (the reason for such a fancy name will become clear later).

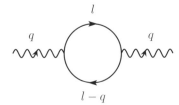

**Fig. 38.1:** The vacuum polarization loop in QED.

As a prelude to the main subject of this chapter, let us examine some simple general properties of the contribution of the closed loop in Fig. 38.1. In the preceding chapter we have arrived at its integral representation, which reads

$$
i\Pi_{\mu\nu}(q) = \int \frac{\mathrm{d}^4 l}{(2\pi)^4} \, \mathrm{Tr}\left(\gamma_\mu \frac{1}{\slashed{l} - \slashed{q} - m}\gamma_\nu \frac{1}{\slashed{l} - m}\right) = \int \frac{\mathrm{d}^4 l}{(2\pi)^4} \frac{\mathrm{Tr}\left[\gamma_\mu(\slashed{l} - \slashed{q} + m)\gamma_\nu(\slashed{l} + m)\right]}{\left[(l-q)^2 - m^2\right](l^2 - m^2)}.
$$
(38.1)

It is not difficult to find out that $\Pi_{\mu\nu}(q)$ is symmetric, i.e. $\Pi_{\mu\nu}(q) = \Pi_{\nu\mu}(q)$, and is also an even function, $\Pi_{\mu\nu}(-q) = \Pi_{\mu\nu}(q)$. Indeed, these two properties can be verified by means of an appropriate formal shift of the integration variable in (38.1) and by using the trace cyclicity (the reader is encouraged to do this, as a simple exercise). In fact, there is an alternative argument, which is perhaps even more elegant. Whatever definition might be used for the (up to now ill-defined) integral in (38.1), one would like to get a covariant form for $\Pi_{\mu\nu}(q)$; this would mean that

$$
\Pi_{\mu\nu}(q) = A(q^2)g_{\mu\nu} + B(q^2)q_\mu q_\nu,
$$
(38.2)

where $A$ and $B$ are essentially arbitrary functions. It is so, because $\Pi_{\mu\nu}(q)$ should be a 2nd rank Lorentz tensor depending on a single four-vector variable $q$; the expression (38.2) is then obviously its most general form. The Ansatz (38.2) makes the above-mentioned symmetry properties manifest. With these simple facts in mind, it is clear that the contribution of the graph in Fig. 38.1 is the same as for Fig. 38.2. It means that when drawing the vacuum polarization

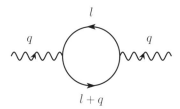

**Fig. 38.2:** An alternative equivalent labelling of the loop momenta for the vacuum polarization graph.

diagram in question, one need not worry about the orientation of the loop (the direction of running around the loop is irrelevant — clockwise or counterclockwise, it doesn't matter).

There is one more property, which may not be so obvious at first sight: $\Pi_{\mu\nu}(q)$ is **transverse** with respect to $q$, i.e.

$$q^{\mu}\Pi_{\mu\nu}(q) = 0 .\qquad(38.3)$$

So, how can one derive Eq. (38.3), at least heuristically? A hand-waving argument may proceed as follows. Multiplying (38.1) by $q^{\mu}$, one has

$$
\begin{aligned}
iq^{\mu}\Pi_{\mu\nu}(q) &= \int \frac{\mathrm{d}^4 l}{(2\pi)^4} \; \mathrm{Tr}\left( \slashed{q}\, \frac{1}{\slashed{l} - \slashed{q} - m} \gamma_{\nu} \frac{1}{\slashed{l} - m} \right) \\
&= \int \frac{\mathrm{d}^4 l}{(2\pi)^4} \; \mathrm{Tr}\left[ \left( (\slashed{l} - m) - (\slashed{l} - \slashed{q} - m) \right) \frac{1}{\slashed{l} - \slashed{q} - m} \gamma_{\nu} \frac{1}{\slashed{l} - m} \right] \\
&= \int \frac{\mathrm{d}^4 l}{(2\pi)^4} \; \mathrm{Tr}\left[ (\slashed{l} - m) \frac{1}{\slashed{l} - \slashed{q} - m} \gamma_{\nu} \frac{1}{\slashed{l} - m} \right] - \int \frac{\mathrm{d}^4 l}{(2\pi)^4} \; \mathrm{Tr}\left( \gamma_{\nu} \frac{1}{\slashed{l} - m} \right) .
\end{aligned}
$$

Now, using the trace cyclicity in the first term, one has

$$iq^{\mu}\Pi_{\mu\nu}(q) = \int \frac{\mathrm{d}^4 l}{(2\pi)^4} \; \mathrm{Tr}\left( \gamma_{\nu} \frac{1}{\slashed{l} - \slashed{q} - m} \right) - \int \frac{\mathrm{d}^4 l}{(2\pi)^4} \; \mathrm{Tr}\left( \gamma_{\nu} \frac{1}{\slashed{l} - m} \right) .\qquad(38.4)$$

Then, upon the shift $l - q \to l$ in the first integral, the first and the second term on the right-hand side of Eq. (38.4) become identical and Eq. (38.3) is thereby proved. Of course, we have used here manipulations with badly divergent integrals, so our "proof" has been rather sloppy, certainly not rigorous.

On the other hand, the transversality property (38.3) is highly desirable for ensuring gauge independence of physical $S$-matrix elements. Such a strong statement clearly calls for an explanation. So, as an instructive example, let us consider the two-loop "sausage-like" diagram contributing (at 6th order) e.g. to our earlier sample process $e + \mu \to e + \mu$, which is shown in Fig. 38.3. If one uses photon propagators in the form (E.16) (see Appendix E), i.e. in the general covariant $\alpha$-gauge, the contribution of the diagram in Fig. 38.3 would certainly be $\alpha$-independent if (38.3) holds. However, if Eq. (38.3) were invalid, the $\alpha$-dependence in the contribution of graph in Fig. 38.3 would persist, because of the propagator 2 in the middle. Notice also that the propagators 1 and 3 are harmless in this context because of familiar identities $\bar{u}(p_1')\slashed{q}u(p_1) = 0$, $\bar{u}(p_2')\slashed{q}u(p_2) = 0$.

So, any definition of the tensor $\Pi_{\mu\nu}(q)$ based on an explicit regularization should maintain the identity (38.3). Employing the general form of $\Pi_{\mu\nu}(q)$ according to (38.2), the relation (38.3) obviously amounts to

$$A(q^2) = -q^2 B(q^2) .\qquad(38.5)$$

227

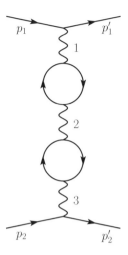

**Fig. 38.3:** An example of a multiloop QED diagram, whose contribution is sensitive to the gauge dependence of photon propagators and to the transversality of the vacuum polarization loop.

Thus, the general form of $\Pi_{\mu\nu}(q)$, satisfying Eq. (38.3), can be written in terms of a single "form factor". In a universally accepted convention it reads

$$\Pi_{\mu\nu}(q) = \Pi(q^2)(q^2 g_{\mu\nu} - q_\mu q_\nu). \tag{38.6}$$

Note also that the form factor $\Pi(q^2)$ is dimensionless, as it should be clear from the integral representation (38.1).

There is another important point that should be mentioned here in connection with the transversality property (38.3). It can be shown that $\Pi_{\mu\nu}(q)$, as given by the expression (38.1), is in fact the Fourier transform of the vacuum expectation value of the T-product of two currents,

$$\langle 0|T\big(J_\mu(x)J_\nu(0)\big)|0\rangle, \tag{38.7}$$

with $J_\mu = \bar\psi\gamma_\mu\psi$. Multiplication of $\Pi_{\mu\nu}(q)$ by $q^\mu$ is thus equivalent to differentiating the expression (38.7) with respect to $x$. However, the current $J_\mu$ is conserved, $\partial^\mu J_\mu = 0$. So, one may write (on the heuristic level we still stick to)

$$\partial^\mu J_\mu(x) = 0 \implies q^\mu \Pi_{\mu\nu}(q) = 0. \tag{38.8}$$

Technical details of the calculation confirming the result (38.8) are left as a non-trivial exercise for an interested reader.

In QFT, identities that are related to the gauge invariance (manifested here in the current conservation) are generally called the **Ward identities**, in honour of the British theorist John C. Ward (1924–2000). So, the transversality relation (38.3) is one of the whole set of QED Ward identities. Note that, historically, the first identity of such a type, discovered by Ward, was a simple relation between quantities $\Sigma$ and $\Gamma_\mu$ introduced in Chapter 37 (see (37.16) and (37.17)); we will discuss it later on, in Chapter 42.

Let us now proceed to the main theme of this chapter, the UV regularization. A regularization of UV divergences in an expression like (38.1) means that the corresponding divergent integral is replaced by a convergent one, at the price of introducing an auxiliary (regularization) parameter, in such a way that by removing it one returns to the original form. This is admittedly rather vague statement, so now we are going to make it explicit. There are essentially three possible strategies for the UV regularization. First, one can replace the infinite integration domain by a finite one, simply by introducing a finite upper bound for the integration over the loop four-momentum. This type of regularization is usually called **momentum cut-off**, for obvious reasons. Such a straightforward procedure is too simple so as to guarantee special relations like e.g. (38.3), i.e. to maintain the symmetry properties of the QFT model in question, but we will see that it can be useful as an auxiliary regularization within a more complicated scheme.

Second, taking into account that the UV divergence is also due to a bad behaviour of the integrand at infinity, one may improve such an asymptotics by modifying appropriately the integrand. For instance, the propagator $1/(l^2 - m^2)$ may be replaced by a subtracted form,

$$\frac{1}{l^2 - m^2} \longrightarrow \frac{1}{l^2 - m^2} - \frac{1}{l^2 - M^2}, \tag{38.9}$$

where $M$ is a "regulator mass". Obviously, the expression (38.9) behaves like $1/(l^2)^2$ for $l \to \infty$ in contrast with the $1/l^2$ behaviour of the original propagator. So, here the regularization parameter is the auxiliary mass $M$ and removing such a regularization means, of course, taking $M \to \infty$. An example of this regularization strategy is the so-called **Pauli–Villars method** (to be discussed in detail later on).

Third, another source of UV divergences is the integration in the four-dimensional space. The integrals in question would, obviously, have better convergence properties in lower-dimensional spaces. To implement this idea, one may derive formulae for the relevant integrals in dependence on the dimensionality $n$ of the integration space. Then one can use an analytic continuation of these formulae to non-integer (or even complex) values of the parameter $n$. The original UV divergence is then revealed as a singularity of such an $n$-dependent regularized expression in the limit $n \to 4$. This last mentioned method is called simply the **dimensional regularization**. It had been invented independently by the Dutch theorists Gerardus 't Hooft and Martinus Veltman, and Argentinians Carlos Bollini and Juan Giambiagi in the early 1970s (but usually it is attributed only to 't Hooft and Veltman, who became much more famous and also received the Nobel Prize in 1999 for their work on gauge theories in particle physics). For the original papers, see [41] and [42]. In fact, the dimensional regularization (DR) is the least intuitive method introduced here, and for a newcomer in QFT it might represent a sort of "shock therapy". Nevertheless, from the point of view of practical calculations, in most cases it is a most efficient and flexible procedure, so we will employ it as the first example of how to deal with the UV divergence appearing in $\Pi_{\mu\nu}(q)$, the contribution of the vacuum polarization graph.

Before implementing the relevant steps of the DR method, let us introduce a general calculational tool (used in any computation of Feynman diagrams), namely the so-called **Feynman parametrization**. Such a trick amounts to replacing a product of propagator denominators by a power of a single expression, at the price of introducing some new integrations over finite domain of "Feynman parameters". The simplest example is the elementary formula

$$\frac{1}{ab} = \int_0^1 dx \, \frac{1}{[ax + b(1-x)]^2}$$

$$\left( = \int_0^1 dx \, \frac{1}{[(a-b)x + b]^2} \right). \tag{38.10}$$

A generalization of (38.10) reads

$$\frac{1}{a_1 a_2 \ldots a_n} = (n-1)! \int_0^1 dx_1 \int_0^{1-x_1} dx_2 \ldots \int_0^{1-x_1-\ldots-x_{n-2}} dx_{n-1}$$

$$\times \frac{1}{\left[(a_1 - a_n)x_1 + \cdots + (a_{n-1} - a_n)x_{n-1} + a_n\right]^n}. \qquad (38.11)$$

A formal proof of the identity (38.11) can be found in many places and it is also an appropriate topic for a tutorial.

A remark concerning Eq. (38.10) is in order here (in fact it could be extended to the general case (38.11) as well). Proving the identity (38.10) is elementary. But one might wonder how can it hold in case that $a$, $b$ are real, and $ab < 0$, while the integral in the right-hand side is apparently positive. This is a right question and the answer is as follows. By $a$ and $b$ one means propagator denominators that have, as we know, the form $l^2 - m^2 + i\epsilon$, $\epsilon > 0$. It turns out that it is precisely the tiny piece $i\epsilon$, which saves the consistency of the relation (38.10). For simplicity, let us consider an almost trivial example, where $a = 1$ and $b = -1$, i.e. in fact

$$a = 1 + i\epsilon, \quad b = -1 + i\epsilon. \qquad (38.12)$$

Then, obviously, the left-hand side of (38.10) is $-1$. On the right-hand side one has

$$I(\epsilon) = \int_0^1 dx \, \frac{1}{(2x - 1 + i\epsilon)^2}. \qquad (38.13)$$

Now it is clear what is the role of $i\epsilon$: without it, the integrand in (38.13) has a singularity at $x = \frac{1}{2}$ and the integral is divergent. In the presence of $i\epsilon$, the integrand is equal to $-1/\epsilon^2$ for $x = \frac{1}{2}$ (so it becomes highly negative!). The evaluation of the integral (38.13) is of course elementary; using the substitution $2x - 1 = y$ for convenience, one gets

$$I(\epsilon) = \frac{1}{2} \int_{-1}^1 \frac{1}{(y + i\epsilon)^2} \, dy = -\frac{1}{2} \left( \frac{1}{1 + i\epsilon} - \frac{1}{-1 + i\epsilon} \right) = -\frac{1}{1 + \epsilon^2},$$

so that $\lim_{\epsilon \to 0} I(\epsilon) = -1$.

Thus, the $i\epsilon$ prescription is welcome for avoiding singularities of integrands consisting of products of field propagators — a typical situation in the evaluation of Feynman diagrams. Note also that there are some further generalizations of the Feynman-parametric formulae (38.10) or (38.11) that can be found easily in the current literature.

Let us now start the evaluation of the $\Pi_{\mu\nu}(q)$ within dimensional regularization. We will consider the diagram in Fig. 38.2, so one may write, integrating formally in $n$ dimensions:

$$i\Pi_{\mu\nu}^{DR}(q) = \mu^{4-n} \int \frac{d^n l}{(2\pi)^n} \, \text{Tr}\left( \gamma_\mu \frac{1}{l - m} \gamma_\nu \frac{1}{l + q - m} \right). \qquad (38.14)$$

Note that we have introduced an additional overall factor $\mu^{4-n}$, where $\mu$ is an arbitrary mass scale; this is done for preserving the right dimensionality of $\Pi_{\mu\nu}(q)$. So, the integral on the right-hand side of (38.14) is

$$I_{\mu\nu}(q) = \int \frac{d^n l}{(2\pi)^n} \, \frac{\text{Tr}[\gamma_\mu(l + m)\gamma_\nu(l + q + m)]}{[(l + q)^2 - m^2](l^2 - m^2)}. \qquad (38.15)$$

Introducing the Feynman parametrization according to (38.10), the expression (38.15) becomes

$$I_{\mu\nu}(q) = \int\limits_0^1 dx \int \frac{d^n l}{(2\pi)^n} \frac{\mathrm{Tr}[\gamma_\mu(\slashed{l}+m)\gamma_\nu(\slashed{l}+\slashed{q}+m)]}{\left([(l+q)^2 - l^2]x + l^2 - m^2\right)^2}$$

$$= \int\limits_0^1 dx \int \frac{d^n l}{(2\pi)^n} \frac{\mathrm{Tr}[\gamma_\mu(\slashed{l}+m)\gamma_\nu(\slashed{l}+\slashed{q}+m)]}{\left[(2l\cdot q + q^2)x + l^2 - m^2\right]^2} . \tag{38.16}$$

The next step is an appropriate shift of the integration variable: since we assume implicitly that the integral (38.16) is now convergent (regularized), a shift of $l$ is legitimate. For this purpose the expression (38.16) is recast as

$$I_{\mu\nu}(q) = \int\limits_0^1 dx \int \frac{d^n l}{(2\pi)^n} \frac{\mathrm{Tr}[\gamma_\mu(\slashed{l}+m)\gamma_\nu(\slashed{l}+\slashed{q}+m)]}{\left[(l+xq)^2 - x^2 q^2 + xq^2 - m^2\right]^2} . \tag{38.17}$$

Now, the relevant shift is $l' = l + xq$, i.e. $l = l' - xq$. Performing this in (38.17) and renaming $l'$ again as $l$, one has

$$I_{\mu\nu}(q) = \int\limits_0^1 dx \int \frac{d^n l}{(2\pi)^n} \frac{\mathrm{Tr}[\gamma_\mu(\slashed{l}-x\slashed{q}+m)\gamma_\nu(\slashed{l}+(1-x)\slashed{q}+m)]}{(l^2 - C)^2} , \tag{38.18}$$

with

$$C = m^2 - x(1-x)q^2 \tag{38.19}$$

(one should keep in mind that $m^2$ in the expression (38.19) is in fact $m^2 - i\epsilon$).

# Chapter 39

# Accomplishing dimensional regularization of $\Pi_{\mu\nu}(q)$

Our starting point is now the integral (38.18) along with the formula (38.19) from the preceding chapter. A great advantage of the form (38.18) is that the denominator of the integrand is an even function of $l$. This enables one to eliminate immediately terms in the numerator that are odd (i.e. linear in $l$). So, working out the matrix trace under the integral in (38.18) in a straightforward way and discarding the odd terms, one gets

$$
I_{\mu\nu}(q) = \int_0^1 dx \int \frac{d^n l}{(2\pi)^n} \frac{\mathrm{Tr}(\gamma_\mu \slashed{l} \gamma_\nu \slashed{l}) - x(1-x)\,\mathrm{Tr}(\gamma_\mu \slashed{q} \gamma_\nu \slashed{q}) + m^2\,\mathrm{Tr}(\gamma_\mu \gamma_\nu)}{(l^2 - C)^2}
$$

$$
= 4 \int_0^1 dx \int \frac{d^n l}{(2\pi)^n} \frac{2l_\mu l_\nu - l^2 g_{\mu\nu} - x(1-x)(2q_\mu q_\nu - q^2 g_{\mu\nu}) + m^2 g_{\mu\nu}}{(l^2 - C)^2}.
$$
(39.1)

A remark is in order here: The overall factor 4 comes from the familiar formulae for traces of products of $\gamma$-matrices in four dimensions (recall that it is the trace of unit 4×4 matrix). In fact, in $n$ dimensions, one should work with $2^{n/2} \times 2^{n/2}$ matrices, so the trace factor would be then $2^{n/2}$. However, such an overall factor does not play any significant role in practical calculations, so usually one employs, as an "operational prescription", the trace factor equal to four.

Now, the first term in the numerator of the integrand in (39.1) can be simplified by using a rule for "symmetric integration", namely by means of the replacement

$$
l_\alpha l_\beta \longrightarrow \frac{1}{n} l^2 g_{\alpha\beta}.
$$
(39.2)

How can one justify the "trick" (39.2)? It is clear that the integral

$$
\int d^n l \, \frac{l_\alpha l_\beta}{(l^2 - C)^2}
$$
(39.3)

must have the tensor form $f(C)g_{\alpha\beta}$. Taking the trace over the indices $\alpha$, $\beta$ and realizing that $g_\alpha{}^\alpha = n$ in $n$-dimensional space, one has

$$
\int d^n l \, \frac{l^2}{(l^2 - C)^2} = f(C)n,
$$

i.e.

$$
f(C) = \frac{1}{n} \int d^n l \, \frac{l^2}{(l^2 - C)^2}.
$$
(39.4)

Returning to the expression (39.3) and using (39.4), one thus has the result

$$\int d^n l \, \frac{l_\alpha l_\beta}{(l^2 - C)^2} = f(C)g_{\alpha\beta} = \frac{1}{n} g_{\alpha\beta} \int d^n l \, \frac{l^2}{(l^2 - C)^2} \, ,$$

and the rule (39.2) is thereby proved.

In this way, the integral (39.1) becomes

$$I_{\mu\nu}(q) = 4 \int_0^1 dx \int \frac{d^n l}{(2\pi)^n} \, \frac{(\frac{2}{n} - 1)l^2 g_{\mu\nu} - 2x(1-x)q_\mu q_\nu + x(1-x)q^2 g_{\mu\nu} + m^2 g_{\mu\nu}}{(l^2 - C)^2} \, . \qquad (39.5)$$

Now, we would like to prove that our regularized expression satisfies the transversality condition (38.3), i.e. that the integral $I_{\mu\nu}(q)$ (which is simply proportional to $\Pi_{\mu\nu}(q)$) has the form (38.6). To this end, let us separate in (39.5) the transverse part and then try to prove that the rest is zero. Thus, the expression (39.5) may be recast as

$$I_{\mu\nu}(q) = 4 \int_0^1 dx \int \frac{d^n l}{(2\pi)^n} \, \frac{2x(1-x)(q^2 g_{\mu\nu} - q_\mu q_\nu) + \left( (\frac{2}{n} - 1)l^2 - x(1-x)q^2 + m^2 \right) g_{\mu\nu}}{(l^2 - C)^2} \, ,$$

$$(39.6)$$

and we would like to prove that

$$\int \frac{d^n l}{(2\pi)^n} \, \frac{[(\frac{2}{n} - 1)l^2 - x(1-x)q^2 + m^2]}{(l^2 - C)^2} = 0 \, . \qquad (39.7)$$

For this purpose, one needs a general formula suitable for the type of integrals appearing in (39.7), which in fact represents the very core of the method of dimensional regularization. It reads

$$\int \frac{d^n l}{(2\pi)^n} \, \frac{(l^2)^r}{(l^2 - C + i\epsilon)^s} = \frac{i}{(4\pi)^{\frac{n}{2}}} (-1)^{r-s} \, C^{r + \frac{n}{2} - s} \, \frac{\Gamma(r + \frac{n}{2}) \Gamma(s - r - \frac{n}{2})}{\Gamma(\frac{n}{2}) \Gamma(s)} \, , \qquad (39.8)$$

where $\Gamma$ denotes the Euler gamma function. Note that the integral in (39.8) is convergent for $s - r > n/2$. Notice also that we have marked explicitly the contribution $i\epsilon$ in the denominator on the left-hand side so as to emphasize the way of avoiding a possible singularity of the integrand. Because of the key position of the relation (39.8) in the whole regularization scheme, it may be aptly dubbed the "master formula" of the DR method. We will prove the formula (39.8) at the end of this chapter, now let us utilize it for proving the identity (39.7). Taking into account that $m^2 - x(1-x)q^2 = C$ (see Eq. (38.19)), the integral in Eq. (39.7) is

$$I = \int \frac{d^n l}{(2\pi)^n} \, \frac{[(\frac{2}{n} - 1)l^2 + C]}{(l^2 - C)^2} \, . \qquad (39.9)$$

Then, using the formula (39.8), one has

$$I = \left( \frac{2}{n} - 1 \right) \frac{i}{(4\pi)^{\frac{n}{2}}} (-1) C^{\frac{n}{2} - 1} \frac{\Gamma(1 + \frac{n}{2}) \Gamma(1 - \frac{n}{2})}{\Gamma(\frac{n}{2}) \Gamma(2)} + \frac{i}{(4\pi)^{\frac{n}{2}}} C \cdot C^{\frac{n}{2} - 2} \frac{\Gamma(\frac{n}{2}) \Gamma(2 - \frac{n}{2})}{\Gamma(\frac{n}{2}) \Gamma(2)} \, . \qquad (39.10)$$

The last expression can be simplified by means of the well-known identity

$$z\,\Gamma(z) = \Gamma(1 + z) \, . \qquad (39.11)$$

233

One thus gets, after a simple manipulation,

$$I = \frac{i}{(4\pi)^{\frac{n}{2}}} \, C^{\frac{n}{2}-1} \Gamma\left(1 - \frac{n}{2}\right) \left[-\left(\frac{2}{n} - 1\right)\frac{n}{2} + \left(1 - \frac{n}{2}\right)\right] = 0 \, .$$

There is another interesting aspect of the above calculation that is worth mentioning here. Along with the elimination of the non-transverse part of the expression (39.6), the original quadratic divergence drops out (note that such a divergence would correspond to the part of the integrand involving $l^2/(l^2 - C)^2$). Thus, the remaining transverse term in Eq. (39.6) is only logarithmically divergent, according to our earlier preliminary classification.

In this way, the integral (39.6) has been reduced to

$$I_{\mu\nu}(q) = 8 \, (q^2 g_{\mu\nu} - q_\mu q_\nu) \int_0^1 dx \, x(1 - x) \int \frac{d^n l}{(2\pi)^n} \frac{1}{(l^2 - C)^2} \, . \tag{39.12}$$

Getting back to the original quantity $\Pi_{\mu\nu}(q)$, this means (cf. (38.6) and (38.14)) that we have

$$i\Pi_{\mu\nu}^{\text{DR}}(q) = i\Pi^{\text{DR}}(q^2)(q^2 g_{\mu\nu} - q_\mu q_\nu) \, ,$$

with

$$i\Pi^{\text{DR}}(q^2) = \mu^{4-n} \, 8 \int_0^1 dx \, x(1 - x) \int \frac{d^n l}{(2\pi)^n} \frac{1}{(l^2 - C)^2} \, . \tag{39.13}$$

One may now carry out the integration over $d^n l$ with the help of the formula (39.8) and we thus get, after a simple manipulation,

$$\Pi^{\text{DR}}(q^2) = 8 \, \frac{1}{(4\pi)^{\frac{n}{2}}} \, \Gamma\left(2 - \frac{n}{2}\right) \int_0^1 dx \, x(1 - x) \left(\frac{C}{\mu^2}\right)^{\frac{n}{2}-2} \, . \tag{39.14}$$

Removing the regularization would mean, formally, performing the limit $n \to 4$, so it makes sense to expand the expression (39.14) in the vicinity of $n = 4$. Its form suggests that it is natural to introduce a parameter

$$\epsilon = 2 - \frac{n}{2} \, , \tag{39.15}$$

i.e. $4 - n = 2\epsilon$ (hopefully, such a notation will not cause confusion as regards the symbol $\epsilon$ appearing in any propagator denominator). So, (39.14) can be recast as

$$\Pi^{\text{DR}}(q^2) = \frac{1}{2\pi^2} \, (4\pi)^\epsilon \, \Gamma(\epsilon) \int_0^1 dx \, x(1 - x) \left(\frac{C}{\mu^2}\right)^{-\epsilon} \, . \tag{39.16}$$

In view of the definition (39.15), the limit $n \to 4$ corresponds to $\epsilon \to 0$. The Euler gamma function $\Gamma(\epsilon)$ has a simple pole at $\epsilon = 0$, and we may utilize the well-known form of Laurent expansion around $\epsilon = 0$:

$$\Gamma(\epsilon) = \frac{1}{\epsilon} - \gamma_{\text{E}} + \mathcal{O}(\epsilon) \, , \tag{39.17}$$

where $\gamma_{\text{E}} \approx 0.577$ is the Euler–Mascheroni constant (note that $\gamma_{\text{E}} = -\Gamma'(1)$). The remaining $\epsilon$-dependent factors in (39.16) can be expanded in Taylor series around $\epsilon = 0$. One has

$$(4\pi)^\epsilon = 1 + \epsilon \ln 4\pi + \mathcal{O}(\epsilon^2) \, ,$$

$$\left(\frac{C}{\mu^2}\right)^{-\epsilon} = 1 - \epsilon \ln \frac{C}{\mu^2} + \mathcal{O}(\epsilon^2) \, . \tag{39.18}$$

234

Putting all this together, one gets finally[24]

$$\Pi^{DR}(q^2) = \frac{1}{2\pi^2} \left[ \frac{1}{6} \left( \frac{1}{\epsilon} - \gamma_E + \ln 4\pi \right) - \int_0^1 dx\, x(1-x) \ln \frac{m^2 - x(1-x)q^2}{\mu^2} + \mathcal{O}(\epsilon) \right]. \quad (39.19)$$

In this way, we have isolated the UV divergence as the pole term $1/\epsilon$. The general strategy of the dimensional regularization should now be clear: we use the master formula (39.8) that is strictly valid for some values of parameters $n$, $r$, $s$ ($s - r > n/2$) and then one can employ its analytic continuation in the parameter $n$. This is quite straightforward, because of the known simple properties of $\Gamma$ function.

As a technical appendix to our conceptual exposition of basic DR techniques, let us now prove the "master formula" (39.8). So, we would like to compute the integral

$$I = \int \frac{d^n l}{(2\pi)^n} \frac{(l^2)^r}{(l^2 - C + i\epsilon)^s}. \quad (39.20)$$

The integration variable $l$ lives in the space with pseudo-Euclidean metric, so that

$$l^2 = l_0^2 - |\vec{l}|^2 = l_0^2 - (l_1^2 + \cdots + l_{n-1}^2). \quad (39.21)$$

In (39.20) one may integrate first over the components of $\vec{l}$ and subsequently over $l_0$. In this last integration along the real axis one avoids potential singularities (poles) thanks to the presence of $i\epsilon$. One may then rotate the real interval $(-\infty, +\infty)$ for $l_0$ to the imaginary axis in complex plane (this should be understandable for anybody with at least basic knowledge of complex analysis). Such a transformation then means that $l_0 = i l_n$, $l_n \in (-\infty, +\infty)$ and $l^2$ in (39.21) becomes

$$l^2 = l_0^2 - |\vec{l}|^2 = -l_E^2, \quad (39.22)$$

with

$$l_E^2 = l_1^2 + \cdots + l_{n-1}^2 + l_n^2 \quad (39.23)$$

(the index E stands here for Euclidean, for obvious reasons). Such a trick is quite common in QFT and is called **Wick rotation** (due to the same G. C. Wick as before). The substitution $l_0 = i l_n$ in the integral of course means that

$$d^n l \longrightarrow i d^n l_E. \quad (39.24)$$

So, we have, as a first step,

$$I = i(-1)^{r-s} \int \frac{d^n l_E}{(2\pi)^n} \frac{(l_E^2)^r}{(l_E^2 + C)^s}. \quad (39.25)$$

For simplicity, let us assume that $C > 0$ throughout our calculation. A most convenient way of the evaluation of the integral (39.25) is based on the exponentiation of the rational function in

---

[24]Note that the factor $1/6$ arises here from the integration over the Feynman parameter, as

$$\int_0^1 dx\, x(1-x) = \frac{1}{6}.$$

the integrand, with the help of the identity

$$\int_0^\infty t^k e^{-tA}\, dt = \frac{k!}{A^{k+1}}$$

(39.26)

$$\left(= \frac{\Gamma(k+1)}{A^{k+1}}\right).$$

Thus, we may write (omitting the index E)

$$I = i(-1)^{r-s} \int \frac{d^n l}{(2\pi)^n} (l^2)^r \frac{1}{\Gamma(s)} \int_0^\infty t^{s-1} e^{-t(l^2+C)}\, dt$$

$$= \frac{i(-1)^{r-s}}{\Gamma(s)} \frac{1}{(2\pi)^n} \int_0^\infty dt\, t^{s-1} e^{-tC} \int d^n l\, (l^2)^r e^{-tl^2}.$$

(39.27)

The integral over $l$ is expressed easily in terms of the Gaussian integral, since it holds, obviously,

$$\int d^n l\, (l^2)^r e^{-tl^2} = (-1)^r \left(\frac{d}{dt}\right)^r \int d^n l\, e^{-tl^2},$$

(39.28)

and

$$\int d^n l\, e^{-tl^2} = \int d^n l\, e^{-t(l_1^2+\cdots+l_n^2)} = \left(\frac{\sqrt{\pi}}{\sqrt{t}}\right)^n.$$

(39.29)

Differentiating the last expression as in (39.28) one gets, after some simple manipulations (and using some well-known properties of the $\Gamma$ function),

$$\int d^n l\, (l^2)^r e^{-tl^2} = \pi^{\frac{n}{2}} \frac{\Gamma(\frac{n}{2}+r)}{\Gamma(\frac{n}{2})} t^{-\frac{n}{2}-r}.$$

(39.30)

Inserting (39.30) into (39.27), one has

$$I = i(-1)^{r-s} \frac{1}{\Gamma(s)} \frac{1}{(2\pi)^n} \pi^{\frac{n}{2}} \frac{\Gamma(\frac{n}{2}+r)}{\Gamma(\frac{n}{2})} \int_0^\infty dt\, t^{s-r-\frac{n}{2}-1} e^{-tC}.$$

(39.31)

Using the familiar definition

$$\Gamma(z) = \int_0^\infty dt\, t^{z-1} e^{-t},$$

(39.32)

one gets easily, via a substitution $tC = u$ in (39.31),

$$I = i(-1)^{r-s} \frac{1}{(4\pi)^{\frac{n}{2}}} C^{r+\frac{n}{2}-s} \frac{\Gamma(r+\frac{n}{2})\Gamma(s-r-\frac{n}{2})}{\Gamma(\frac{n}{2})\Gamma(s)},$$

(39.33)

and the proof is thereby completed.

# Chapter 40

# Pauli–Villars regularization

As we have already indicated in Chapter 38, the Pauli–Villars regularization relies on the strategy of modifying ("deforming") the integrand of an integral representing a closed-loop Feynman diagram (while staying firmly in four dimensions). The simplest variant of such a scheme can be implemented via a propagator subtraction shown in (38.9). Such a procedure is good enough e.g. for the triangular loop in Fig. 37.2 or the diagram in Fig. 37.3, where one can consider the above-mentioned modification of the photon propagator. However, for the purely fermion loop appearing in the vacuum polarization graph one has to employ a more sophisticated approach. The reason is that one would like to maintain the transversality property (38.3); performing a subtraction in just a single propagator in the loop would introduce terms involving different masses in the internal lines attached to the same vertex, and this would certainly destroy the identity (38.3) — it should be clear from our heuristic derivation of the condition in question. For an additional argument, one may realize that the appearance of two different masses in such a loop would mean that it is made of non-conserved currents (recall that the four-divergence of a vector current made of Dirac fields with unequal masses is proportional to their difference).

Wolfgang Pauli and Felix Villars came up with the right (and very natural) solution of this problem in the early days of modern era of QED, in 1949 (for the original paper see [43]). They proposed to subtract the whole fermion loop involving an auxiliary regulator mass — then, obviously, the transversality (38.3) is preserved automatically. More generally, it may happen that more than one subtraction of such a type is needed; as we are going to see immediately, this is precisely the case of the quantity $\Pi_{\mu\nu}(q)$, which is currently our toy object for testing various regularization methods.

So, let us consider the contribution of the familiar loop in Fig. 40.1 and try to regularize its contribution using the above-mentioned basic idea of the Pauli–Villars (PV) method. It amounts to replacing the ill-defined contribution of the diagram in Fig. 40.1 by a *bona fide* convergent

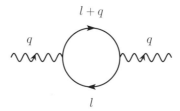

**Fig. 40.1:** Vacuum polarization loop reproduced here for reader's convenience.

integral

$$i\Pi_{\mu\nu}^{\mathrm{PV}}(q) = \int \frac{\mathrm{d}^4 l}{(2\pi)^4} \left[ J(m) - C_1 J(M_1) - C_2 J(M_2) \right], \tag{40.1}$$

where we have denoted

$$J(m) = \frac{\mathrm{Tr}\left[ \gamma_\mu (\slashed{l} + m) \gamma_\nu (\slashed{l} + \slashed{q} + m) \right]}{(l^2 - m^2)[(l + q)^2 - m^2]}, \tag{40.2}$$

and similarly for the other two terms in the integrand of the expression (40.1). We will see that in the considered case, two PV subtractions are sufficient. Thus, the prescription (40.1) can be pictured symbolically as shown in Fig. 40.2.

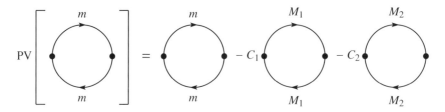

**Fig. 40.2:** Pictorial representation of the Pauli–Villars regularization of the vacuum polarization loop.

One may employ now the Feynman parametrization (38.10) for each term in (40.1) and then shift the integration variable $l$ as before, in Chapter 38 ($l' = l + xq$). Let us stress that such a shift is legitimate, since we work with a convergent integral. From $J(m)$ one then gets (preserving the original notation)

$$J(m) = \frac{\mathrm{Tr}\left[ \gamma_\mu (\slashed{l} - x\slashed{q} + m) \gamma_\nu (\slashed{l} + (1 - x)\slashed{q} + m) \right]}{\left( l^2 - f(m) \right)^2}, \tag{40.3}$$

with

$$f(m) = m^2 - x(1 - x)q^2 \tag{40.4}$$

(cf. (38.19)) and similarly for $J(M_1)$ and $J(M_2)$. Working out the trace in (40.3) and carrying out the symmetric integration (see Chapter 39), the relevant integrand in (40.1) becomes, after some simple manipulations,

$$J = J(m) - C_1 J(M_1) - C_2 J(M_2), \tag{40.5}$$

where

$$J(m) = 4 \left[ -\frac{1}{2} l^2 g_{\mu\nu} - 2x(1 - x)q_\mu q_\nu + x(1 - x)q^2 g_{\mu\nu} + m^2 g_{\mu\nu} \right] \frac{1}{\left( l^2 - f(m) \right)^2}, \tag{40.6}$$

and analogously for $J(M_1)$ and $J(M_2)$.

Consider now the power-like behaviour of the individual terms in the integrand $J$ for $l \to \infty$ and the potential UV divergences they could produce. Obviously, the $l^2$ term in the numerators could lead to a quadratic divergence; so, let us examine first a condition for its elimination. The part of the integrand (40.5) responsible for such a would-be leading UV divergence is proportional to

$$X_{\mathrm{q.\ div.}} = \frac{l^2}{\left( l^2 - f(m) \right)^2} - C_1 \frac{l^2}{\left( l^2 - f(M_1) \right)^2} - C_2 \frac{l^2}{\left( l^2 - f(M_2) \right)^2}. \tag{40.7}$$

238

Let us denote, for brevity,

$$f(m) = f_0, \quad f(M_1) = f_1, \quad f(M_2) = f_2. \tag{40.8}$$

The expression (40.7) can then be written as

$$X_{q.\ div.} = l^2 \frac{(l^2 - f_1)^2(l^2 - f_2)^2 - C_1(l^2 - f_0)^2(l^2 - f_2)^2 - C_2(l^2 - f_0)^2(l^2 - f_1)^2}{(l^2 - f_0)^2(l^2 - f_1)^2(l^2 - f_2)^2}. \tag{40.9}$$

The denominator in (40.9) is of the order $\mathcal{O}(l^{12})$ for $l \to \infty$, so it is clear that a UV divergence can arise from the terms in the numerator that are at least of the order $\mathcal{O}(l^8)$. In other words, contributions of the type $l^2(l^2)^2 f_1^2$, $l^2(l^2)^2 f_1 f_2$, etc. are harmless (i.e. lead to UV convergent integrals). Let us start with the worst possible case, namely the power $l^2(l^2)^2(l^2)^2$ appearing in the numerator. Clearly, the condition for eliminating these terms reads $1 - C_1 - C_2 = 0$, i.e.

$$C_1 + C_2 = 1. \tag{40.10}$$

Next, some logarithmic (next-to-leading) UV divergences would originate in terms of the type $l^2(l^2)^2(-2l^2 f)$. There are plenty of them, a full list is as follows:

$$\begin{aligned}
& l^2(l^2)^2(-2l^2 f_2) + l^2(l^2)^2(-2l^2 f_1) \\
& - C_1\left[l^2(l^2)^2(-2l^2 f_2) + l^2(l^2)^2(-2l^2 f_0)\right] \\
& - C_2\left[l^2(l^2)^2(-2l^2 f_1) + l^2(l^2)^2(-2l^2 f_0)\right].
\end{aligned} \tag{40.11}$$

Now, the factors $f_0$, $f_1$, $f_2$ have the structure (40.4), so the mass-independent terms get cancelled because of the relation (40.10). On the other hand, the mass-dependent terms are proportional to

$$\begin{aligned}
& M_2^2 - C_1 M_2^2 - C_2 M_1^2 \\
& + M_1^2 - C_1 m^2 - C_2 m^2 \\
& = M_1^2 + M_2^2 - C_1 M_2^2 - C_2 M_1^2 - (C_1 + C_2)m^2.
\end{aligned} \tag{40.12}$$

Using now (40.10), the last expression becomes $C_1 M_1^2 + C_2 M_2^2 - m^2$, and the condition for the elimination of the considered next-to-leading terms reads

$$C_1 M_1^2 + C_2 M_2^2 = m^2. \tag{40.13}$$

Completing this discussion, it is not difficult to see that due to the conditions (40.10) and (40.13), also the $l$-independent terms in (40.6) do not contribute to potential UV divergences.

To summarize our results: We have found out that UV divergences disappear from our expression (40.1) if the parameters $C_1$, $C_2$ and $M_1$, $M_2$ satisfy the conditions

$$\begin{aligned}
C_1 + C_2 &= 1, \\
C_1 M_1^2 + C_2 M_2^2 &= m^2.
\end{aligned} \tag{40.14}$$

Obviously, there are many ways of how to satisfy the relations (40.14); one particular choice could be

$$\begin{aligned}
C_1 &= 2, & M_1^2 &= m^2 + \overline{M}^2, \\
C_2 &= -1, & M_2^2 &= m^2 + 2\overline{M}^2.
\end{aligned} \tag{40.15}$$

The conditions (40.14) also show clearly that just a single PV subtraction would not be enough; indeed, such a setting would correspond to $C_2 = 0$ and, consequently, $C_1 = 1$ along with $M_1 = m$, which of course would be useless. Thus, we have verified explicitly the well-known fact that the number of PV subtractions depends generally on the degree of UV divergence of the Feynman diagram in question.

The preceding discussion was, in a sense, just a warm-up exercise, showing that the proposal of Pauli and Villars does work. Now we would like to evaluate explicitly the regularized contribution of $\Pi^{\text{PV}}_{\mu\nu}(q)$ and establish its dependence on the employed regularization parameters.

To this end, it is very helpful to use an auxiliary "intermediate" regularization for the evaluation of the individual integrals appearing in the whole expression for $\Pi^{\text{PV}}_{\mu\nu}$; the reason for such a "side step" (following the mythical Czech genius Jára Cimrman) is that a direct evaluation of the complete integral in Eq. (40.1) using the integrand given by Eqs. (40.5), (40.6) would be a hopeless task. A needed auxiliary tool could be dimensional regularization, but we would rather stick to our original commitment to stay firmly in four dimensions. Thus, we are going to use a direct "momentum cut-off" mentioned earlier as one of the possible regularization strategies. We will need two integrals, namely

$$\int \frac{\mathrm{d}^4 l}{(2\pi)^4} \frac{l^2}{(l^2 - f)^2}, \tag{40.16}$$

and

$$\int \frac{\mathrm{d}^4 l}{(2\pi)^4} \frac{1}{(l^2 - f)^2}, \tag{40.17}$$

with an appropriately chosen integration upper limit. The integrals (40.16) and (40.17) can be computed easily with the help of the Wick rotation (sketched at the end of Chapter 39). A relevant cut-off $\Lambda$ can be then defined as the radius of a hypersphere in four-dimensional Euclidean space, which represents the boundary of the finite integration region. Details of the calculation are left to the reader as an exercise; here let us only show the relevant results, up to inessential terms vanishing for $\Lambda \to \infty$. One gets

$$\int_{(\Lambda)} \frac{\mathrm{d}^4 l}{(2\pi)^4} \frac{l^2}{(l^2 - f)^2} = -\frac{i}{16\pi^2} \left[ \Lambda^2 - 2f \ln \frac{\Lambda^2}{f} + f + \mathscr{O}\left(\frac{1}{\Lambda^2}\right) \right], \tag{40.18}$$

$$\int_{(\Lambda)} \frac{\mathrm{d}^4 l}{(2\pi)^4} \frac{1}{(l^2 - f)^2} = \frac{i}{16\pi^2} \left[ \ln \frac{\Lambda^2}{f} - 1 + \mathscr{O}\left(\frac{1}{\Lambda^2}\right) \right]. \tag{40.19}$$

Now we can employ these formulae in the expression for $\Pi^{\text{PV}}_{\mu\nu}$ in its present form (cf. (40.5), (40.6)). Instead of writing an unbearably long expression, let us try to find out, which contributions will drop out (in fact, we expect some significant cancellations). First of all, it is clear that the terms from (40.18) proportional to $\Lambda^2$ will disappear because of $1 - C_1 - C_2 = 0$ (i.e. the quadratic divergence is cancelled, similarly to the case of dimensional regularization). Analogously, terms $f$ from (40.18) get cancelled owing to the relations (40.14) and the constant $(-1)$ from (40.19) also drops out due to $1 - C_1 - C_2 = 0$. Thus, it is sufficient to take into account the logarithmic terms only. One then gets, returning to the original notation for $f$'s:

$$i\Pi_{\mu\nu}^{\mathrm{PV}} = 4 \int_0^1 dx \left( \left\{ -\frac{1}{2}g_{\mu\nu}\left(-\frac{i}{16\pi^2}\right)\left(-2f(m)\ln\frac{\Lambda^2}{f(m)}\right) \right. \right.$$

$$+ \left[ -2x(1-x)q_\mu q_\nu + x(1-x)q^2 g_{\mu\nu} + m^2 g_{\mu\nu} \right]\frac{i}{16\pi^2}\ln\frac{\Lambda^2}{f(m)} \Big\}$$

$$\left. - C_1\{m \to M_1\} - C_2\{m \to M_2\} \right)$$

$$= \frac{i}{4\pi^2} \int_0^1 dx \left( \left\{ \left[ -(m^2 - x(1-x)q^2)g_{\mu\nu} \right. \right. \right.$$

$$- 2x(1-x)q_\mu q_\nu + x(1-x)q^2 g_{\mu\nu} + m^2 g_{\mu\nu} \Big] \ln\frac{\Lambda^2}{f(m)} \Big\}$$

$$\left. - C_1\{m \to M_1\} - C_2\{m \to M_2\} \right)$$

$$= \frac{i}{2\pi^2} \int_0^1 dx\, x(1-x)(q^2 g_{\mu\nu} - q_\mu q_\nu)\left[ \ln\frac{\Lambda^2}{f(m)} - C_1\ln\frac{\Lambda^2}{f(M_1)} - C_2\ln\frac{\Lambda^2}{f(M_2)} \right].$$

(40.20)

For arriving at a final neat form of the expression $\Pi_{\mu\nu}^{\mathrm{PV}}$, it is useful to write

$$\ln\frac{\Lambda^2}{f(m)} = \ln\frac{\Lambda^2}{m^2} + \ln\frac{m^2}{f(m)}, \qquad \ln\frac{\Lambda^2}{f(M_1)} = \ln\frac{\Lambda^2}{m^2} + \ln\frac{m^2}{f(M_1)}, \qquad \text{etc.}$$

Then,

$$\ln\frac{\Lambda^2}{m^2} - C_1\ln\frac{\Lambda^2}{m^2} - C_2\ln\frac{\Lambda^2}{m^2} = 0,$$

because of $1 - C_1 - C_2 = 0$. In this way, one may say that the auxiliary cut-off has fulfilled its role and disappears from the scene; of course, we are left with a non-trivial dependence on the PV regulator masses $M_1$, $M_2$. Another reassuring feature of (40.20) is that we have recovered the transverse tensor structure — since we know *a priori* that the PV recipe should obey such a rule, we thus have a consistency check of the explicit calculation.

So, we have

$$i\Pi_{\mu\nu}^{\mathrm{PV}}(q) = \frac{i}{2\pi^2}(q^2 g_{\mu\nu} - q_\mu q_\nu)$$

$$\times \int_0^1 dx\, x(1-x)\left[ \ln\frac{m^2}{m^2 - x(1-x)q^2} - C_1\ln\frac{m^2}{M_1^2 - x(1-x)q^2} - C_2\ln\frac{m^2}{M_2^2 - x(1-x)q^2} \right].$$

(40.21)

The regulator masses $M_1$, $M_2$ are supposed to be, in principle, arbitrarily large. It means that one may write e.g.

$$\ln\frac{m^2}{M_1^2 - x(1-x)q^2} = \ln\frac{m^2}{M_1^2} - \ln\left[ 1 - x(1-x)\frac{q^2}{M_1^2} \right]$$

$$= \ln\frac{m^2}{M_1^2} + \mathcal{O}\left(\frac{1}{M_1^2}\right).$$

So, we are allowed to replace the second and third logarithmic terms in (40.21) by $\ln(m^2/M_1^2)$ and $\ln(m^2/M_2^2)$, respectively, and recast (40.21) as

$$
i\Pi_{\mu\nu}^{\text{PV}}(q) = \frac{i}{2\pi^2}\,(q^2 g_{\mu\nu} - q_\mu q_\nu)
$$

$$
\times \int_0^1 dx\, x(1-x)\left[\ln\frac{m^2}{m^2 - x(1-x)q^2} + C_1\ln\frac{M_1^2}{m^2} + C_2\ln\frac{M_2^2}{m^2} + \mathscr{O}\!\left(\frac{1}{M^2}\right)\right], \qquad (40.22)
$$

where $M$ is a generic large mass scale. The two large logarithms involving $M_1$ and $M_2$ can be combined into one term, i.e. one can define a new mass parameter $\overline{M}$ according to

$$
\ln\frac{\overline{M}^2}{m^2} = C_1\ln\frac{M_1^2}{m^2} + C_2\ln\frac{M_2^2}{m^2} \qquad (40.23)
$$

(the reader is recommended to use in the expression (40.23) the earlier choice (40.15) and find an approximate relation between $M$ and $\overline{M}$).

Thus, finally,

$$
\Pi_{\mu\nu}^{\text{PV}}(q) = \Pi^{\text{PV}}(q^2)(q^2 g_{\mu\nu} - q_\mu q_\nu)\,,
$$

with

$$
\Pi^{\text{PV}}(q^2) = \frac{1}{2\pi^2}\int_0^1 dx\, x(1-x)\left[\ln\frac{\overline{M}^2}{m^2} - \ln\frac{m^2 - x(1-x)q^2}{m^2}\right],
$$

i.e.

$$
\Pi^{\text{PV}}(q^2) = \frac{1}{2\pi^2}\left[\frac{1}{6}\ln\frac{\overline{M}^2}{m^2} - \int_0^1 dx\, x(1-x)\ln\frac{m^2 - x(1-x)q^2}{m^2}\right]. \qquad (40.24)
$$

Notice that this result is quite similar to our previous expression obtained within the DR scheme: there is a simple correspondence between the UV divergences in both schemes, namely

$$
\ln\frac{\overline{M}^2}{m^2} \longleftrightarrow \frac{1}{\epsilon}\,, \qquad (40.25)
$$

or, if you want,

$$
\ln\frac{\overline{M}}{m} \longleftrightarrow \frac{1}{4-n}\,. \qquad (40.26)
$$

It was a lot of work, a real *tour de force*, but the result (40.24) is quite elegant and in fact gratifying. One may now also appreciate the flexibility and technical simplicity of the dimensional regularization scheme. Nevertheless, in some situations the DR scheme may encounter specific difficulties. In particular, the $\gamma_5$ matrix needs special treatment within the DR scheme, since, as we know, the identities for traces of products of $\gamma$-matrices involving $\gamma_5$ do not have a simple uniform structure as those without $\gamma_5$. Fortunately, in QED we are spared those subtle problems.

In closing this chapter, an additional general remark is in order. Apart from the examples we have discussed explicitly up to now, there are in fact infinitely many UV regularization schemes with the desirable properties. In particular, one may consider a class of regularization devised in 1980s by Oleg I. Zavialov, which are implemented as "continuous superpositions of the PV cut-offs" (see [44] and references therein). An individual scheme belonging to this class can be defined by means of an integration over the variable PV mass parameter, involving an appropriate wight function (whose form is rather general). A specific peculiar choice of such a weight function enables one to imitate the dimensional regularization (sic!). This amusing example may thus be dubbed aptly "DR in four dimensions". Anyway, the straightforward DR scheme remains to be the most efficient regularization method in perturbative QFT.

# Chapter 41

# $\Sigma(p)$ and all that

---

We are going to examine the other one-loop QED diagrams that we have already encountered in Chapter 37 (as well as some extra species that are yet to be identified). Let us start with the loop shown in Fig. 41.1 (we have reproduced here Fig. 37.3 for reader's convenience). As we have noted in Chapter 37, it is commonly known as the fermion **self-energy diagram** — the origin of such a name will become clear later.

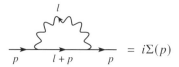

**Fig. 41.1:** Electron self-energy loop.

The original integral representation (37.17) for $\Sigma(p)$ can now be written within the dimensional regularization scheme as

$$i\Sigma^{\text{DR}}(p) = \mu^{4-n} e^2 \int \frac{d^n l}{(2\pi)^n} \gamma_\alpha \frac{1}{\not{l} + \not{p} - m} \gamma_\beta \frac{-g^{\alpha\beta}}{l^2 - \lambda^2} . \tag{41.1}$$

Note that we have included here also the coupling factor $e^2$ (of course, in the same way we might complete our previous result for the vacuum polarization graph). Another important point is to be noticed here: In the photon propagator we have included an auxiliary "fictitious" mass squared $\lambda^2$, since we anticipate later appearance of the **infrared divergence** due to massless photon; this aspect will be clarified when we arrive at the relevant explicit formulae for $\Sigma(p)$. In subsequent manipulations with the expression (41.1) we will set, temporarily, $e = 1$ and restore the coupling factor at the very end of the calculation.

From (41.1) one gets first

$$i\Sigma^{\text{DR}}(p) = -\mu^{4-n} \int \frac{d^n l}{(2\pi)^n} \gamma_\mu \left(\not{l} + \not{p} + m\right) \gamma^\mu \frac{1}{[(l+p)^2 - m^2](l^2 - \lambda^2)} . \tag{41.2}$$

Introducing Feynman parametrization, shifting appropriately the integration variable $l$ ($l' = l + xp$) and utilizing symmetric integration, one has

$$i\Sigma^{\text{DR}}(p) = \mu^{4-n} \int_0^1 dx\ [(n-2)(1-x)\not{p} - nm] \int \frac{d^n l}{(2\pi)^n} \frac{1}{(l^2 - C)^2} , \tag{41.3}$$

where

$$C = xm^2 + (1-x)\lambda^2 - x(1-x)p^2 . \tag{41.4}$$

The reader is encouraged to verify independently the results (41.3) and (41.4). Note that in arriving at (41.3), we have used the identities

$$\gamma_\mu(\slashed{l} + \slashed{p})\gamma^\mu = (2-n)(\slashed{l} + \slashed{p}) ,$$
$$\gamma_\mu\gamma^\mu = n \tag{41.5}$$

valid in $n$ dimensions. One should also mention that the superficial linear divergence apparent in the original expression (41.1) disappears upon the symmetric integration and one is thus left with a logarithmic UV divergence only.

Next, employing the master formula (39.8) and introducing the parameter $\epsilon = 2 - n/2$ instead of $n$, one gets

$$\Sigma(p) = \frac{1}{8\pi^2} (4\pi)^\epsilon \Gamma(\epsilon) \int_0^1 dx \left[ (1-\epsilon)(1-x)\slashed{p} - (2-\epsilon)m \right] \left( \frac{C}{\mu^2} \right)^{-\epsilon} . \tag{41.6}$$

The result thus can be expressed in the form

$$\Sigma^{\mathrm{DR}}(p) = X(p^2) + Y(p^2)\slashed{p} \tag{41.7}$$

(where, in general, $p^2 \neq m^2$), with

$$X(p^2) = -\frac{1}{8\pi^2} m\, (4\pi)^\epsilon\, \Gamma(\epsilon)(2-\epsilon) \int_0^1 dx \left( \frac{C}{\mu^2} \right)^{-\epsilon} ,$$

$$Y(p^2) = \frac{1}{8\pi^2} (4\pi)^\epsilon\, \Gamma(\epsilon)(1-\epsilon) \int_0^1 dx\, (1-x) \left( \frac{C}{\mu^2} \right)^{-\epsilon} ,$$

and, using $\epsilon\Gamma(\epsilon) = \Gamma(1+\epsilon)$, this can be conveniently rewritten as

$$X(p^2) = -\frac{1}{4\pi^2} m\, (4\pi)^\epsilon\, \Gamma(\epsilon) \int_0^1 dx \left( \frac{C}{\mu^2} \right)^{-\epsilon}$$
$$+ \frac{1}{8\pi^2} m\, (4\pi)^\epsilon\, \Gamma(1+\epsilon) \int_0^1 dx \left( \frac{C}{\mu^2} \right)^{-\epsilon} , \tag{41.8}$$

$$Y(p^2) = \frac{1}{8\pi^2} (4\pi)^\epsilon\, \Gamma(\epsilon) \int_0^1 dx\, (1-x) \left( \frac{C}{\mu^2} \right)^{-\epsilon}$$
$$- \frac{1}{8\pi^2} (4\pi)^\epsilon\, \Gamma(1+\epsilon) \int_0^1 dx\, (1-x) \left( \frac{C}{\mu^2} \right)^{-\epsilon} . \tag{41.9}$$

Expanding now the expressions (41.8), (41.9) around $\epsilon = 0$, one obtains

$$X(p^2) = -\frac{1}{4\pi^2} m \left[ \Delta_{UV} - \frac{1}{2} - \int_0^1 dx \, \ln \frac{C}{\mu^2} + \mathcal{O}(\epsilon) \right], \qquad (41.10)$$

$$Y(p^2) = \frac{1}{16\pi^2} \left[ \Delta_{UV} - 1 - 2 \int_0^1 dx \, (1-x) \ln \frac{C}{\mu^2} + \mathcal{O}(\epsilon) \right], \qquad (41.11)$$

where

$$\Delta_{UV} = \frac{1}{\epsilon} - \gamma_E + \ln 4\pi .$$

In general, the loop in Fig. 41.1 may emerge as an insertion into the fermion propagator, for instance in the 4th order diagram for Compton scattering like the one in Fig. 41.2. In other

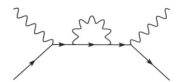

**Fig. 41.2:** A fourth-order contribution to the Compton scattering, involving a one-loop correction to the electron propagator.

words, the quantity $\Sigma(p)$ may play the role of a correction to the fermion propagator (we will discuss this in more detail later on). With this in mind, it is natural to consider an expansion of $\Sigma(p)$ in powers of $\not{p} - m$, writing

$$\Sigma(p) = A + B(\not{p} - m) + C(\not{p} - m)^2 + \dots , \qquad (41.12)$$

where $A, B, \dots$ are constants (i.e. independent of $p^2$). The most economical way of obtaining an expansion like (41.12) is to employ, quite formally, the corresponding Taylor expansion with the coefficients $A, B$, etc. being determined by derivatives of the function $\Sigma(p)$ with respect to $\not{p}$, taken at $\not{p} = m$ (although we know that there is no $p$ such that $\not{p} = m$!). In doing this, one should also take into account that $p^2 = \not{p}\not{p}$. Then, using the decomposition (41.7), one gets

$$A = \Sigma(\not{p})\big|_{\not{p}=m} = X(m^2) + m \, Y(m^2) , \qquad (41.13)$$

$$\begin{aligned} B = \frac{\partial \Sigma}{\partial \not{p}}\bigg|_{\not{p}=m} &= X'(p^2 = m^2) \, 2\not{p}\big|_{\not{p}=m} \\ &\quad + Y'(p^2 = m^2) \, 2\not{p} \cdot \not{p}\big|_{\not{p}=m} \\ &\quad + Y(p^2 = m^2) \cdot \mathbb{1} \\ &= Y(m^2) + 2m \, X'(m^2) + 2m^2 Y'(m^2) . \end{aligned} \qquad (41.14)$$

A watchful reader, who would consider our calculational trick with $\not{p} = m$ suspect, may arrive at the same results by expanding $X(p^2)$ and $Y(p^2)$ around $p^2 = m^2$ and using $p^2 = \not{p}\not{p}$ wherever necessary (in particular, $p^2 - m^2 = (\not{p} - m)(\not{p} + m)$, etc.). In any case, it is also important to notice that the coefficients $A$ and $B$ contain the UV divergence, while $C$ is UV finite (as well as the higher terms). Indeed, for computing the coefficient $C$ one has to take the second derivative

246

of $\Sigma(p)$ with respect to $p$ and the resulting expression then involves only derivatives of $X$ and $Y$; from (41.10), (41.11) it is clear that these are UV finite (since $\Delta_{UV}$ is manifestly $p$-independent). The evaluation of $A$ and $B$ by means of the relations (41.13), (41.14), (41.10) and (41.11) is straightforward; one gets

$$A = -\frac{1}{4\pi^2} m \left( \Delta_{UV} - \frac{1}{2} - \int_0^1 dx \, \ln \frac{(1-x)\lambda^2 + x^2 m^2}{\mu^2} \right)$$

$$+ \frac{1}{16\pi^2} m \left( \Delta_{UV} - 1 - 2 \int_0^1 dx \, (1-x) \ln \frac{(1-x)\lambda^2 + x^2 m^2}{\mu^2} \right) \qquad (41.15)$$

and

$$B = \frac{1}{16\pi^2} \left( \Delta_{UV} - 1 - 2 \int_0^1 dx \, (1-x) \ln \frac{(1-x)\lambda^2 + x^2 m^2}{\mu^2} \right)$$

$$- \frac{1}{2\pi^2} m^2 \int_0^1 dx \, x(1-x) \, \frac{1}{(1-x)\lambda^2 + x^2 m^2}$$

$$+ \frac{1}{4\pi^2} m^2 \int_0^1 dx \, x(1-x)^2 \, \frac{1}{(1-x)\lambda^2 + x^2 m^2} \, . \qquad (41.16)$$

In (41.15) the infrared (IR) regulator $\lambda$ can be removed ($\lambda \to 0$) and the result is simplified considerably. Restoring also the coupling factor $e^2 = 4\pi\alpha$, one has finally

$$A = -\frac{3\alpha}{4\pi} m \left( \Delta_{UV} - \ln \frac{m^2}{\mu^2} + \frac{4}{3} \right) \qquad (41.17)$$

(the reader is encouraged to verify this independently). In the expression (41.16) for the coefficient $B$ the IR divergence persists (the second and third integrals obviously diverge for $\lambda = 0$), but one may simplify the result by using an appropriate expansion in the vicinity of $\lambda = 0$, and discarding the terms that vanish for $\lambda \to 0$. One thus gets, after somewhat tedious manipulations (including then $e^2 = 4\pi\alpha$ as well)

$$B = \frac{\alpha}{4\pi} \left( \Delta_{UV} - \ln \frac{m^2}{\mu^2} + 2 \ln \frac{\lambda^2}{m^2} + 4 \right) . \qquad (41.18)$$

A detailed derivation of the result (41.18) is left as a challenge for a hard-working reader.

Note that such a calculation can be carried out also by using the Pauli–Villars regularization. The results can be found e.g. in the textbook [6]; it is quite remarkable that the UV divergence is reproduced in much the same way as in the case of the vacuum polarization graph. In particular, there is a one-to-one correspondence

$$\Delta_{UV} - \ln \frac{m^2}{\mu^2} \quad \longleftrightarrow \quad \ln \frac{M^2}{m^2} \, .$$

On the other hand, the constant terms are different; for $A$, instead of $4/3$ one has $1/2$ in the PV scheme, and for $B$, 4 is replaced by $9/4$. There is nothing wrong with such a discrepancy; it corresponds to a common experience with regularization procedures of different types.

All this was again quite a long and tedious procedure, but our goal was to exhibit a new feature of the loop calculations in QED, namely the possible appearance of the IR divergence, and this requires a detailed calculation. Moreover, the expansion of $\Sigma(p)$ shown in (41.3) and the coefficients $A$, $B$ will play an important role in our future discussion of the renormalization program in QED.

Let us now proceed to another example of a closed-loop diagram, namely the triangle shown in Fig. 41.3. In this case (usually called, for obvious reasons, the **vertex correction**),

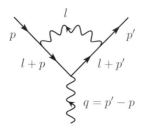

**Fig. 41.3:** One-loop vertex correction in QED.

the UV finite part is quite complicated, and we will not discuss it now; later on, we will evaluate in detail at least a part of it in connection with the famous QED application — the so-called Schwinger correction to the electron magnetic moment. Anyway, the diagram in Fig. 41.3 provides a good opportunity to demonstrate the power and efficiency of the dimensional regularization method for extracting the UV divergence — it turns out that many calculational details can be simply ignored and the road to the UV divergent part of the loop contribution is surprisingly short and straightforward. So, using musical terminology, while the tempo of our previous calculations was "andante", now it will be "allegro", or "allegro moderato". Let us start with the DR form of the contribution of Fig. 41.3 (cf. (37.16)). Setting again provisionally $e = 1$, one has, after a trivial manipulation,

$$i\Gamma_\mu^{\mathrm{DR}}(p',p) = \mu^{4-n} \int \frac{\mathrm{d}^n l}{(2\pi)^n} \frac{\gamma_\alpha(\slashed{l} + \slashed{p}' + m)\gamma_\mu(\slashed{l} + \slashed{p} + m)\gamma^\alpha}{[(l+p')^2 - m^2][(l+p)^2 - m^2](l^2 - \lambda^2)}. \tag{41.19}$$

The relevant Feynman parametrization now reads (cf. (38.11))

$$\frac{1}{abc} = 2 \int_0^1 \mathrm{d}x \int_0^{1-x} \mathrm{d}y \frac{1}{[(a-c)x + (b-c)y + c]^3}.$$

Using this, and performing an appropriate shift of the integration variable in (41.19) (note that such a shift is a linear combination of $p$ and $p'$ involving Feynman parameters $x$, $y$), one is left with an expression that has rather complicated structure for UV finite terms, but the part of the integrand leading to UV divergence is very simple: it is $\gamma_\alpha \slashed{l} \gamma_\mu \slashed{l} \gamma^\alpha$ and nothing more. So, one has

$$i\Gamma_\mu^{\mathrm{DR}}(p',p) = 2\mu^{2\epsilon} \int \frac{\mathrm{d}^n l}{(2\pi)^n} (\gamma_\alpha \slashed{l} \gamma_\mu \slashed{l} \gamma^\alpha + \dots) \int_0^1 \mathrm{d}x \int_0^{1-x} \mathrm{d}y \frac{1}{(l^2 - f)^3}, \tag{41.20}$$

where $f$ is a function of $p^2$, $p'^2$, $p \cdot p'$, $x$, $y$.

Now comes a crucial observation: The integration using our master formula (39.8) produces $\Gamma(2 - n/2) = \Gamma(\epsilon)$, and this is multiplied by factors that are finite for $n = 4$, some of them appearing under the integral over Feynman parameters. Thus, for extracting the UV divergence, i.e. the pole term $1/\epsilon$, one can set $n = 4$ in the factors surrounding $\Gamma(\epsilon)$. Employing in (41.20) the trick of symmetric integration and an identity for $\gamma$-matrices (see (41.5)), one thus has

$$i\Gamma_\mu^{\mathrm{DR}}(p', p) = 2 \cdot (-2)^2 \cdot \frac{1}{4}\, \gamma_\mu \int_0^1 dx \int_0^{1-x} dy \int \frac{d^n l}{(2\pi)^n} \frac{l^2}{(l^2 - f)^3} + \ldots$$

$$= 2\gamma_\mu \frac{i}{16\pi^2} \int_0^1 dx \int_0^{1-x} dy\, f^{\frac{n}{2}-2} \frac{\Gamma(1 + \frac{n}{2})\Gamma(2 - \frac{n}{2})}{\Gamma(\frac{n}{2})\Gamma(3)} + \ldots$$

$$= 2\gamma_\mu \frac{i}{16\pi^2} \Gamma\left(2 - \frac{n}{2}\right) \int_0^1 dx \int_0^{1-x} dy + \ldots$$

$$= \frac{i}{16\pi^2} \Gamma(\epsilon)\gamma_\mu + \ldots$$

(41.21)

In all preceding expressions, the ellipses stand for the UV finite contributions. Notice that a highly welcome simplification of our procedure is due to the fact that $f^{n/2-2}$ is trivialized for $n = 4$, so there are no complications related to the integration over $x$ and $y$.

Thus, the calculation carried out here has led to a remarkably simple result for the UV divergent part of the vertex loop, namely

$$\Gamma_\mu^{\mathrm{DR}}(p', p) = \frac{1}{16\pi^2} \frac{1}{\epsilon} \gamma_\mu + \ldots$$

(41.22)

The tempo was perhaps too fast for a beginner, so I recommend the reader to go through the calculation "da capo al fine" and try to master firmly the DR method, which is so useful for QFT practitioners. In particular, it is useful for anybody to work out the details that we did not need for obtaining the results (41.22), e.g. to find the explicit expression for the shift of the loop momentum $l$ in (41.19), as well as the function $f$ in (41.20). Good luck!

# Chapter 42

# More about QED loops

Before proceeding further, let us make more precise our notation conventions for contributions of QED loops we are dealing with. As it is practical to work sometimes with the relevant quantities stripped of the corresponding coupling factors, it is also useful to introduce special symbols for them: we will use tilde to distinguish such "truncated" contributions from those involving the full coupling factor (i.e. $\widetilde{\Sigma}$ instead of $\Sigma$, with $\Sigma = e^2\widetilde{\Sigma}$, etc.).

Let us now come back to our results for the self-energy and the vertex correction. We have seen that $\widetilde{\Sigma}$ can be expanded like

$$\widetilde{\Sigma}(p) = \widetilde{A} + \widetilde{B}(\not{p} - m) + \text{UV finite terms}, \qquad (42.1)$$

where, among other things,

$$\widetilde{B} = \frac{1}{16\pi^2}\frac{1}{\epsilon} + \dots \qquad (42.2)$$

(note that the coefficient of the pole term $1/\epsilon$ is obtained from (41.18) by dividing it by $4\pi\alpha = e^2$). For the $\widetilde{\Gamma}_\mu(p', p)$ we have

$$\widetilde{\Gamma}_\mu(p', p) = \frac{1}{16\pi^2}\frac{1}{\epsilon}\gamma_\mu + \dots \qquad (42.3)$$

Now, one might think that the coincidence of expressions (42.2) and (42.3) is purely accidental. But it is not. It turns out that it has a deeper origin — it is a simple consequence of the so-called **Ward identity**, as we are going to explain below.

The identity we have in mind reads

$$\frac{\partial}{\partial p^\mu}\widetilde{\Sigma}(p) = \widetilde{\Gamma}_\mu(p, p) \qquad (42.4)$$

(as we have already noted in Chapter 38, it had been first observed by J. C. Ward in 1950, see the paper [45]). A proof of the identity (42.4) is not difficult; it is based on a simple formula for the differentiation of the propagator of Dirac field, namely

$$\frac{\partial}{\partial p^\mu}\frac{1}{\not{p} - m} = -\frac{1}{\not{p} - m}\gamma_\mu\frac{1}{\not{p} - m}. \qquad (42.5)$$

So, let us first verify the relation (42.5). It is elementary, one has just to keep in mind that the derivative of a matrix $M$ does not, in general, commute with $M$. One must therefore start with the definition of inverse matrix

$$(\not{p} - m)(\not{p} - m)^{-1} = 1. \qquad (42.6)$$

250

Differentiating (42.6) one gets

$$\gamma_\mu (\not p - m)^{-1} + (\not p - m) \frac{\partial}{\partial p^\mu} (\not p - m)^{-1} = 0,$$

and thus

$$\frac{\partial}{\partial p^\mu} (\not p - m)^{-1} = -(\not p - m)^{-1} \gamma_\mu (\not p - m)^{-1},$$

which is precisely the identity (42.5).

Now one may recall our previous expressions for $\widetilde{\Sigma}(p)$ and $\widetilde{\Gamma}_\mu(p, p)$, namely

$$i\widetilde{\Sigma}^{\mathrm{DR}}(p) = \mu^{4-n} \int \frac{d^n l}{(2\pi)^n} \gamma_\alpha \frac{1}{\not l + \not p - m} \gamma_\beta \frac{-g^{\alpha\beta}}{l^2 - \lambda^2}, \tag{42.7}$$

$$i\widetilde{\Gamma}_\mu^{\mathrm{DR}}(p, p) = -\mu^{4-n} \int \frac{d^n l}{(2\pi)^n} \gamma_\alpha \frac{1}{\not l + \not p - m} \gamma_\mu \frac{1}{\not l + \not p - m} \gamma_\beta \frac{-g^{\alpha\beta}}{l^2 - \lambda^2} \tag{42.8}$$

(just to be sure: let us remind the reader that the minus sign in (42.8) is $i^6$ coming from three vertices and three propagators, while in (42.7) the corresponding "bookkeeping factor" is $i^4 = 1$). Taking into account the elementary identity (42.5), the validity of (42.4) is obvious, since the differentiation of $(\not l + \not p - m)^{-1}$ in (42.7) with respect to $p^\mu$ leads precisely to the matrix chain in (42.8), including the necessary minus sign.

So, how can one utilize the Ward identity (42.4) for elucidating the coincidence of the pole terms in (42.2) and (42.3)? In fact, it is quite obvious now. Differentiating the expression (42.1), one gets

$$\frac{\partial}{\partial p^\mu} \widetilde{\Sigma}(p) = \widetilde{B} \gamma_\mu + \text{UV finite terms}, \tag{42.9}$$

and so, according to (42.4), one also has

$$\widetilde{\Gamma}_\mu(p, p) = \widetilde{B} \gamma_\mu + \text{UV finite terms}. \tag{42.10}$$

Since we know that the UV divergence in $\widetilde{\Gamma}_\mu(p', p)$ does not depend on the external momenta $p$, $p'$, the validity of the results (42.9) and (42.10) is sufficient for our argument.

Let us remark that our derivation of the Ward identity (42.4) in fact does not depend on an explicit use of dimensional regularization. It is clear that it would be equally valid within the conventional Pauli–Villars scheme (or any other that does not deform fermion propagators). Therefore we may work formally with the expressions for $\widetilde{\Sigma}$ and $\widetilde{\Gamma}_\mu$, in which the regularization operation is suppressed.

The discussion of the Ward identity can be extended so as to get it in a form that involves the full vertex function $\widetilde{\Gamma}_\mu(p', p)$ with $p' \neq p$ in general. The resulting relation was derived first by Yoshio Takahashi in 1957 (see [46]) and thus carries the name **Ward–Takahashi identity**. Let us demonstrate how such an identity emerges from the structure of one-loop QED diagrams we are dealing with.

After a trivial manipulation, $\widetilde{\Gamma}_\mu(p', p)$ has the form (as we have stressed, the dimension of the integration space is now irrelevant)

$$i\widetilde{\Gamma}_\mu(p', p) = \int \frac{d^4 l}{(2\pi)^4} \gamma_\alpha \frac{1}{\not l + \not p' - m} \gamma_\mu \frac{1}{\not l + \not p - m} \gamma^\alpha \frac{1}{l^2 - \lambda^2}, \tag{42.11}$$

and for $\widetilde{\Sigma}(p)$ we have

$$i\widetilde{\Sigma}(p) = - \int \frac{d^4 l}{(2\pi)^4} \gamma_\alpha \frac{1}{\not l + \not p - m} \gamma^\alpha \frac{1}{l^2 - \lambda^2}. \tag{42.12}$$

251

Now, multiplying the expression (42.11) by $q^\mu$, where $q = p' - p$, one gets

$$iq^\mu \widetilde{\Gamma}_\mu(p', p) = \int \frac{d^4 l}{(2\pi)^4} \gamma_\alpha \frac{1}{\slashed{l} + \slashed{p}' - m} (\slashed{p}' - \slashed{p}) \frac{1}{\slashed{l} + \slashed{p} - m} \gamma^\alpha \frac{1}{l^2 - \lambda^2} . \qquad (42.13)$$

One may now write

$$\slashed{p}' - \slashed{p} = (\slashed{l} + \slashed{p}' - m) - (\slashed{l} + \slashed{p} - m),$$

and utilizing this in Eq. (42.13), the propagator denominators can be partially cancelled, so that one gets

$$iq^\mu \widetilde{\Gamma}_\mu(p', p) = \int \frac{d^4 l}{(2\pi)^4} \left( \gamma_\alpha \frac{1}{\slashed{l} + \slashed{p} - m} \gamma^\alpha - \gamma_\alpha \frac{1}{\slashed{l} + \slashed{p}' - m} \gamma^\alpha \right) \frac{1}{l^2 - \lambda^2} . \qquad (42.14)$$

Thus, taking into account (42.12), the identity (42.14) can be obviously recast as

$$q^\mu \widetilde{\Gamma}_\mu(p', p) = \widetilde{\Sigma}(p') - \widetilde{\Sigma}(p) . \qquad (42.15)$$

The relation (42.15) is just the famous Ward–Takahashi (WT) identity. It is not difficult to find out that the original Ward identity (42.4) can be obtained from the WT identity by means of the Taylor expansion of both sides of Eq. (42.15) in powers of $q$. Indeed, on the left-hand side one then has

$$q^\mu \widetilde{\Gamma}_\mu(p', p) = q^\mu \widetilde{\Gamma}_\mu(p, p) + \mathcal{O}(q^2) , \qquad (42.16)$$

while the right-hand side of (42.15) gives

$$\widetilde{\Sigma}(p') - \widetilde{\Sigma}(p) = \frac{\partial}{\partial p^\mu} \widetilde{\Sigma}(p) \, q^\mu + \mathcal{O}(q^2) . \qquad (42.17)$$

So, comparing the terms of the first order in $q^\mu$ one gets immediately the relation (42.4).

One more remark is in order here. If the vertex triangular loop is a part of a Feynman graph with external lines carrying on-shell four-momenta $p$ and $p'$ (and the plane-wave amplitudes $u(p)$, $\bar{u}(p')$), the matrix function $\widetilde{\Gamma}_\mu(p', p)$ is then sandwiched between $\bar{u}(p')$ and $u(p)$; the multiplication by $q^\mu$ then gives, according to the WT identity,

$$q^\mu \bar{u}(p') \widetilde{\Gamma}_\mu(p', p) u(p) = \bar{u}(p') \left[ \widetilde{\Sigma}(p') - \widetilde{\Sigma}(p) \right] u(p) . \qquad (42.18)$$

Next, using the decomposition (42.1) for $\widetilde{\Sigma}(p')$ and $\widetilde{\Sigma}(p)$, one has

$$\begin{aligned} \widetilde{\Sigma}(p') - \widetilde{\Sigma}(p) &= \widetilde{B}(\slashed{p}' - m) - \widetilde{B}(\slashed{p} - m) \\ &+ \widetilde{C}(\slashed{p}' - m)^2 - \widetilde{C}(\slashed{p} - m)^2 + \dots \end{aligned} \qquad (42.19)$$

So, the right-hand side of (42.18) obviously vanishes, because of $\bar{u}(p')(\slashed{p}' - m) = 0$, $(\slashed{p} - m)u(p) = 0$, and one is left with the relation

$$q^\mu \bar{u}(p') \widetilde{\Gamma}_\mu(p', p) u(p) = 0 , \qquad (42.20)$$

which looks like "current conservation" in momentum representation (in this context, recall the familiar identity $q^\mu \bar{u}(p') \gamma_\mu u(p) = \bar{u}(p') \slashed{q} u(p) = 0$ that we used repeatedly in our previous calculations). The identity (42.20) will be useful later, in the calculation of the celebrated Schwinger correction to the spin magnetic moment of electron.

252

As we have already noted before, the transversality of the vacuum polarization tensor $\Pi_{\mu\nu}(q)$ is an example of a Ward identity, in general sense. So, now we have two examples of this kind, at the level of one-loop QED diagrams. A general systematic analysis of WT identities and their connection with gauge invariance of QED would go beyond the scope of the present text; an interested reader can find the relevant deeper exposition of the subject e.g. in the book [6] (Chapter 8, section 8.4.1 therein).

Up to now we have discussed in detail three one-loop diagrams that indeed belong to the standard "QED household", and we have used them primarily for practicing techniques of the UV regularization. Now, a legitimate question could be: How about other possible one-loop diagrams? In particular, can one encounter also UV finite loops? In closing this chapter, let us mention briefly at least two new specific examples; the remaining more substantial stuff will be a subject of the next chapter.

In chapters 36 and 37, we have in fact missed the opportunity to notice a curious closed loop that could, naïvely, also be present within the set of 4th order graphs contributing to our sample scattering process $e + \mu \rightarrow e + \mu$. The example in question may be represented by the funny picture shown in Fig. 42.1.

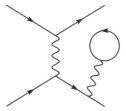

**Fig. 42.1:** A would-be tadpole contribution to the 4th order matrix element for $e–\mu$ scattering.

When saying "naïvely", one means a mechanical usage of the "pictographic script" of Feynman diagrams, involving internal lines, external lines and familiar vertices specified in Chapter 35. In fact, a diagram like that in Fig. 42.1 is excluded automatically owing to the Wick's theorem T3. Indeed, one should keep in mind that in our definition of the QED interaction Lagrangian, the current has the form : $\bar{\psi}(x)\gamma_\mu\psi(x)$ : and the loop in Fig. 42.1 would involve a contraction of field operators at the same spacetime point; however, they occur inside the normal product and such contractions are omitted, according to the cited Wick's theorem. It is quite amusing that the contribution of such a loop would, in fact, vanish even without the intervention of the Wick's theorem (the reader is encouraged to prove this statement independently; hint: don't forget that for the evaluation of a closed fermion loop, matrix trace is involved). The funny loop in Fig. 42.1 has its special name in the QFT literature: it is called **tadpole** because of its similarity with the junior (larval) stage of a frog.[25]

The second example to be mentioned is the box diagram shown in Fig. 42.2. It is easy to guess that its contribution is UV finite. Indeed, taking into account the asymptotic behaviour of the propagators, the relevant integrand corresponding to the loop in Fig. 42.2 behaves like

$$l^3 \, dl \cdot l^{-1} \cdot l^{-1} \cdot l^{-2} \cdot l^{-2}$$

for $l \rightarrow \infty$, i.e. summarily we have $dl \cdot l^{-3}$ (so, in a sense, the integral over $l$ is "more convergent than necessary"). Later on, in our discussion of the renormalization program in QED, we will appreciate greatly the UV finiteness of the loop in Fig. 42.2, which involves four external fermion

---

[25] In Czech: pulec, in Slovak: žubrienka.

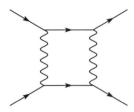

**Fig. 42.2:** A QED box diagram that is manifestly UV finite.

lines. In the next chapter, we will discover more sophisticated examples of UV finite diagrams, namely the higher purely fermion loops.

# Chapter 43

# Fate of higher fermionic loops

In chapters 38, 39 and 40, we have processed the by now familiar vacuum polarization graph in rather detailed manner and in various ways (perhaps almost *ad nauseam*). Now it is straight-forward to introduce other close relatives of the bubble shown in Fig. 40.1, namely the purely fermionic loops with more than two vertices — a triangle, a square ("box"), etc. This is what is meant by "higher fermionic loops" in the title of the present chapter. It is also easy to imagine how such loops could emerge within Feynman diagrams for physical scattering amplitudes. For instance, let us consider the two-photon annihilation of an electron–positron pair. In the second order, there are the familiar tree-level diagrams shown in Fig. 43.1.

(a)             (b)

**Fig. 43.1:** Tree-level QED graphs for the process $e^+e^- \to \gamma\gamma$.

In the 4th order, there is a lot of diagrams; among them, one could consider the one in Fig. 43.2, where "cross." means crossing of the photon lines in analogy with Fig. 43.1.

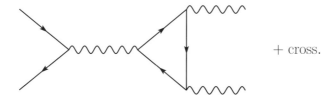

$+ \text{ cross.}$

**Fig. 43.2:** A fourth-order QED graph contributing to the process $e^+e^- \to \gamma\gamma$.

Further, within the 4th order of QED one could contemplate possible diagrams for a process involving four real photons. Obviously, there is only one type (topology) of such a diagram, namely the box shown in Fig. 43.3. A detailed analysis of the 4th order $S$-matrix contribution shows that one must include all six permutations of three external photon lines, while keeping the fourth one fixed. Similarly, one could consider the process $\gamma\gamma \to \gamma\gamma\gamma$, which would proceed via a pentagon loop, in the 5th order in QED. So, there is a hierarchy of fermionic

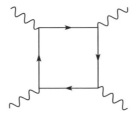

**Fig. 43.3:** The fourth-order QED graph describing the elastic photon–photon scattering.

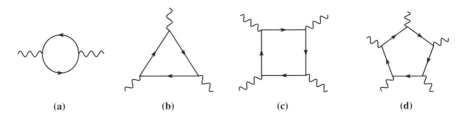

(a)  (b)  (c)  (d)

**Fig. 43.4:** Examples of purely fermionic QED loops with external photon lines.

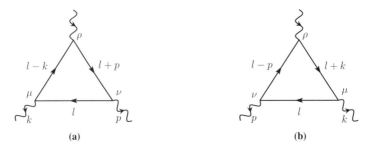

(a)  (b)

**Fig. 43.5:** Pure fermionic triangles in QED.

loops shown in Fig. 43.4, and our goal is to explore properties of the other members of the family appearing there, beyond the familiar vacuum polarization bubble.

Let us start with the triangle. Considering the 4th order annihilation diagram in Fig. 43.2, it is not difficult to realize that the part of its contribution involving the loop in question amounts to $T_{\rho\mu\nu}(k,p)\epsilon^{*\mu}(k)\epsilon^{*\nu}(p)$, where $k$, $p$ denote photon four-momenta and $T_{\rho\mu\nu}(k,p)$ is, due to the crossing, given by the sum of two graphs, namely those shown in Fig. 43.5. It means that

$$T_{\rho\mu\nu}(k,p) = \Gamma_{\rho\mu\nu}(k,p) + \Gamma_{\rho\nu\mu}(p,k)\,, \tag{43.1}$$

where, symbolically (i.e. without invoking an explicit regularization),

$$\Gamma_{\rho\mu\nu}(k,p) = \int d^4l \ \mathrm{Tr}\left(\frac{1}{\slashed{l}-\slashed{k}-m}\gamma_\mu \frac{1}{\slashed{l}-m}\gamma_\nu \frac{1}{\slashed{l}+\slashed{p}-m}\gamma_\rho\right), \tag{43.2}$$

so that

$$\Gamma_{\rho\nu\mu}(p,k) = \int d^4l \ \mathrm{Tr}\left(\frac{1}{\slashed{l}-\slashed{p}-m}\gamma_\nu \frac{1}{\slashed{l}-m}\gamma_\mu \frac{1}{\slashed{l}+\slashed{k}-m}\gamma_\rho\right). \tag{43.3}$$

Note that the crossing operation (43.1) is usually called "Bose symmetrization". Let us also remark that throughout this chapter we set $e = 1$, since the coupling factors are irrelevant for our

256

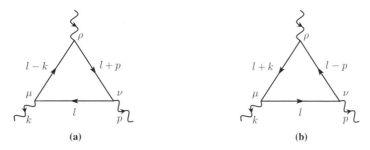

**Fig. 43.6:** Topologically inequivalent triangle loops with mutually reversed orientation.

present purpose. In the same way, we suppress the overall factor (-1) that otherwise belongs to any closed fermionic loop. For the sake of brevity, we also write simply $d^4l$ instead of the usual $d^4l/(2\pi)^4$. Concerning the convergence properties of the considered loops, they are obviously linearly divergent, since the integrand behaves like $l^3 dl/(l^{-1})^3 = dl$.

It is easy to see that the expression $\Gamma_{\rho\nu\mu}(p, k)$ can equivalently be obtained by reversing the direction of the loop momenta flowing through the first diagram in Fig. 43.5. In other words, instead of the pair from Fig. 43.5 we can use the one shown in Fig. 43.6 (note that, conventionally, we read the contribution of the loops by running against the arrows in fermion internal lines).

Let us now compare the contributions of loops (a) and (b) in Fig. 43.6. We have

$$\Gamma_{\rho\mu\nu}(k, p) = \int d^4l \, \frac{\text{Tr}[(\slashed{l} - \slashed{k})\gamma_\mu \slashed{l}\gamma_\nu (\slashed{l} + \slashed{p})\gamma_\rho] + \dots}{[(l - k)^2 - m^2](l^2 - m^2)[(l + p)^2 - m^2]}, \tag{43.4}$$

$$\Gamma_{\rho\nu\mu}(p, k) = \int d^4l \, \frac{\text{Tr}[(\slashed{l} - \slashed{p})\gamma_\nu \slashed{l}\gamma_\mu (\slashed{l} + \slashed{k})\gamma_\rho] + \dots}{[(l + k)^2 - m^2](l^2 - m^2)[(l - p)^2 - m^2]}. \tag{43.5}$$

In the last two expressions, the ellipsis represents terms proportional to $m^2$, which involve just four $\gamma$-matrices inside the trace; for the moment, we are going to deal with the leading mass-independent terms only. The structure of the expression (43.5) suggests that the substitution $l \to -l$ might be useful. So, doing this and using trace cyclicity, one gets

$$\Gamma_{\rho\nu\mu}(p, k) = \int d^4l \, \frac{(-1)^3 \, \text{Tr}[\gamma_\rho (\slashed{l} + \slashed{p})\gamma_\nu \slashed{l}\gamma_\mu (\slashed{l} - \slashed{k})] + \dots}{[(l - k)^2 - m^2](l^2 - m^2)[(l + p)^2 - m^2]}. \tag{43.6}$$

To proceed further, let us recall a remarkable "palindromic" identity for the trace of a product of $\gamma$-matrices, namely

$$\text{Tr}(\gamma_\alpha \gamma_\beta \dots \gamma_\tau \gamma_\omega) = \text{Tr}(\gamma_\omega \gamma_\tau \dots \gamma_\beta \gamma_\alpha), \tag{43.7}$$

i.e. the trace is not changed when the order of $\gamma$-matrices is reversed (see (C.10) in Appendix C, or Chapter 5; its proof follows (5.44)).

So, with the identity (43.7) at hand, the expression (43.6) can be recast as

$$\Gamma_{\rho\nu\mu}(p, k) = -\int d^4l \, \frac{\text{Tr}[(\slashed{l} - \slashed{k})\gamma_\mu \slashed{l}\gamma_\nu (\slashed{l} + \slashed{p})\gamma_\rho]}{[(l - k)^2 - m^2](l^2 - m^2)[(l + p)^2 - m^2]} + \dots, \tag{43.8}$$

which makes it clear that the leading term in $\Gamma_{\rho\nu\mu}(p, k)$ is exactly opposite to its counterpart in $\Gamma_{\rho\mu\nu}(k, p)$. In other words, the leading terms (involving six $\gamma$-matrices) in the sum (43.1) are mutually cancelled.

In fact, one can proceed in much the same way in the case of terms that are proportional to $m^2$ and find out that they are cancelled in the sum (43.1) as well (the reader is encouraged to verify independently such a statement).

It is important to notice that our analysis admits a generalization for any closed fermionic loop with an odd number of vertices. Indeed, there have been two main ingredients in our calculation: first, the substitution $l \to -l$ that brings about the factor $(-1)^n$, with $n$ being the number of the internal lines in the loop (and this of course coincides with the number of vertices). Second, the $\gamma$-matrix identity (43.7). In this way, one can prove that the contribution of a fermionic loop with an odd number of vertices is exactly cancelled by its counterpart with the reverse orientation of the loop momenta. There are some further generalizations of such a statement, but we will not need them now. In QED, the exact cancellation of odd fermionic loops that we have just observed is called the **Furry's theorem**, in honour of Wendell Furry (1907–1984) who discovered it in 1937.

To conclude this part of the story, the main message of the Furry's theorem is that the odd members of the family in Fig. 43.4 can be discarded completely. Among other things, the elimination of the triangular loops that, at first sight, could generate linear UV divergences, plays an important role in the implementation of the renormalization program in QED.

Let us now proceed to the box diagram sketched in Fig. 43.3. As we have already noted, for its full contribution one has to take into account six permutations of three external momenta, with a fourth one fixed. Let us display the box diagram once again, now equipped with all labels necessary for the explicit calculation. So, we will select as a "basic configuration" the picture shown in Fig. 43.7. Conventionally, we have taken all external momenta as outgoing, so that $k_1 + k_2 + k_3 + k_4 = 0$.

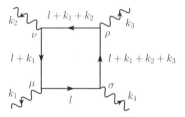

**Fig. 43.7:** A detailed labelling of the four-momenta appearing in the fermionic box diagram.

Our primary interest is a possible UV divergence that might originate in these box diagrams. Obviously, the diagram in Fig. 43.7 is *a priori* logarithmically divergent, since the relevant integrand behaves like $l^3 \, dl \, (l^{-1})^4 = dl \, l^{-1}$ for $l \to \infty$. So, our approach will consist in computing the UV divergence descending from the diagram in Fig. 43.7 for the given setting of vertex indices, and then perform the "Bose symmetrization" with respect to indices $\nu, \rho, \sigma$.

We are going to consider the fermionic loop itself, i.e. stripped of the external photon lines (we also suppress all irrelevant overall factors, similarly as in the preceding example of the triangle loop). Thus, we start with

$$i\Gamma_{\mu\nu\rho\sigma}(k_1, k_2, k_3, k_4) = \int \frac{d^n l}{(2\pi)^n} \frac{\mathrm{Tr}[(\slashed{l} + m)\gamma_\mu(\slashed{l} + \slashed{p}_1 + m)\gamma_\nu(\slashed{l} + \slashed{p}_2 + m)\gamma_\rho(\slashed{l} + \slashed{p}_3 + m)\gamma_\sigma]}{(l^2 - m^2)[(l + p_1)^2 - m^2][(l + p_2)^2 - m^2][(l + p_3)^2 - m^2]} ,$$
$$(43.9)$$

where $p_1 = k_1$, $p_2 = k_1 + k_2$, $p_3 = k_1 + k_2 + k_3$. For the evaluation of the anticipated UV divergence we choose the tempo "allegro moderato", in analogy with our calculation of the vertex function in Chapter 41. It means that we are going to preserve only those parts of the full contribution, which are absolutely necessary for extracting the pole factor $1/\epsilon$ within the DR scheme. In particular, we will set $n = 4$ in factors multiplying $1/\epsilon$ whenever possible.

To begin with, we introduce Feynman parametrization and carry out a pertinent shift of the loop momentum $l$; as a result of performing these routine steps one gets first (suppressing also the factor $1/(2\pi)^n$ in the integrand)

$$i\Gamma_{\mu\nu\rho\sigma} = \int d^n l \int dX \ \mathrm{Tr}(l\gamma_\mu l\gamma_\nu l\gamma_\rho l\gamma_\sigma + \ldots) \frac{1}{(l^2 - f)^4}, \qquad (43.10)$$

where we have used an abbreviation $\int dX$ for the full integration over Feynman parameters, namely

$$\int dX = 3! \int_0^1 dx \int_0^{1-x} dy \int_0^{1-x-y} dz,$$

and the ellipsis denotes, as usual, the irrelevant terms. The quantity $f$ in the denominator is a function of the mass $m$, external momenta and Feynman parameters, but we will not need its explicit value (however, hard-working readers are encouraged to strain their muscles to evaluate $f$, at least for on-shell photon momenta $k_i^2 = 0$, $i = 1, 2, 3, 4$).

The next step is the symmetric integration; one has to find out what becomes of

$$I^{\alpha\beta\iota\kappa} = \int d^n l \ l^\alpha l^\beta l^\iota l^\kappa \frac{1}{(l^2 - f)^4}. \qquad (43.11)$$

In fact, it is not difficult to get the answer. The integral $I^{\alpha\beta\iota\kappa}$ is a completely symmetric tensor that can be expressed solely in terms of components of the metric tensor (there is not any other vector or tensor at play; of course, $f$ is a scalar). So, denoting for brevity the denominator in the expression (43.11) as $D(l^2)$, we may write first

$$I^{\alpha\beta\iota\kappa} = F(g^{\alpha\beta}g^{\iota\kappa} + g^{\alpha\iota}g^{\beta\kappa} + g^{\alpha\kappa}g^{\beta\iota}), \qquad (43.12)$$

and contracting the indices $\alpha$, $\beta$, one gets

$$\int d^n l \ \frac{l^2 l^\iota l^\kappa}{D(l^2)} = F(ng^{\iota\kappa} + g^{\iota\kappa} + g^{\iota\kappa}) \qquad (43.13)$$

$$= (n+2)Fg^{\iota\kappa}.$$

Contracting now $\iota$ and $\kappa$ in (43.13), one has

$$\int d^n l \ \frac{(l^2)^2}{D(l^2)} = n(n+2)F,$$

so that

$$F = \frac{1}{n(n+2)} \int d^n l \ \frac{(l^2)^2}{D(l^2)}. \qquad (43.14)$$

Thus, we have finally

$$\int d^n l \ \frac{l^\alpha l^\beta l^\iota l^\kappa}{D(l^2)} = \frac{1}{n(n+2)} \int d^n l \ \frac{(l^2)^2}{D(l^2)} (g^{\alpha\beta}g^{\iota\kappa} + g^{\alpha\iota}g^{\beta\kappa} + g^{\alpha\kappa}g^{\beta\iota}).$$

In other words, a rule for the symmetric integration in the tensor expression (43.11) reads

$$l_\alpha l_\beta l_\iota l_\kappa \longrightarrow \frac{1}{n(n+2)} (l^2)^2 (g_{\alpha\beta}g_{\iota\kappa} + g_{\alpha\iota}g_{\beta\kappa} + g_{\alpha\kappa}g_{\beta\iota}). \qquad (43.15)$$

Of course, in our accelerated process of evaluating the UV divergence one can replace the factor $1/n(n+2)$ with $1/24$. Using the rule (43.15) in (43.10) we have

$$
\begin{aligned}
i\Gamma_{\mu\nu\rho\sigma} \\
&= \frac{1}{24} \int dX \int d^n l \, \frac{(l^2)^2}{(l^2-f)^4} \, \mathrm{Tr}(\gamma_\alpha\gamma_\mu\gamma_\beta\gamma_\nu\gamma_\iota\gamma_\rho\gamma_\kappa\gamma_\sigma)(g^{\alpha\beta}g^{\iota\kappa}+g^{\alpha\iota}g^{\beta\kappa}+g^{\alpha\kappa}g^{\beta\iota}) + \ldots \\
&= \frac{1}{24} \int dX \int d^n l \, \frac{(l^2)^2}{(l^2-f)^4} \\
&\quad \times \Big[ \mathrm{Tr}(\gamma_\alpha\gamma_\mu\gamma^\alpha\gamma_\nu\gamma_\iota\gamma_\rho\gamma^\iota\gamma_\sigma) + \mathrm{Tr}(\gamma_\alpha\gamma_\mu\gamma_\beta\gamma_\nu\gamma^\alpha\gamma_\rho\gamma^\beta\gamma_\sigma) + \mathrm{Tr}(\gamma_\alpha\gamma_\mu\gamma_\beta\gamma_\nu\gamma^\beta\gamma_\rho\gamma^\alpha\gamma_\sigma) \Big] + \ldots
\end{aligned}
\tag{43.16}
$$

One may now use the identities for chains of $\gamma$-matrices, namely

$$
\begin{aligned}
\gamma_\alpha\gamma_\mu\gamma^\alpha &= (2-n)\gamma_\mu \,, \\
\gamma_\alpha\gamma_\mu\gamma_\nu\gamma^\alpha &= 4g_{\mu\nu} + (n-4)\gamma_\mu\gamma_\nu \,, \\
\gamma_\alpha\gamma_\mu\gamma_\nu\gamma_\rho\gamma^\alpha &= -2\gamma_\rho\gamma_\nu\gamma_\mu + (n-4)\gamma_\mu\gamma_\nu\gamma_\rho \,,
\end{aligned}
$$

in their simplified form for $n = 4$; note that in the last trace in (43.16), also the palindromic identity (43.7) is eventually utilized. The integration over $l$ with the help of the master formula (39.8) obviously generates $\Gamma(\epsilon)$ and $f$ appears in the factor of $f^{n/2-2}$. Thus, ignoring all inessential factors, one may write down the UV divergence in question as

$$
i\Gamma_{\mu\nu\rho\sigma}\Big|_{\text{UVdiv}} = \text{const.} \, \frac{1}{\epsilon} \Big[ (-2)^2 \, \mathrm{Tr}(\gamma_\mu\gamma_\nu\gamma_\rho\gamma_\sigma) + (-2) \, \mathrm{Tr}(\gamma_\nu\gamma_\beta\gamma_\mu\gamma_\rho\gamma^\beta\gamma_\sigma) + (-2)^2 \, \mathrm{Tr}(\gamma_\mu\gamma_\nu\gamma_\rho\gamma_\sigma) \Big] \,,
$$

so that finally

$$
\Gamma_{\mu\nu\rho\sigma}\Big|_{\text{UVdiv}} = \text{const.} \, \frac{1}{\epsilon}(g_{\mu\nu}g_{\rho\sigma} + g_{\mu\sigma}g_{\nu\rho} - 2g_{\mu\rho}g_{\nu\sigma})
\tag{43.17}
$$

(needless to say, "const." here is different from the preceding one). The aforementioned Bose symmetrization means adding to the expression (43.17) the other five terms involving permutations of the indices $\nu$, $\rho$, $\sigma$. For better transparency, let us denote $\mu \equiv 1$, $\nu \equiv 2$, $\rho \equiv 3$, $\sigma \equiv 4$. The permutations of the expression in parentheses in (43.17) are collected in Table 43.1 and we have shown there the cancellations among the terms in the sum of UV divergent contributions of the form (43.17). Thus, it turns out that the box diagram is effectively UV *finite*.

| permutation | expression in parentheses (43.17) |
|:---:|:---:|
| 2 3 4 | $g_{12}\,g_{34} + g_{14}\,g_{23} - 2g_{13}\,g_{24}$ |
| 3 4 2 | $g_{13}\,g_{42} + g_{12}\,g_{34} - 2g_{14}\,g_{32}$ |
| 4 2 3 | $g_{14}\,g_{23} + g_{13}\,g_{42} - 2g_{12}\,g_{43}$ |
| 3 2 4 | $g_{13}\,g_{24} + g_{14}\,g_{32} - 2g_{12}\,g_{34}$ |
| 2 4 3 | $g_{12}\,g_{43} + g_{13}\,g_{24} - 2g_{14}\,g_{23}$ |
| 4 3 2 | $g_{14}\,g_{32} + g_{12}\,g_{43} - 2g_{13}\,g_{42}$ |

**Table 43.1:** Cancellations among UV divergent terms in permutations of (43.17).

This is again a favourable circumstance from the point of view of the QED renormalization program, as we will see later. Moreover, it means that e.g. the process of photon–photon elastic scattering is **calculable** in QED. More about this later.

# Chapter 44

# Index of UV divergence of 1PI diagram

Up to now, we have discussed one-loop diagrams in QED. Our conclusions can be summarized as follows. There are three types of UV divergent loops, namely those shown in Fig. 44.1, while

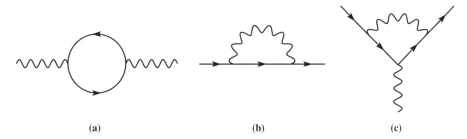

**(a)**                **(b)**                **(c)**

**Fig. 44.1:** "Canonical" UV divergent one-loop diagrams in QED.

some other potentially suspicious species eventually turn out to be UV finite or even trivial. In particular, it is the case of pure fermionic loops: the triangle drops out completely (due to the Furry's theorem) and for the box (square) diagrams the UV divergent parts vanish (while the finite part survives and gives a calculable contribution to the intriguing process of photon–photon scattering). Needless to say, even higher fermionic loops (hexagon, etc.) are obviously UV convergent. This valuable knowledge is for good and we will utilize it amply later on.

However, one might tackle a more ambitious goal: to explore the convergence properties of higher-order diagrams, in principle multiloop ones, as well. An analysis of such a problem is the main topic of this chapter; note that we follow here closely the exposition of the book [6].

To begin with, we must introduce a basic definition: A **one-particle-irreducible (1PI) Feynman diagram** is such that it does not fall into two disjoint pieces by cutting one internal line. For an illustration: the diagram shown in Fig. 44.2 is obviously 1PI, while the one depicted in Fig. 44.3 is one-particle-reducible (to see this, please cut the middle wavy line).

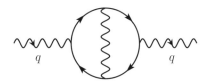

$q$                                    $q$

**Fig. 44.2:** 1PI two-loop vacuum polarization graph in QED.

261

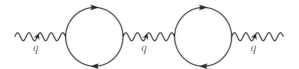

**Fig. 44.3:** An example of a one-particle-reducible graph.

Let us now consider the contribution of a general 1PI graph $G$. It is represented by an integral

$$\Gamma = \int \mathrm{d}^4 l_1 \dots \mathrm{d}^4 l_L \, J(l_1, \dots, l_L; p_{\mathrm{ext}}, m) \,, \tag{44.1}$$

where the integration variables are the independent loop momenta, $p_{\mathrm{ext}}$ denotes collectively the external momenta and $m$ stands for masses in propagators. The index $L$ is, by definition, the number of closed loops. A terminological remark is perhaps in order here: Looking at the diagram in Fig. 44.2, one might be tempted to say that it involves three closed loops (and in a purely geometrical sense it is true). However, we classify it as a two-loop graph, since there are only two independent integration variables (loop momenta) for its contribution: one may choose them e.g. as belonging to the internal photon line and one of the internal fermion lines — the other loop momenta are then already fixed in terms of these two and the external momentum $q$. It should also be clear that if we work with 1PI diagrams, integration variables are present in any internal line (an obvious counterexample is provided by the picture shown in Fig. 44.3). Precisely this is the crucial feature of 1PI diagrams, as regards the analysis we are going to carry out.

Our previous experience with convergence properties of one-loop graphs can be extended and reformulated in the following way: In the UV asymptotic region of the integration variables $l_1, \dots, l_L$ the integrand in (44.1) is a homogeneous function, since the dependence on external momenta and masses can be neglected there. Including also the integration volume elements $\mathrm{d}^4 l_1 \dots \mathrm{d}^4 l_L$, one may employ the degree of homogeneity of the whole expression as an indicator of UV convergence properties of the integral in Eq. (44.1). We will denote such a number as $\omega(G)$ and call it the **index of divergence** of the graph $G$ (note that in many textbooks it is also called "superficial degree of divergence"). Our previous results for one-loop diagrams suggest that $\omega(G) < 0$ would correspond to a convergent integral, and $\omega(G) \geq 0$ to a divergent one; for $\omega(G) = 0$ the UV divergence is logarithmic, for $\omega(G) = 1$ it is linear, etc. (let us remind the reader that for a one-loop logarithmically divergent graph the basic structure of the integrand for $l \to \infty$ is $\mathrm{d}l/l$, etc.).

Now, let us consider a general QFT model involving bosons and fermions with polynomial interaction terms (including, in general, also field derivatives). For simplicity, we are going to assume first that all boson propagators behave like $1/l^2$ for $l \to \infty$; a straightforward generalization will follow later on. So, what is the value of $\omega(G)$? Just to be sure, let us recall that the degree of homogeneity is obtained from rescaling the relevant variables by a factor, say, $\lambda$ and finding the corresponding power of $\lambda$. In the case of the integrand in (44.1) one takes into account that each boson propagator contributes $\lambda^{-2}$, each fermion (Dirac) propagator gives $\lambda^{-1}$, any $\mathrm{d}^4 l_i$, $i = 1, \dots, L$ will produce $\lambda^4$, and one must also consider positive powers of loop momenta generated possibly by derivatives acting in some interaction vertices. With all this in mind, we can write down the value of $\omega(G)$ as

$$\omega(G) = 4L - 2I_{\mathrm{B}} - I_{\mathrm{F}} + \sum_v \delta_v \,, \tag{44.2}$$

where $I_B$ is the number of boson internal lines (propagators), $I_F$ means the same for fermions and there is a sum over vertices $v$, with $\delta_v$ denoting the power of a loop momentum generated by derivatives operating in the vertex $v$ (i.e. acting on the internal lines attached to such a vertex). Further, it holds

$$L = I - V + 1 , \tag{44.3}$$

where $I$ is the total number of internal lines, $I = I_F + I_B$, and $V$ is the number of vertices of the diagram. The relation (44.3) means that there are $V$ constraints (four-momentum conservation) on the loop momenta appearing in the internal lines, but one such constraint represents just the overall conservation of the external momenta. Thus, the expression (44.2) can be recast, after a trivial manipulation, as

$$\omega(G) - 4 = 3I_F + 2I_B + \sum_v (\delta_v - 4) . \tag{44.4}$$

Now, $I_F$ and $I_B$ can be expressed as

$$I_F = \frac{1}{2} \sum_v f_v ,$$
$$I_B = \frac{1}{2} \sum_v b_v , \tag{44.5}$$

where $f_v$ is the number of fermion lines attached to the vertex $v$, and $b_v$ means the same for boson lines (note that the factor $1/2$ in (44.5) is included there so as to avoid double counting — every internal line has two ends inside the diagram). So, we have

$$\omega(G) - 4 = \sum_v \left( \frac{3}{2} f_v + b_v + \delta_v - 4 \right) . \tag{44.6}$$

Next, $f_v$, $b_v$ and $\delta_v$ can be written as

$$f_v = n_{F;v} - E_{F;v} ,$$
$$b_v = n_{B;v} - E_{B;v} , \tag{44.7}$$
$$\delta_v = n_{D;v} - E_{D;v} ,$$

where $n_{F;v}$ means the full number of fermion lines attached to the vertex $v$ (of course, it is the number of fermion fields appearing in the interaction term corresponding to the vertex $v$) and $E_{F;v}$ is the number of external fermion lines attached to the vertex $v$. An analogous notation holds for boson lines and, concerning the last line in (44.7), $n_{D;v}$ denotes the number of derivatives in the interaction term corresponding to vertex $v$ and $E_{D;v}$ means the number of derivatives "acting on external lines" (this amounts to the factorization of a power of external momenta). From (44.6) and (44.7) we thus get

$$\omega(G) - 4 = \sum_v (\omega_v - 4) - \left( \frac{3}{2} E_F + E_B + E_D \right) , \tag{44.8}$$

where the "index of interaction vertex" $\omega_v$ is

$$\omega_v = \frac{3}{2} n_{F;v} + n_{B;v} + n_{D;v} \tag{44.9}$$

and $E_F = \sum_v E_{F;v}$ is the total number of external fermion lines; $E_B$ means the same for boson lines and $E_D$ is the complete power of external momenta factorized in the diagram contribution due to the operation of derivatives appearing the interaction terms.

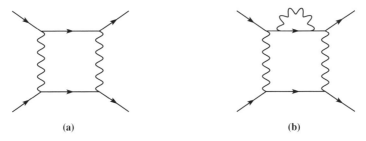

**Fig. 44.4:** (a) UV convergent box diagram, (b) the box diagram involving an UV divergent subdiagram.

The results (44.8), (44.9) are quite remarkable, but there is also one caveat: in the interpretation of the value of $\omega(G)$ as the criterion for UV convergence or divergence we have, in fact, tacitly assumed that the diagram in question does not involve UV divergent subdiagrams. This can be illustrated on an example of the QED diagrams shown in Fig. 44.4. In QED, obviously, $\omega_v = 4$, since $n_F = 2$, $n_B = 1$ and $n_D = 0$. Thus, according to the formula (44.8) we get for both diagrams in Fig. 44.4 the value $\omega(G) = -2$. While for the diagram in Fig. 44.4(a) the conclusion about UV convergence is correct, for the diagram (b) we observe a UV divergence hidden in the self-energy subdiagram inserted in an internal fermion line.[26]

Thus, for the integral representing the diagram (b) it is more appropriate to use the term "conditional convergence", which is quite common in the literature (cf. e.g. the book [6]).

In any case, the formulae (44.8), (44.9) are very valuable for providing us with the information about the possible number of types of UV divergent 1PI diagrams. In this connection, the value $\omega_v = 4$ is in a sense critical. Indeed, if for a given QFT model one has at least one interaction term with $\omega_v > 4$, it is clear that for any chosen configuration of external lines (i.e. for chosen values of $E_F$ and $E_B$) one can get an UV divergent graph by going to a sufficiently high order of perturbation expansion. On the other hand, if for any $v$ one has $\omega_v = 4$, or even $\omega_v < 4$, the number of types of UV divergent graphs is finite.

Another point is supposed to be obvious: the simple "power counting" leading to the formula (44.8) is dealing just with individual diagrams and cannot reflect e.g. the cancellation of some UV divergences within groups of diagrams (cf. the case of the fermion box in QED).

One more remark is in order here. The value of $\omega_v$ is seen to coincide with the "mass dimension" of the corresponding term in the interaction Lagrangian, stripped of the coupling constant (i.e. simply the dimension of the relevant monomial in participating fields). Indeed, we know that the dimension of Dirac field is $3/2$, for any boson field it is 1 and the derivative has the dimension of inverse length, which is also 1. Complete Lagrangian density, i.e. also the interaction terms (including coupling constants) has the dimension four; thus, the critical value $\omega_v = 4$ corresponds to dimensionless coupling constant.

The derivation of the formulae (44.8), (44.9) can be generalized to the case where bosons are represented by a massive vector (Proca) field. In such a case, the boson propagator has

---

[26]It is fair to admit that our simple analysis of the convergence properties of Feynman graphs, which has led us to the key quantity $\omega(G)$, has been somewhat sloppy. A detailed treatment of the problem and a relevant rigorous theorem concerning the convergence issue in question can be found e.g. in the book [6] (see the Section 8.1.3 therein). In any case, an important result of the rigorous analysis is that a non-negative value of $\omega(G)$ definitely signifies UV divergent 1PI diagram.

different scaling properties: its form is

$$D_{\mu\nu}(l) = \frac{-g_{\mu\nu} + \dfrac{l_\mu l_\nu}{M^2}}{l^2 - M^2} \,, \tag{44.10}$$

so that for $l \to \infty$ it behaves like $\mathcal{O}(1)$, i.e. the zeroth power of $l$. Thus, for $\omega(G)$ one may now write, instead of (44.2),

$$\omega(G) = 4L - I_F + \sum_\nu \delta_\nu \,. \tag{44.11}$$

The following steps are the same as before and one ends up with the formula

$$\omega(G) - 4 = \sum_\nu (\omega_\nu - 4) - \left( \frac{3}{2} E_F + 2 E_B + E_D \right), \tag{44.12}$$

where now

$$\omega_\nu = \frac{3}{2} n_{F;\nu} + 2 n_{B;\nu} + n_{D;\nu} \,. \tag{44.13}$$

In the most general case, where both types of bosons are present, one gets

$$\omega_\nu = \frac{3}{2} n_{F;\nu} + n_{B;\nu}^{(1)} + 2 n_{B;\nu}^{(2)} + n_{D;\nu} \,, \tag{44.14}$$

where (1) and (2) refer to the bosons of the 1st type (like the photon or a scalar) and 2nd type (Proca fields), respectively.

Finally, for an illustration of our general formulae, let us discuss some explicit examples.

1) **QED**

As we have already noted, here $\omega_\nu = 4$, so that for any 1PI graph one has, using the formula (44.8), $\omega(G) = 4 - \frac{3}{2} E_F - E_B$. The readers are encouraged to identify a full set of configurations $(E_F, E_B)$ for which $\omega(G) \geq 0$, and confront their findings with our previous knowledge concerning one-loop diagrams.

2) **QED involving Pauli term**

As we have noted in Chapter 13, if one wants to describe a spin-$1/2$ particle with an essentially arbitrary magnetic moment, one has to include the so-called Pauli term in the Dirac equation. In field theory, this amounts to adding the (gauge invariant) interaction term

$$\mathscr{L}_{\text{Pauli}} = f \bar{\psi} \sigma^{\mu\nu} \psi F_{\mu\nu} \tag{44.15}$$

to the basic QED Lagrangian. It is easy to see that the coupling constant $f$ has the dimension of an inverse mass. The numbers relevant for the evaluation of the corresponding index of interaction vertex (44.9) are $n_F = 2$, $n_B = 1$ and $n_D = 1$ so that $\omega_\nu = 5$. This indicates that within such an "extended QED" one may anticipate an unlimited proliferation of types of UV divergent graphs in higher orders of perturbation expansion. It turns out that it is indeed so — their number becomes infinite.

3) **QED with massive photon**

In such a case, one should use the formulae (44.12), (44.13). Thus, one gets $\omega_\nu = 5$ and this suggests that the number of types of UV divergent graphs is infinite. However, it turns out (it is a non-trivial finding) that there are many cancellations and the resulting picture is essentially the same as in ordinary QED.

265

**Fig. 44.5:** Interaction vertices in the scalar QED.

4) **Scalar QED**

This describes the interaction of a charged scalar and the photon. Here we have two types of interaction vertices, namely those shown in Fig. 44.5.

For the vertex (a) one has $n_B = 2$, $n_D = 1$ (the interaction term $ie\varphi^* \overset{\leftrightarrow}{\partial^\mu} \varphi A_\mu$ involving the derivative), while for (b) $n_B = 4$, $n_D = 0$ (the "seagull" term $e^2\varphi^* \varphi A_\mu A^\mu$) and, of course, for both cases $n_F = 0$. Thus, according to (44.9), both for (a) and (b) one has $\omega_v = 4$.

5) **Direct four-fermion interaction**

Schematically, take $\mathcal{L}_{int} = G(\bar{\psi}_1 \psi_2)(\bar{\psi}_2 \psi_1)$ (which is a simplified prototype of the weak interaction of the electron and neutrino). Here $n_F = 4$, $n_B = 0$, $n_D = 0$, so that $\omega_v = 6$. In this case, the number of types of UV divergent graphs is indeed infinite; the types of diagrams involving UV divergences proliferate with increasing order of perturbation expansion (i.e. with the increasing number of vertices in the sum appearing in (44.8)).

6) **Standard model of electroweak interactions**

One encounters there, among other things, self-interactions of massive vector bosons ($W^\pm$, $Z$) and these are of two types — trilinear (involving one derivative) and quadrilinear (without derivative). Moreover, charged vector bosons $W^\pm$ also interact with photons. For such interactions one then has, according to our formulae:

$$\omega_{WW\gamma} = 2 \cdot 2 + 1 + 1 = 6,$$
$$\omega_{WWZ} = 2 \cdot 3 + 1 \qquad = 7,$$
$$\omega_{WWWW} = 2 \cdot 4 \qquad = 8,$$
$$\omega_{WW\gamma\gamma} = 2 \cdot 2 + 2 \qquad = 6.$$

Thus, it might seem that the number of types of UV divergences is infinite. However, there are delicate cancellations due to the underlying gauge invariance, and when the dust settles, the resulting situation is similar as in QED. An alternative way of understanding the remarkable phenomenon of "miraculous" cancellations is a sophisticated reformulation of the theory in such a way that the propagators of massive vector bosons also behave like $1/l^2$ for $l \to \infty$, so that the simple formula (44.8) is applicable (for details, see e.g. [22]).

In summary, one might be tempted to say now that a subtitle for this chapter could read "UV divergences wherever you look". In such a situation, Eric Cartman from South Park would probably exclaim: "Screw you guys! I'm going home." But we persist. In the forthcoming chapters, we will try to cope with the problem of UV divergences by developing the program of the **renormalization** in QED.

# Chapter 45

# Renormalization in QED: preliminary considerations

Let us come back to our sample scattering process $e(k) + \mu(p) \rightarrow e(k') + \mu(p')$. The second order matrix element corresponding to the familiar tree-level diagram has the form

$$\mathcal{M}^{(2)} = e^2 \big[\bar{u}(p')\gamma_\alpha u(p)\big] \frac{g^{\alpha\beta}}{q^2} \big[\bar{u}(k')\gamma_\beta u(k)\big], \tag{45.1}$$

where $q = k' - k \ (= p - p')$. The 4th order correction due to the vacuum polarization loop corresponds to the diagram shown in Fig. 45.1. Using our conventional notation, its contribution

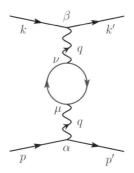

**Fig. 45.1:** Vacuum polarization insertion into the photon propagator in the matrix element for $e$–$\mu$ scattering.

can be written as

$$\mathcal{M}_a^{(4)} = e^2 \big[\bar{u}(p')\gamma_\alpha u(p)\big] \frac{g^{\alpha\mu}}{q^2} \big(-\Pi_{\mu\nu}(q)\big) \frac{g^{\nu\beta}}{q^2} \big[\bar{u}(k')\gamma_\beta u(k)\big], \tag{45.2}$$

where another factor of $e^2$ is included in $\Pi_{\mu\nu}(q)$. Taking into account the transversality property of $\Pi_{\mu\nu}(q)$, i.e.

$$\Pi_{\mu\nu}(q) = \Pi(q^2)(q^2 g_{\mu\nu} - q_\mu q_\nu), \tag{45.3}$$

and using the identity $\bar{u}(p')\slashed{q}u(p) = 0$, or $\bar{u}(k')\slashed{q}u(k) = 0$ (due to the Dirac equation) one gets, after some simple manipulations,

$$\mathcal{M}_a^{(4)} = \mathcal{M}^{(2)} \cdot \big(-\Pi(q^2)\big). \tag{45.4}$$

267

So, it is surprisingly simple, isn't it? Nevertheless, the main problem lies ahead. We know how to regularize the UV divergent form factor $\Pi(q^2)$, but the question is what is the right candidate for a "true" contribution of the expression like (45.4); obviously, a physical scattering amplitude depending on a regularization parameter would be unsatisfactory.

For the exploration of such a problem it will be useful to utilize our earlier "tilde" notation for quantities stripped of the coupling factors (see chapter 42). Then, we may write

$$\mathscr{M}^{(2)} + \mathscr{M}_a^{(4)} = e^2\left(1 - e^2\widetilde{\Pi}(q^2)\right).\widetilde{\mathscr{M}}^{(2)}. \tag{45.5}$$

Obviously, the UV divergence may be singled out in the form factor $\widetilde{\Pi}(q^2)$ for any particular value of $q^2$, e.g. $q^2 = 0$. Splitting $\widetilde{\Pi}(q^2)$ into $\widetilde{\Pi}(0)$ and the UV finite part $\widetilde{\Pi}(q^2) - \widetilde{\Pi}(0)$, Eq. (45.5) may be recast as

$$\begin{aligned}\mathscr{M}^{(2)} + \mathscr{M}_a^{(4)} &= e^2\left(1 - e^2\widetilde{\Pi}(0)\right).\widetilde{\mathscr{M}}^{(2)} \\ &\quad - e^4\left(\widetilde{\Pi}(q^2) - \widetilde{\Pi}(0)\right).\widetilde{\mathscr{M}}^{(2)}.\end{aligned} \tag{45.6}$$

Now we are ready to make a crucial step forward. The first line in (45.6) suggests that we might try to reinterpret the original coupling constant $e$ in such a way that the whole coefficient multiplying $\widetilde{\mathscr{M}}^{(2)}$ is identified with the physical coupling constant squared, i.e. to define a "renormalized" coupling constant $e_R$ by

$$e_R^2 = e^2\left(1 - e^2\widetilde{\Pi}(0)\right). \tag{45.7}$$

Thus, $e$ is now considered to be an unphysical parameter, usually called **bare coupling constant** (in Czech: **holá vazbová konstanta**). Our trick of introducing $e_R$ instead of $e$ is therefore a reinterpretation of the relevant parameter in the interaction Lagrangian and the subsequent reparametrization of the expression for the scattering matrix element in question.

Let us now return to the relation (45.7). One would like to express $e^2$ in terms of $e_R^2$. It is almost trivial; but, just to be sure, we will do it explicitly here.

So, denoting $e^2$ for brevity as $x$, one has to solve the quadratic equation

$$\widetilde{\Pi}(0)x^2 - x + e_R^2 = 0. \tag{45.8}$$

Of course, it has two solutions, namely

$$x_{1,2} = \frac{1}{2\widetilde{\Pi}(0)}\left(1 \pm \sqrt{1 - 4e_R^2\widetilde{\Pi}(0)}\right). \tag{45.9}$$

It is clear how to select the right solution. We wish, naturally, that $e^2 = 0$ also means $e_R^2 = 0$ and vice versa. So, we choose

$$x = \frac{1}{2\widetilde{\Pi}(0)}\left(1 - \sqrt{1 - 4e_R^2\widetilde{\Pi}(0)}\right). \tag{45.10}$$

A remark is in order here. Having in mind that $\widetilde{\Pi}(0)$ contains the UV divergence, one might worry about a negative number under the square root sign, resulting in complex values of $x$. Please, don't be afraid! In our calculation, we always keep the regularization parameter fixed, and its value can be taken such that $e_R^2\widetilde{\Pi}(0) \ll 1$. So, without further misgivings, we may expand the square root in (45.10) to get, after some simple manipulations,

$$e^2 = e_R^2\left(1 + e_R^2\widetilde{\Pi}(0)\right) + \mathcal{O}(e_R^6). \tag{45.11}$$

Next, in the second line of (45.6) one may now simply replace $e^4$ with $e_R^4$ and one thus has, to the order $\mathcal{O}(e_R^4)$,

$$\mathcal{M}^{(2)} + \mathcal{M}_a^{(4)} = e_R^2 \cdot \widetilde{\mathcal{M}}^{(2)} - e_R^4 \big( \widetilde{\Pi}(q^2) - \widetilde{\Pi}(0) \big) \cdot \widetilde{\mathcal{M}}^{(2)} + \mathcal{O}(e_R^6). \tag{45.12}$$

What we have done up to now is just a first hint of the renormalization procedure. Of course, the diagram in Fig. 45.1 is not the only 4th order contribution to the considered process, so we cannot yet make a full statement about the meaning of $e_R$ as the physical measurable coupling constant. Nevertheless, our first step into the area of renormalization theory gives us useful insight into the conceptual foundations of the whole method.

Next, we are going to discuss one more example of this kind, namely the renormalization of fermion mass. For this purpose, we will consider the fermion (e.g. electron) self-energy loop inserted into the propagator in a 4th order Feynman diagram contributing to the Compton scattering, as shown in Fig. 41.2. In this picture, the self-energy loop is sandwiched between two Dirac field propagators, so that the interior of the diagram in Fig. 41.2 is proportional to $(\not{p} - m)^{-1} \Sigma(p) (\not{p} - m)^{-1}$. Before proceeding further, let us recall a matrix identity that will come in handy in our calculation. The identity in question is as follows. Suppose that $X, Y$ are non-singular matrices such that $X - Y$ is non-singular as well; then

$$(X - Y)^{-1} = X^{-1} + X^{-1} Y X^{-1} + X^{-1} Y X^{-1} Y X^{-1} + \dots \tag{45.13}$$

The proof of Eq. (45.13) is easy. Indeed, one may write

$$(X - Y)^{-1} = \Big( X(1 - X^{-1}Y) \Big)^{-1} = (1 - X^{-1}Y)^{-1} X^{-1}$$
$$= (1 + X^{-1}Y + X^{-1}Y X^{-1}Y + \dots) X^{-1}, \tag{45.14}$$

and there you are. Note that in arriving at the last line in (45.14) we have utilized a standard formula for geometric series.

When the diagram in Fig. 41.2 is added to the familiar 2nd order (tree-level) graph, one is dealing with "corrected propagator"

$$= \frac{i}{\not{p} - m} + \frac{i}{\not{p} - m} i\Sigma(p) \frac{i}{\not{p} - m}$$

$$= i \left( \frac{1}{\not{p} - m} + \frac{1}{\not{p} - m} (-\Sigma(p)) \frac{1}{\not{p} - m} \right),$$

and this, using the identity (45.13), can be written as

$$i \frac{1}{\not{p} - m + \Sigma(p)} + \mathcal{O}(e^4) \tag{45.15}$$

(the point is that $\Sigma(p)$ is of the order $\mathcal{O}(e^2)$; for the time being we do not distinguish here between bare and renormalized couplings).

The quantity $\Sigma(p)$ has been examined in detail in chapters 41 and 42, where we have used the expansion

$$\Sigma(p) = A + B(\not{p} - m) + \dots, \tag{45.16}$$

with the remainder "$\dots$" that can be written, for convenience, as

$$C(\not{p} - m)^2 + \dots = C(p)(\not{p} - m) \tag{45.17}$$

(so that $C(p) = \mathcal{O}(\not{p} - m)$ and this is, as we know, UV finite). Thus, the expression for the corrected propagator in (45.15) can be recast as

$$\frac{1}{\not{p} - m + \Sigma(p)} = \frac{1}{\not{p} - m + A + B(\not{p} - m) + C(p)(\not{p} - m)} . \tag{45.18}$$

The structure of the denominator in (45.18) suggests a reinterpretation of the original mass parameter $m$: one may now denote $m$ as an unphysical "bare mass" and identify the combination $m - A \equiv m - \delta m$ as the physical ("renormalized") mass. The contribution $-\delta m = -A$, originating in $\Sigma(p)$, may be intuitively understood as an "electromagnetic self-energy" (since the corresponding closed loop looks like an interaction of the electron with its own electromagnetic field, represented by the virtual photon in Fig. 41.2). Thus, the term "self-energy graph" for $\Sigma(p)$ is clarified.

In any case, the idea of the **mass renormalization** is embodied in the relation

$$m - A = m_{\text{phys.}} . \tag{45.19}$$

As we know from Chapter 41, the constant $A$ depends on a regularization parameter; in particular, in the Pauli–Villars (PV) scheme one has for the UV divergent part

$$A_{\text{div}} = -\frac{3\alpha}{4\pi} m \ln \frac{M^2}{m^2} \tag{45.20}$$

(see the result (41.17) and its PV counterpart shown e.g. in the book [6]). So, $\delta m = A$ is an extra (negative) ingredient of the bare mass, in addition to (positive) $m_{\text{phys.}}$. Both the bare mass and $\delta m$ depend on the regularization (cut-off) parameter, while $m_{\text{phys.}}$ should be the (measurable) physical mass. Notice also a curious feature of the relation (45.19) along with (45.20): for $M \to \infty$ (which corresponds to $\epsilon \to 0$ in dimensional regularization) one would get negative value for the bare mass. However, one need not worry about a "paradox" like this, since we always have in mind a fixed finite value of the regularization parameter; the limit of removed cut-off is thus of purely academic interest (this is similar to our preceding discussion around Eq. (45.10) for the coupling).

From now on, we will write simply $m$ instead of $m_{\text{phys.}}$. Since the coefficients $A$, $B$ and $C(p)$ in (45.16), (45.17) are of the order $\mathcal{O}(e^2)$, the expression (45.18) can be rewritten in terms of the renormalized mass as

$$\frac{1}{\not{p} - m + B(\not{p} - m) + C(p)(\not{p} - m)} , \tag{45.21}$$

if one neglects terms of the order $\mathcal{O}(e^4)$. We may make another step forward and recast the expression (45.21) as

$$\frac{1}{\not{p} - m} \frac{1}{1 + B + C(p)} = \frac{1}{\not{p} - m} \left(1 - B - C(p)\right) + \mathcal{O}(e^4)$$

$$= \frac{1}{\not{p} - m} \left(1 - B\right)\left(1 - C(p)\right) + \mathcal{O}(e^4) . \tag{45.22}$$

The idea of mass renormalization outlined above is, hopefully, quite plausible. But, in (45.22) there is still an UV divergent remnant; the fermion propagator with the one-loop correction incorporates, among other things, a constant UV divergent factor $1 - B$. We know that the propagator is defined by means of a product of fields. So, one might come up with a crazy idea to renormalize, with the help of an appropriate factor, the quantized field itself. Well, such an idea is crazy only at first sight. We will see soon that its implementation is feasible, but for such a purpose it is convenient to develop a novel technique, based on the concept of the so-called renormalization **counterterms**. This will be the main topic of the next chapter.

The contents of the present chapter may be summarized as follows. We have introduced the idea of **renormalization**,[27] which means *reinterpretation* of some parameters in the basic Lagrangian and subsequent *reparametrization* of the scattering amplitudes in terms of those renormalized constants (which are supposed to be measurable quantities). The procedure highlights the concept of "bare parameters" along with the renormalized ones. The UV divergences coming from the closed-loop diagrams are eliminated by renormalization, as they are "absorbed" in the redefinition of parameters (like the coupling constant and the mass) when passing from bare to renormalized ones. In other words, UV divergences become part of the unphysical bare parameters.

In several forthcoming chapters, we are going to describe a basic practical technique of the renormalization in perturbative QED. In fact, the theme of renormalization of QFT is wide-ranging and its various aspects are treated in many books and review papers (see e.g. [2, 6, 8, 10–14, 19, 24, 25, 27–31, 55]. To paraphrase Woody Allen, in the aforementioned literature there is (almost) "everything you always wanted to know about renormalization (but were afraid to ask)".

---

[27]Note that, historically, the term "renormalization" (or rather "to renormalize") had been used apparently for the first time by Robert Serber in the paper [48].

# Chapter 46

# Renormalization counterterms

The QFT model we are going to employ in what follows is the QED of electrons, positrons and photons; a single fermion species is sufficient for the discussion of methods we would like to develop here. The relevant Lagrangian can be written, schematically, as

$$\mathscr{L}_{\text{QED}} = \mathscr{L}^{(1)} + \mathscr{L}^{(2)} + \mathscr{L}^{(3)} , \tag{46.1}$$

where

$$
\begin{aligned}
\mathscr{L}^{(1)} &= i\bar{\psi}\,\partial\!\!\!/\,\psi - m\bar{\psi}\psi , \\
\mathscr{L}^{(2)} &= -\frac{1}{4}F_{\mu\nu}F^{\mu\nu} , \quad F_{\mu\nu} = \partial_\mu A_\nu - \partial_\nu A_\mu , \\
\mathscr{L}^{(3)} &= e\bar{\psi}\gamma^\mu\psi A_\mu .
\end{aligned}
\tag{46.2}
$$

In the preceding chapter we have observed, besides other things, that there is a reasonable motivation for introducing the concept of bare mass for the electron. The point is that a part of the UV divergent contribution of the electron self-energy loop can be buried in such an unphysical (regularization-dependent) parameter and one is left with the physical renormalized mass that defines the pole of the corresponding field propagator. Thus, $\mathscr{L}^{(1)}$ in (46.2) may be rewritten, tentatively, as

$$\mathscr{L}^{(1)} = i\bar{\psi}\,\partial\!\!\!/\,\psi - m_0\bar{\psi}\psi , \tag{46.3}$$

where $m_0$ is the bare mass. Let us denote the difference between $m_0$ and the physical mass $m$ as $\delta m$, so that

$$m_0 = m + \delta m . \tag{46.4}$$

The Lagrangian (46.3) is thus

$$\mathscr{L}^{(1)} = i\bar{\psi}\,\partial\!\!\!/\,\psi - m\bar{\psi}\psi - \delta m\bar{\psi}\psi . \tag{46.5}$$

Now we make a basic strategic step. Since we would like to define the free part of the relevant Lagrangian in terms of the physical mass $m$, the additional term $-\delta m\bar{\psi}\psi$ will be incorporated in the interaction part of the QED Lagrangian. Of course, up to now we have been used to identify as "interactions" the terms involving at least three fields, but in fact there is no fundamental principle that would forbid quadratic field monomials as interactions — from our point of view it is just a choice of an organizational principle for the perturbation expansion. The full correction to the electron propagator then looks, pictorially, as shown in Fig. 46.1. Using the familiar expansion

$$\Sigma(p) = A + B(\not{p} - m) + \dots ,$$

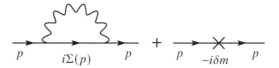

**Fig. 46.1:** Electron self-energy loop and the corresponding mass counterterm.

the sum corresponding to diagrams in Fig. 46.1 is proportional to

$$A + B(\not p - m) + \cdots + (-\delta m) \,, \tag{46.6}$$

and an obvious choice for the cancellation is

$$\delta m = A \,. \tag{46.7}$$

Comparing this with the relation (46.4), one sees that the mass renormalization introduced in the preceding chapter is thereby recovered.

Thus, the funny interaction term $-\delta m \bar\psi \psi$ plays the role of a compensation of a part of the self-energy loop in Fig. 46.1; precisely this is the reason why we call it the **counterterm**: against the UV divergent loop term, a counterterm is invoked.[28]

So, we have reproduced, in an almost trivial way, our previous picture of the mass renormalization. However, having done it, there is still an UV divergent remnant in the sum (46.6). Once we have launched the program relying on counterterms, one may cast off all inhibitions and continue such a game. In particular, the coefficient $B$ multiplies $\not p$ and this indicates that a structure involving a pair of Dirac fields and a derivative could come in handy. Having this in mind and recalling the elementary result for the mass renormalization, one may consider a pair of counterterms incorporated in the Lagrangian structure

$$\mathscr{L}_{\mathrm{CT}}^{(1)} = iK_2 \bar\psi \not\partial \psi - K_1 \bar\psi \psi \,, \tag{46.8}$$

designed to take care of the UV divergent terms coming from $\Sigma(p)$. Thus, one may say that the Dirac-like structure (46.8) is the counterterm Lagrangian induced by the electron self-energy function $\Sigma(p)$, corresponding to the loop diagram in Fig. 46.1. Now, the corrected electron propagator up to the order $\mathcal{O}(e^2)$, including contributions of the counterterms (46.8), is, pictorially, shown in Fig. 46.2. In the last diagram we have indicated the Feynman rule

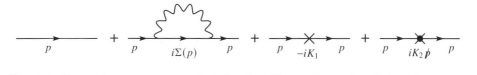

**Fig. 46.2:** Corrected electron propagator including the self-energy loop and two independent counterterms.

---

[28]In Czech, it reads **kontrčlen**, though it is an ugly hybrid word. A linguistic remark addressed to potential polyglots: In English, the word "counterterm" is obviously quite consistent. In Czech, so as to avoid the hybrid word "kontrčlen", one should in fact use a funny but consistent term "protičlen". But this is clearly not viable. Some people tried to say "kontračlen", but this also did not take root. Most probably, the word "kontrčlen" migrated to the Czech language from Russian term "контрчлен", which is constructed by using the prefix "contre" of French origin (see e.g. the Russian original of the book [16]).

corresponding to the first term in (46.8). It is not difficult to find out that such a rule is indeed valid, but we will not discuss it now (it is also a good topic for a tutorial, not time-consuming). Thus, the corrections shown in Fig. 46.2 amount to the sum

$$\Sigma(p) - K_1 + K_2 \not{p} . \tag{46.9}$$

Again, using the familiar expansion of $\Sigma(p)$, namely

$$\Sigma(p) = A + B(\not{p} - m) + \overline{\Sigma}(p) , \tag{46.10}$$

where $\overline{\Sigma}(p)$ is a shorthand notation for the UV finite part, one may set the values of the counterterm constants $K_1$, $K_2$ so as to cancel the first two terms in (46.10) completely. It means

$$\begin{aligned} A - Bm - K_1 &= 0 , \\ B + K_2 &= 0 , \end{aligned} \tag{46.11}$$

i.e.

$$\begin{aligned} K_1 &= A - Bm , \\ K_2 &= -B . \end{aligned} \tag{46.12}$$

Let us now see what one gets for the sum of the free Dirac Lagrangian and $\mathscr{L}_{\text{CT}}^{(1)}$ (recall that now the free Lagrangian is simply $i\overline{\psi}\not{\partial}\psi - m\overline{\psi}\psi$ with $m$ being the physical mass). We have

$$\mathscr{L}^{(1)} + \mathscr{L}_{\text{CT}}^{(1)} = i(1 + K_2)\overline{\psi}\not{\partial}\psi - (m + K_1)\overline{\psi}\psi . \tag{46.13}$$

The form of the expression (46.13) suggests that one may introduce a rescaled Dirac field according to

$$\psi_0 = Z_2^{1/2}\psi , \tag{46.14}$$

with $Z_2 = 1 + K_2$, i.e., using (46.12),

$$Z_2 = 1 - B . \tag{46.15}$$

Note that the choice of notation in (46.14) is a standard convention. The rescaling (46.14) is in fact an implementation of our previous hint at a "crazy" idea to renormalize the field itself. In the spirit of previous considerations we will call $\psi_0$ the **bare field** and the sum $\mathscr{L}^{(1)} + \mathscr{L}_{\text{CT}}^{(1)}$ can be denoted as $\mathscr{L}_{\text{bare}}^{(1)}$. If we express this in terms of the bare field (46.14) one gets, using (46.12) and (46.13),

$$\mathscr{L}_{\text{bare}}^{(1)} = i\overline{\psi}_0\not{\partial}\psi_0 - (m + A - Bm)Z_2^{-1}\overline{\psi}_0\psi_0 . \tag{46.16}$$

Then, up to the lowest non-trivial order in the coupling constant $e$, (46.16) can be recast as

$$\mathscr{L}_{\text{bare}}^{(1)} = i\overline{\psi}_0\not{\partial}\psi_0 - (m + A)\overline{\psi}_0\psi_0 . \tag{46.17}$$

Note that in arriving at (46.17) we have used $Z_2^{-1} = (1 - B)^{-1} = 1 + B + \mathscr{O}(e^4)$ and then we have discarded terms proportional to $AB$ or $B^2$, since these are of the order $\mathscr{O}(e^4)$. Thus, we have

$$\mathscr{L}_{\text{bare}}^{(1)} = i\overline{\psi}_0\not{\partial}\psi_0 - m_0\overline{\psi}_0\psi_0 , \tag{46.18}$$

with

$$m_0 = m + A ,$$

274

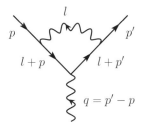

**Fig. 46.3:** QED vertex one-loop correction reproduced here for reader's convenience.

which is just our earlier relation between the bare and physical mass (cf. (45.19)).

As a next step, let us consider the vertex correction corresponding to the diagram shown in Fig. 46.3. We know from our previous treatment of the vertex function $\Gamma_\mu(p', p)$ (see chapters 41, 42) that its UV divergent part resides e.g. in $\Gamma_\mu(p, p)$. In other words, if we split $\Gamma_\mu(p', p)$ as

$$\Gamma_\mu(p', p) = \Gamma_\mu(p, p) + \left(\Gamma_\mu(p', p) - \Gamma_\mu(p, p)\right), \tag{46.19}$$

the term in parentheses is UV finite. Further, an important information is provided by the Ward identity (42.4), which tells us that, employing the "tilde notation", it holds

$$\widetilde{\Gamma}_\mu(p, p) = \widetilde{B}\gamma_\mu + \text{UV finite terms}. \tag{46.20}$$

Including the coupling factors, $\Gamma_\mu(p, p)$ is proportional to $e^3$, while $e^2\widetilde{B} = B$. So, one may write, instead of (46.20),

$$\Gamma_\mu(p, p) = eB\gamma_\mu + \dots \tag{46.21}$$

Now, it is clear what could be the relevant counterterm compensating the UV divergent term in (46.21). Since the latter is proportional to $\gamma_\mu$ (which appears in the basic QED interaction term), it is natural to choose the counterterm with the structure of the interaction Lagrangian itself, i.e.

$$\mathscr{L}_{\text{CT}}^{(3)} = K_v \bar{\psi}\gamma^\mu \psi A_\mu. \tag{46.22}$$

Of course, the field composition of the counterterm (46.22) is dictated by the structure of the diagram in Fig. 46.3, which contains two external fermion lines and one external photon line. For a compensation of the UV divergent term in (46.21) one must obviously set

$$K_v = -eB. \tag{46.23}$$

Thus, our result for the "vertex counterterm" reads

$$\mathscr{L}_{\text{CT}}^{(3)} = -eB\bar{\psi}\gamma^\mu \psi A_\mu, \tag{46.24}$$

so that

$$\begin{aligned}\mathscr{L}^{(3)} + \mathscr{L}_{\text{CT}}^{(3)} &= e(1 - B)\bar{\psi}\gamma^\mu \psi A_\mu \\ &= eZ_1\bar{\psi}\gamma^\mu \psi A_\mu.\end{aligned} \tag{46.25}$$

Here we have introduced the conventional notation for the renormalization constant

$$Z_1 = 1 - B. \tag{46.26}$$

Thus, we see that owing to the Ward identity one has

$$Z_1 = Z_2 \qquad (46.27)$$

(cf. (46.15)).

Finally, let us examine the familiar vacuum polarization bubble shown in Fig. 40.1. Let us recall that, conventionally, we denote its contribution as $(-i\Pi_{\mu\nu}(q))$ and we know that in any good regularization scheme the tensor $\Pi_{\mu\nu}(q)$ is transverse, i.e.

$$\Pi_{\mu\nu}(q) = \Pi(q^2)(q^2 g_{\mu\nu} - q_\mu q_\nu) \,.$$

From our earlier results it is clear that if $\Pi(q^2)$ is split as e.g.

$$\Pi(q^2) = \Pi(0) + \left(\Pi(q^2) - \Pi(0)\right), \qquad (46.28)$$

then the term in parentheses is UV finite. This gives us a hint for finding a corresponding counterterm. It is clear that it should be quadratic in the electromagnetic fields (because there are two photon lines attached to the fermion loop in Fig. 40.1) and it is expected to involve two derivatives (because of the quadratic $q$-dependence of $\Pi(0)(q^2 g_{\mu\nu} - q_\mu q_\nu)$). It turns out that the right answer is

$$\mathcal{L}_{CT}^{(2)} = K_3 \left(-\frac{1}{4} F_{\mu\nu} F^{\mu\nu}\right). \qquad (46.29)$$

So, the counterterm in question is proportional to the Maxwell Lagrangian in (46.2)! This might seem quite surprising at first sight, but our comments preceding (46.29) should make it plausible. In any case, a detailed examination of the contribution of the counterterm (46.29) leads to the conclusion that the corresponding Feynman rule is

$$= -iK_3(q^2 g_{\mu\nu} - q_\mu q_\nu) \qquad (46.30)$$

(verifying this is left to a diligent reader as an instructive exercise). So, following the decomposition (46.28), one may set

$$K_3 = -\Pi(0) \,. \qquad (46.31)$$

Thus,

$$\mathcal{L}^{(2)} + \mathcal{L}_{CT}^{(2)} = -\frac{1}{4}(1 + K_3) F_{\mu\nu} F^{\mu\nu} \,, \qquad (46.32)$$

and this leads us naturally to rescaling the photon field $A_\mu$, in analogy with what we have done before for the Dirac field. So, let us define the bare photon field by

$$A_0^\mu = Z_3^{1/2} A^\mu \,, \qquad (46.33)$$

where

$$Z_3 = 1 + K_3 = 1 - \Pi(0) \,. \qquad (46.34)$$

Thus, the original Maxwell term $\mathcal{L}^{(2)}$ in the QED Lagrangian is reproduced, with $A^\mu$ being replaced by the bare field $A_0^\mu$.

Now we may express the interaction term (46.25) in terms of the bare fields. One has

$$\begin{aligned} \mathcal{L}^{(3)} + \mathcal{L}_{CT}^{(3)} &= e Z_1 \bar{\psi} \gamma_\mu \psi A^\mu \\ &= e Z_1 Z_2^{-1} Z_3^{-1/2} \bar{\psi}_0 \gamma_\mu \psi_0 A_0^\mu \\ &= e Z_3^{-1/2} \bar{\psi}_0 \gamma_\mu \psi_0 A_0^\mu \,, \end{aligned} \qquad (46.35)$$

276

where we have used (46.14) and the Ward identity $Z_1 = Z_2$, displayed in (46.27). Thus, the coefficient multiplying the field monomial in (46.35) may be identified with the bare coupling constant, i.e.

$$e_0 = Z_3^{-1/2} e \,, \qquad (46.36)$$

and this in turn means that

$$e = Z_3^{1/2} e_0 \,. \qquad (46.37)$$

It is worth noticing that in such a way we have recovered our earlier preliminary relation between the renormalized and bare coupling constant (cf. (45.7)). The point is that, owing to the Ward identity, the only contribution to the renormalization of the coupling constant ("charge renormalization") comes from the vacuum polarization loop. By the way, from (46.34) and (46.37) it is seen that $e^2 < e_0^2$; we will comment on this at the end of Chapter 48.

So, we have arrived at a rather remarkable conclusion. The original QED Lagrangian, when supplemented with all necessary counterterms, has the same form as before, except that one employs here bare parameters and bare fields. The salient point is that the necessary counterterms have the form of terms already present in the basic Lagrangian. In this sense, QED is a **renormalizable theory** (we have verified it here at the level of one-loop Feynman diagrams). We can also see now, why it is so important that the fermion triangle and box diagrams do not require counterterms. Indeed, these would have to involve three or four photon fields, respectively, in a gauge invariant form; thus, the triangle would induce a counterterm of the type, schematically, $FFF$ and the box would produce a counterterm structure $FFFF$. But those are monomials with mass dimension six and eight, respectively, and according to our earlier formula for the index of divergence of 1PI diagrams (see (44.8), (44.9)) they would lead then to an infinite number of types of UV divergences in higher orders. Similarly beneficial for our purpose is the finiteness of the box diagram with four external fermion lines (see Fig. 42.2). Thus, it is reassuring that we have been able to avoid potential destructive complications that could be induced by interaction terms (counterterms) with dimension greater than four. Let us also recall that adding the Pauli term (44.15) to the standard QED Lagrangian would lead to the appearance of infinite number of types of UV divergent graphs, i.e. it would spoil the renormalizability.

Two more remarks are in order here. First, an essential point for our success was the simple structure of the UV divergent terms in the loop diagrams in question. Indeed, all three of them, corresponding to functions $\Pi_{\mu\nu}(q)$, $\Sigma(p)$ and $\Gamma_\mu(p', p)$ exhibit UV divergences that have *polynomial* dependence on the external momenta. This enables one to use *local* counterterms (which are polynomials made of fields and involving possibly just a finite number of their derivatives). Second, returning to the formulae (44.8), (44.9), one may say in another way what is so problematic about a field theory model involving an interaction for which $\omega_\nu > 4$: the ensuing infinite number of types of UV divergences would mean the need for an infinite number of renormalization counterterms. In conventional terms, it would mean that such a theory is **non-renormalizable**.

A historical remark is perhaps in order here. The Nobel Prize for the modern formulation of QED including renormalization techniques was awarded to Richard Feynman, Julian Schwinger and Sin-itiro Tomonaga in 1965 for their work published in the late 1940s. In fact, another crucial contribution is due to Freeman Dyson (see, in particular, his paper [40]), who reformulated the conventional renormalization procedure in the current "textbook" form. In this context, Steven Weinberg once noted that, taking this into account, one may say that Dyson was in fact "fleeced" of the Nobel Prize. Fortunately, he received many other prestigious prizes for his achievements.

# Chapter 47

# Renormalization and radiative corrections

Before proceeding to some simple practical applications of QED at the one-loop level, let us mention briefly also some other examples of QFT models from the point of view of the renormalization theory and the technique of counterterms.

First, let us consider the model with the interaction of Yukawa type, namely

$$\mathscr{L}_{\text{int}} = g\bar{\psi}\psi\varphi, \tag{47.1}$$

where $\psi$ is a Dirac field (mass $m$) and $\varphi$ is a real scalar (Klein–Gordon) field (mass $M$). The free part of the full Lagrangian has the familiar form, so it need not be repeated here. Feynman diagrams corresponding to such a model are topologically quite similar to those of QED. Namely, one-loop graphs contributing to propagator corrections are shown in Fig. 47.1 and there is also

(a)                                                      (b)

**Fig. 47.1:** One-loop graphs representing propagator corrections in a model of Yukawa-type interaction.

**Fig. 47.2:** One-loop vertex correction for a Yukawa-type interaction.

the vertex correction shown in Fig. 47.2. Obviously, the graphs shown in Figs. 47.1 and 47.2 are analogous to the familiar QED diagrams (cf. Fig. 44.1), with the wavy photon line being replaced by a dashed line corresponding to the scalar field. Needless to say, contributions of Feynman graphs within such a model are algebraically simpler than in QED, since there are no $\gamma$-matrices here. Similarly to QED, the diagrams in Figs. 47.1, 47.2 induce counterterms leading to field renormalizations, renormalization of masses (of course, renormalization of the scalar boson mass is an extra ingredient in comparison with QED), and renormalization of the coupling

278

constant $g$. The latter is determined by the $Z$-factors for the fields $\psi$, $\varphi$ and the interaction vertex $\bar{\psi}\psi\varphi$; similarly to QED, this leads to the relation

$$g_0 = g Z_v Z_\psi^{-1} Z_\varphi^{-1/2} \,, \qquad (47.2)$$

where we have employed a notation that is perhaps more instructive than that practiced in QED (here $Z_v$ corresponds to the constant $Z_1$ of QED, $Z_\psi$ to $Z_2$, and $Z_\varphi$ to $Z_3$). However, unlike QED, there is no Ward identity here, so $Z_v \neq Z_\psi$. An explicit evaluation of the renormalization factors entering (47.2) is left as a challenge for a hard-working reader.

In fact, there is another feature, in which the Yukawa-like model differs from QED. Namely, higher fermion loops, in particular the triangle and the box shown in Fig. 47.3, do not

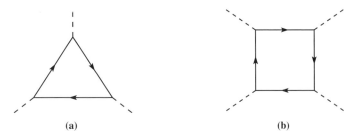

(a)            (b)

**Fig. 47.3:** Fermionic triangle and the box in a Yukawa-type model. These loops yield UV divergent contributions and induce the corresponding counterterms.

drop out of the game. Indeed, as for the triangle, the graph with reverse orientation of the loop momentum does not cancel the contribution of its counterpart, since the Furry's theorem does not work here (because of the absence of $\gamma$-matrices in the triangle vertices). As for the box, its algebraic structure is so simple that there cannot occur any cancellation of UV divergences due to the permutation of external boson lines. It means that two extra counterterms are needed, proportional to $\varphi^3$ and $\varphi^4$, respectively. Nevertheless, both these scalar self-interactions have dimension less than or equal to four (so that $\omega_v \leq 4$ in the formulae (44.8), (44.9)) and the Yukawa-like model supplemented with such counterterms remains to be renormalizable.

A similar situation occurs in the scalar QED (the interaction of charged spin-0 particles with photons, cf. the end of Chapter 44). Vacuum polarization graphs representing the correction to the photon propagator are shown in Fig. 47.4. It is a highly useful and instructive exercise

(a)            (b)

**Fig. 47.4:** Vacuum polarization loop in scalar QED.

to show that both diagrams in Fig. 47.4 are necessary to guarantee the transversality of the corresponding function $\Pi_{\mu\nu}(q)$. The vertex correction is described, at one-loop level, by the diagrams shown in Fig 47.5, and the self-energy graphs for the charged scalar are shown in Fig. 47.6.

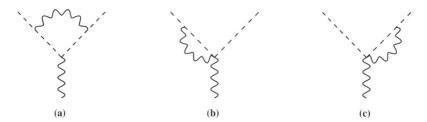

Fig. **47.5:** One-loop vertex correction in scalar QED.

Fig. **47.6:** One-loop corrections to the propagator of the spin-0 charged boson in scalar QED.

The diagrams shown in figures 47.4 through 47.6 contribute to "ordinary" renormalization constants in analogy with spinor QED. However, there are also UV divergent diagrams with four external scalar lines, namely those shown in Fig. 47.7. It is easy to realize that these graphs

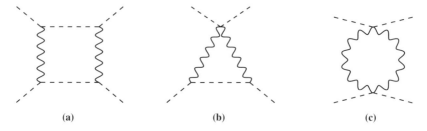

Fig. **47.7:** One-loop graphs generating the counterterm representing a quartic self-interaction of the scalar field.

are logarithmically divergent; so, because of the configuration of the external lines they induce a counterterm proportional to $(\varphi^*\varphi)^2$. On the other hand, diagrams with four external photon lines, namely those shown in Fig. 47.8, give a UV finite contribution (in their sum).

Thus, similarly to the situation in spinor QED, there is no need to introduce a four-photon counterterm (which would be destructive for the renormalization program). The upshot of all this is as follows: Scalar QED with the original interaction terms of the type $\varphi \overset{\leftrightarrow}{\partial} \varphi^* A$ and $\varphi \varphi^* A A$ should be supplemented with the quartic counterterm $(\varphi^*\varphi)^2$, and then it is "closed under renormalization" (i.e. it contains all counterterms necessary for implementing the renormalization program).

Finally, let us mention a curious example of renormalizable field theory model, which involves not only finite number of types of UV divergences, but even a finite number of all UV divergent 1PI diagrams! (such a model is called "superrenomalizable"). The model in question involves the cubic self-interaction $\varphi^3$ alone. We know that such an interaction term is

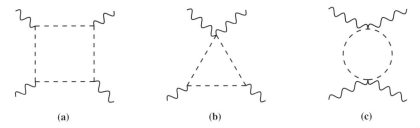

(a)          (b)          (c)

**Fig. 47.8:** One-loop diagrams contributing to the photon–photon scattering in scalar QED; their sum is UV finite.

present within some more complex models; by the way, it also appears in the standard model of electroweak interactions as the (yet untested) self-interaction of the Higgs boson. When one considers the $\varphi^3$ self-interaction by itself, i.e. the model Lagrangian that reads simply

$$\mathscr{L} = \frac{1}{2}\partial_\mu\varphi\partial^\mu\varphi - \frac{1}{2}m^2\varphi^2 + g\varphi^3 , \qquad (47.3)$$

it turns out that there is only one UV divergent 1PI graph (if one ignores tadpoles) namely the self-energy diagram shown in Fig. 47.9. The reader is encouraged to verify such a statement,

**Fig. 47.9:** Scalar self-energy loop in the $\varphi^3$ model.

possibly with the help of formulae (44.8), (44.9). Please notice that the dimension of the interaction term $\varphi^3$ is 3, i.e. the coupling constant $g$ in (47.3) has the dimension of a mass.

Two more remarks addressed to potential QFT enthusiasts: In fact, the model (47.3) is, in a broader perspective, inconsistent, since the corresponding energy density is not bounded from below (precisely because of $\varphi^3$). Of course, such a pathological feature does not prevent us from studying such a model within perturbation theory. Second, when the model (47.3) is considered, academically, in six-dimensional spacetime (see e.g. the book [12]), the coupling constant is dimensionless and the relevant Feynman diagrams are much more interesting than in four dimensions: such a model is renormalizable in the conventional sense, and, moreover, it exhibits a remarkable property, called **asymptotic freedom**, which in four dimensions is reserved just for field theory models involving non-Abelian gauge fields (Yang–Mills fields). The most famous theory of this type in four dimensions is quantum chromodynamics (QCD), the modern theory of strong interactions in particle physics (the Nobel Prize in physics in 2004 was awarded just for this). The end of remarks for QFT enthusiasts.

Let us now come back to our good old QED. We have found that all necessary renormalization counterterms can be incorporated by rewriting the original Lagrangian in terms of bare parameters and bare fields, so that one has

$$\mathscr{L}_{\mathrm{QED}} = i\bar{\psi}_0\slashed{\partial}\psi_0 - m_0\bar{\psi}_0\psi_0 - \frac{1}{4}F_{\mu\nu}^{(0)}F^{(0)\mu\nu} + e_0\bar{\psi}_0\gamma^\mu\psi_0 A_\mu^{(0)} . \qquad (47.4)$$

Thus, the counterterms (which, according to the accepted strategy, belong to interactions) are "wrapped" in an elegant way in the fundamental form (47.4). This Lagrangian could now be "unwrapped" so as to recover all terms that are relevant for the evaluation of renormalized Feynman diagrams, but in fact it is not necessary: we know how we have fixed the counterterms in terms of UV divergent parts of one-loop diagrams and we understand that including the counterterms in the interaction Lagrangian is tantamount to specific subtractions in the regularized functions $\Pi(q^2)$, $\Sigma(p)$ and $\Gamma_\mu(p', p)$. In particular, our renormalization scheme (i.e. our particular choice of counterterms) means that $\Pi(q^2)$ is replaced by

$$\overline{\Pi}(q^2) = \Pi(q^2) - \Pi(0), \qquad (47.5)$$

and from $\Sigma(p)$ one is led to $\overline{\Sigma}(p) = \Sigma(p) - A - B(\not{p} - m)$, i.e.

$$\overline{\Sigma}(p) = \Sigma(p) - \Sigma(p)\big|_{\not{p}=m} - \frac{\partial}{\partial \not{p}} \Sigma(p)\big|_{\not{p}=m}(\not{p} - m). \qquad (47.6)$$

Thus,

$$\overline{\Sigma}(p) = C(\not{p} - m)^2 + \dots \qquad (47.7)$$

As for $\Gamma_\mu(p', p)$, our choice (related to the Ward identity) was (cf. (46.21), (46.24))

$$\overline{\Gamma}_\mu(p', p) = \Gamma_\mu(p', p) - eB\gamma_\mu. \qquad (47.8)$$

It is easy to recast the subtraction in the last expression in a more compact way. Indeed, in arriving at the counterterm in (47.8) we have utilized the Ward identity (42.4) that gives us

$$\begin{aligned}
\widetilde{\Gamma}_\mu(p, p) &= \frac{\partial}{\partial p^\mu} \widetilde{\Sigma}(p) \\
&= \frac{\partial}{\partial p^\mu} \left[ \widetilde{A} + \widetilde{B}(\not{p} - m) + \widetilde{C}(\not{p} - m)^2 + \dots \right] \\
&= \widetilde{B}\gamma_\mu + \widetilde{C}\left(\gamma_\mu(\not{p} - m) + (\not{p} - m)\gamma_\mu\right) + \dots
\end{aligned} \qquad (47.9)$$

Obviously, all terms in the last expression, except $\widetilde{B}\gamma_\mu$, vanish when we set, formally, $\not{p} = m$. Thus, the subtraction in (47.8) can be written as

$$\overline{\Gamma}_\mu(p', p) = \Gamma_\mu(p', p) - \Gamma_\mu(p, p)\big|_{\not{p}=m}. \qquad (47.10)$$

So, the recipe emerging from the above considerations is that the renormalized QED scattering amplitudes can be obtained, at the one-loop level, by replacing contributions of the relevant sub-diagrams (vacuum polarization, fermion self-energy and vertex correction) with the subtracted expressions $\overline{\Pi}(q^2)$, $\overline{\Sigma}(p)$ and $\overline{\Gamma}_\mu(p', p)$ defined in (47.5), (47.6) and (47.10).

Let us now see how it works in practice. In particular, the loop corrections in the external lines deserve special attention. First, let us consider the 4th order diagram for our favourite scattering process $e + \mu \to e + \mu$, shown in Fig. 47.10, where we have denoted by the heavy dot the contribution of the function $\overline{\Sigma}(p)$. Full contribution of the line involving the insertion of $\overline{\Sigma}(p)$ amounts to

$$\frac{1}{\not{p} - m} \overline{\Sigma}(p) u(p). \qquad (47.11)$$

However, according to (47.7), $\overline{\Sigma}(p)$ is proportional to $(\not{p} - m)^2$, so that in the expression (47.11) one certainly gets a factor $(\not{p} - m)u(p)$, which is zero. Of course, the fate of the self-energy insertion in the outgoing line would be the same, because of $\bar{u}(p')(\not{p}' - m) = 0$.

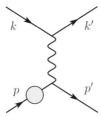

**Fig. 47.10:** A would-be contribution to $e$–$\mu$ scattering involving the self-energy correction in an external line.

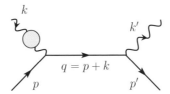

**Fig. 47.11:** A would-be contribution to the Compton scattering, involving the vacuum-polarization correction in an external photon line.

Similarly, one may consider the 4th order diagram for Compton scattering with the insertion of the vacuum polarization bubble in an external photon line, as shown in Fig. 47.11, where the heavy dot denotes the function

$$\overline{\Pi}_{\mu\nu}(k) = \overline{\Pi}(k^2)(k^2 g_{\mu\nu} - k_\mu k_\nu).$$

Now, the full contribution of the line including $\overline{\Pi}_{\mu\nu}(k)$ is proportional to

$$\epsilon^\mu(k)\overline{\Pi}_{\mu\nu}(k)\frac{g^{\nu\rho}}{k^2} = \epsilon^\mu(k)\overline{\Pi}(k^2)(k^2 g_{\mu\nu} - k_\mu k_\nu)\frac{g^{\nu\rho}}{k^2}. \tag{47.12}$$

However, $k_\mu \epsilon^\mu(k) = 0$ for a physical photon polarization, so one is left just with the factor of $\overline{\Pi}(k^2)$, but this is zero, since $k^2 = 0$ and $\overline{\Pi}(0) = 0$ by its definition.

The above two examples show that within our renormalization scheme, corrections on the external lines vanish. In any case, our elementary calculations show that one need not worry about the anticipated singularity of the propagator standing between the vertex and the loop (anticipated because of the value of the physical four-momentum, $p^2 = m^2$ or $k^2 = 0$, imposed by the attached external line).

In this context, a terminological remark is in order. The renormalization scheme we are using is called the **on-shell scheme**, since the choice of counterterms corresponds to subtractions at physical (on-shell) values of the relevant four-momenta. One should come to terms with the fact that the choice of the counterterms (including their UV finite parts) is to some extent arbitrary. For instance, within the dimensional regularization, another popular choice of the counterterms corresponds just to the subtraction of the pole terms $1/\epsilon$; such a renormalization scheme is called **minimal subtraction scheme**, or briefly **MS scheme**.

Apart from the corrections in external lines, which are trivial in the on-shell scheme, in propagators and vertices of Feynman diagrams one, of course, gets non-trivial results. At the one-loop level, the renormalization corrections are of the order $\mathcal{O}(e^2) = \mathcal{O}(\alpha)$ and they are commonly called **radiative corrections**, because they involve, pictorially, virtual particles radiated from some lines of Feynman diagrams. Some examples of radiative corrections will be discussed in the forthcoming chapters.

# Chapter 48

# One-loop vacuum polarization in detail

Renormalized UV finite parts of the contributions of the one-loop QED diagrams, which represent an important portion of the radiative corrections to scattering amplitudes are, in general, quite complicated functions of external momenta. For instance, the expression for the vertex correction $\overline{\Gamma}_\mu(p', p)$ contains also higher transcendental functions like the dilogarithm (Spence's function). On the other hand, the vacuum polarization form factor $\overline{\Pi}(q^2)$ is relatively simple and can be expressed fully in terms of elementary functions. A detailed description of this quantity is the main subject of this chapter.

Let us start with the regularized expressions for $\Pi(q^2)$ that we have obtained in chapters 39 and 40. Including also the coupling factor $e^2 = 4\pi\alpha$ in the formulae (39.19), (40.24), we have

$$\Pi^{DR}(q^2) = \frac{e^2}{2\pi^2}\left[\frac{1}{6}\left(\frac{1}{\epsilon} - \gamma_E + \ln 4\pi - \ln\frac{m^2}{\mu^2}\right) - \int_0^1 dx\, x(1-x)\,\ln\frac{m^2 - x(1-x)q^2}{m^2}\right] \quad (48.1)$$

in the dimensional regularization and

$$\Pi^{PV}(q^2) = \frac{e^2}{2\pi^2}\left[\frac{1}{6}\ln\frac{M^2}{m^2} - \int_0^1 dx\, x(1-x)\,\ln\frac{m^2 - x(1-x)q^2}{m^2}\right] \quad (48.2)$$

in the Pauli–Villars scheme.

Thus, we see that $\overline{\Pi}(q^2) = \Pi(q^2) - \Pi(0)$ does not depend on the regularization scheme. One has, using $4\pi\alpha$ instead of $e^2$,

$$\overline{\Pi}(q^2) = -\frac{2\alpha}{\pi}\int_0^1 dx\, x(1-x)\,\ln\frac{m^2 - x(1-x)q^2}{m^2}. \quad (48.3)$$

A brief inspection of the integrand in the last expression reveals that one should distinguish three regions of the $q^2$ values, namely

$$\text{I)} \qquad q^2 < 0,$$
$$\text{II)} \quad 0 < q^2 \le 4m^2, \quad (48.4)$$
$$\text{III)} \qquad q^2 > 4m^2,$$

with regard to the distinct properties of the quadratic function

$$C(x) = m^2 - x(1-x)q^2 \quad (48.5)$$

285

in these kinematical areas. Indeed, for $q^2 < 0$ the function $C(x)$ is positive for any $x \in (0, 1)$ and has real zeroes outside the interval $(0, 1)$, namely

$$x_\pm = \frac{1}{2}\left(1 \pm \sqrt{1 + \frac{4m^2}{|q^2|}}\right). \tag{48.6}$$

In the region II, $C(x)$ is positive for any $x \in (0, 1)$ and has no real roots whatsoever. The region III is, in a sense, the most interesting case. The function $C(x)$ then has real roots inside the interval $(0, 1)$, namely

$$x_\pm = \frac{1}{2}\left(1 \pm \sqrt{1 - \frac{4m^2}{|q^2|}}\right), \tag{48.7}$$

and thus it changes sign for $x$ between 0 and 1. In particular, $C(x) < 0$ for $x \in (x_-, x_+)$ and $C(x) > 0$ for $x \in (0, x_-) \cup (x_+, 1)$. However, the negative value of $C(x)$ means that the logarithm in the integrand in (48.3) has a **non-trivial imaginary part** for $x \in (x_-, x_+)$. Thus, one may expect that the function $\overline{\Pi}(q^2)$ will be purely real for $q^2 \in (-\infty, 4m^2)$ and complex for $q^2 > 4m^2$!

The evaluation of real parts of the integral in (48.3) in the regions I, II, III is elementary, but somewhat tedious. It is clear that it can be carried out by means of partial integration, which results in integrating a rational function. We leave it to a hard-working reader as an exercise in the elementary calculus, and here we only summarize the relevant results.

I) For $q^2 < 0$, one gets

$$\overline{\Pi}(q^2) = \frac{\alpha}{3\pi}\left[-\frac{1}{3} + \left(1 - \frac{2m^2}{|q^2|}\right)\left(\sqrt{1 + \frac{4m^2}{|q^2|}} \ln \frac{\sqrt{1 + \frac{4m^2}{|q^2|}} - 1}{\sqrt{1 + \frac{4m^2}{|q^2|}} + 1} + 2\right)\right]. \tag{48.8}$$

II) For $0 \le q^2 \le 4m^2$,

$$\overline{\Pi}(q^2) = \frac{\alpha}{3\pi}\left[\frac{1}{3} - 2\left(1 + \frac{2m^2}{q^2}\right)\left(\sqrt{\frac{4m^2}{q^2} - 1}\, \arctan \frac{1}{\sqrt{\frac{4m^2}{q^2} - 1}} - 1\right)\right]. \tag{48.9}$$

III) For $q^2 > 4m^2$,

$$\mathrm{Re}\,\overline{\Pi}(q^2) = \frac{\alpha}{3\pi}\left[-\frac{1}{3} + \left(1 + \frac{2m^2}{q^2}\right)\left(\sqrt{1 - \frac{4m^2}{q^2}} \ln \frac{1 - \sqrt{1 - \frac{4m^2}{q^2}}}{1 + \sqrt{1 - \frac{4m^2}{q^2}}} + 2\right)\right]. \tag{48.10}$$

The evaluation of the imaginary part of $\overline{\Pi}(q^2)$ for $q^2 > 4m^2$ is quite simple, so it is worth doing it here explicitly. To begin with, let us recall what is the origin of the function $C(x)$ in the argument of the logarithm in (48.3). Going back to the expression (38.19) for $C$, one cannot overlook the remark that $m^2$ is to be understood as $m^2 - i\epsilon$, where $\epsilon > 0$ is an infinitesimal constant, ubiquitous in Feynman propagators. Then, if the real part of $C(x)$ is negative, the imaginary part of the logarithm is equal $-i\pi$. An explanatory comment is perhaps in order here: Note that the logarithm of complex variable $z = |z|e^{i\varphi}$ is $\ln z = \ln|z| + i\varphi$ and it has the branch cut on the real axis, extending from $-\infty$ to 0; with the specification of $m^2$ as $m^2 - i\epsilon$ in mind, we are on the lower side of the cut, where $\varphi = -\pi$.

Thus, the evaluation of $\operatorname{Im}\overline{\Pi}(q^2)$ is easy. We have

$$\overline{\Pi}(q^2) = \operatorname{Re}\overline{\Pi}(q^2) + i\operatorname{Im}\overline{\Pi}(q^2),$$

where

$$i\operatorname{Im}\overline{\Pi}(q^2) = -\frac{2\alpha}{\pi}\int\limits_{x_-}^{x_+} dx\, x(1-x)(-i\pi),$$

i.e.

$$\operatorname{Im}\overline{\Pi}(q^2) = 2\alpha\int\limits_{x_-}^{x_+} dx\, x(1-x), \tag{48.11}$$

with $x_\pm$ being given by (48.7) as the solution of the quadratic equation

$$x^2 - x + \frac{m^2}{q^2} = 0. \tag{48.12}$$

So, from (48.11) we have

$$\operatorname{Im}\overline{\Pi}(q^2) = 2\alpha\left[\frac{1}{2}(x_+^2 - x_-^2) - \frac{1}{3}(x_+^3 - x_-^3)\right]. \tag{48.13}$$

Working out the last expression is a refreshing exercise in high school maths. Indeed, one may utilize the elementary identity

$$x_+^3 - x_-^3 = (x_+ - x_-)(x_+^2 + x_-x_+ + x_-^2),$$

as well as the properties of roots of the quadratic equation (48.12), such as

$$x_+ + x_- = 1, \quad x_-x_+ = \frac{m^2}{q^2}, \quad x_+ - x_- = \sqrt{1 - \frac{4m^2}{q^2}},$$

$$x_+^2 + x_-^2 = x_+ + x_- - \frac{2m^2}{q^2} = 1 - \frac{2m^2}{q^2}.$$

For (48.13) one then gets, after a simple manipulation,

$$\operatorname{Im}\overline{\Pi}(q^2) = \frac{\alpha}{3}\left(1 + \frac{2m^2}{q^2}\right)\sqrt{1 - \frac{4m^2}{q^2}}. \tag{48.14}$$

The last result is in fact highly remarkable. If we denote, provisionally, $q^2 = M^2$ and come back to $e^2 = 4\pi\alpha$, the formula (48.14) reads

$$\operatorname{Im}\overline{\Pi}(q^2 = M^2) = \frac{e^2}{12\pi}\left(1 + \frac{2m^2}{M^2}\right)\sqrt{1 - \frac{4m^2}{M^2}}. \tag{48.15}$$

Now it turns out that

$$\operatorname{Im}\overline{\Pi}(q^2 = M^2) = \frac{\Gamma}{M}, \tag{48.16}$$

where $\Gamma$ is the rate of the decay of massive vector boson ("massive photon" with mass $M$) into a pair of fermions, with all particles unpolarized. To appreciate this, the reader is urged to

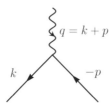

$q = k + p$

$k$

$-p$

**Fig. 48.1:** Tree-level graph for the vector boson decay into a fermion-antifermion pair.

revisit Chapter 24, where we have carried out the calculation of the decay rate $\Gamma$ at the tree level, i.e. using the diagram in Fig. 48.1 (see the result (24.21)). Such a relation, which might seem mysterious at first sight, is in fact not accidental; it is a direct consequence of the $S$-matrix unitarity (a variant of the "optical theorem"). Unfortunately, there is not enough space to discuss this interesting topic in detail now; we may refer the reader to the book [7] for more details. In any case, it is instructive to work out explicitly more examples of this kind, e.g. within a model of Yukawa interaction, etc.

Looking back at the formulae (48.8) through (48.10), it is seen that the analytic form of $\overline{\Pi}(q^2)$ is different in the regions I, II, III, so one would like to have at hand an instructive global picture of this quantity. An exact form of the plot can be found e.g. in the book [15] (see Fig. 5.13 therein; note that an opposite sign convention is used there, so the quantity displayed is $-\overline{\Pi}(q^2)$). Let us mention at least several salient features of the function $\overline{\Pi}(q^2)$. Some of its simple properties can be established rather easily on the basis of the original integral representation (48.3). For instance, it is clear that $\overline{\Pi}(0) = 0$ (by its definition) and $\frac{d}{dq^2}\overline{\Pi}(q^2) > 0$ for any $q^2 \in (-\infty, 4m^2)$. Similarly, $\frac{d}{dq^2} \operatorname{Re} \overline{\Pi}(q^2) < 0$ for $q^2 \in (4m^2, +\infty)$. The function $\overline{\Pi}(q^2)$ is continuous at $q^2 = 4m^2$ and it is not difficult to find out that the corresponding value is

$$\overline{\Pi}(q^2 = 4m^2) = \frac{8}{9}\frac{\alpha}{\pi}. \tag{48.17}$$

On the other hand, the derivative of $\overline{\Pi}(q^2)$ is discontinuous at $q^2 = 4m^2$. It is a refreshing exercise in elementary calculus to show that the left limit of the derivative is infinite, while the right limit is finite. In any case, (48.17) is the maximum value of the real part of $\Pi(q^2)$ and, because of its discontinuous derivative, $\operatorname{Re}\Pi(q^2)$ exhibits a pronounced peak (spike) at $q^2 = 4m^2$; such a type of singularity is commonly called the **cusp**. Let us also notice that the asymptotic behaviour of $\operatorname{Re}\Pi(q^2)$ for $|q^2| \to \infty$ is given by the leading logarithm descending from (48.3), namely $(-\alpha/3\pi)\ln|q^2/m^2|$. The reader is encouraged to verify the above statements explicitly and look up the picturesque plot shown in [15].

Next, we are going to discuss another useful representation of $\overline{\Pi}(q^2)$ that can be obtained by an appropriate transformation of the integration variables in the expression (48.3). We have

$$\overline{\Pi}(q^2) = -\frac{2\alpha}{\pi} I(q^2), \tag{48.18}$$

where

$$I(q^2) = \int_0^1 dx\, x(1-x)\ln\left[1 - x(1-x)\frac{q^2}{m^2}\right]. \tag{48.19}$$

Let us recall again that in the argument of the logarithm, 1 is to be understood as $1 - i\epsilon$. First, we use the simple substitution $x = \frac{1}{2}(1 - u)$, i.e. $1 - x = \frac{1}{2}(1 + u)$, so that $u \in (-1, 1)$. Performing then a partial integration, one gets, after some manipulations,

$$I(q^2) = -\frac{1}{2}\frac{q^2}{4m^2} \int_0^1 du\, u^2 \left(1 - \frac{1}{3}u^2\right) \frac{1}{1 - (1 - u^2)\frac{q^2}{4m^2}}, \tag{48.20}$$

where we have also taken into account that the integrand is an even function of the variable $u$. As a second step, we employ the substitution

$$u = \left(1 - \frac{4m^2}{t}\right)^{1/2}. \tag{48.21}$$

The integral (48.20) is then recast as

$$I(q^2) = -\frac{1}{6}q^2 \int_{4m^2}^{\infty} \frac{dt}{t} \sqrt{1 - \frac{4m^2}{t}} \left(1 + \frac{2m^2}{t}\right) \frac{1}{t - q^2 - i\epsilon}, \tag{48.22}$$

where we have also restored the infinitesimal $i\epsilon$ term implicit in the denominator of the integrand in (48.20). So, returning to the relation (48.18), one has

$$\overline{\Pi}(q^2) = \frac{\alpha}{3\pi} q^2 \int_{4m^2}^{\infty} \frac{dt}{t} \sqrt{1 - \frac{4m^2}{t}} \left(1 + \frac{2m^2}{t}\right) \frac{1}{t - q^2 - i0}, \tag{48.23}$$

where we have used the symbol $i0$ instead of $i\epsilon$. The formula (48.23) represents another highly remarkable result that we have achieved here. Why is it so remarkable? Well, looking back at our result for $\mathrm{Im}\,\overline{\Pi}$, one may notice that (48.23) can be recast as

$$\frac{1}{q^2}\overline{\Pi}(q^2) = \frac{1}{\pi} \int_{4m^2}^{\infty} dt\, \frac{\mathrm{Im}\,\overline{\Pi}(t)}{t} \frac{1}{t - q^2 - i0}. \tag{48.24}$$

It means that the full expression for $\overline{\Pi}(q^2)$ can be obtained from its imaginary part $\mathrm{Im}\,\overline{\Pi}$ by means of a relatively simple integral transformation! (Remember how easy it is to get $\mathrm{Im}\,\overline{\Pi}$ — in principle one could just use the "optical theorem" (48.16).) By the way, $\mathrm{Im}\,\overline{\Pi} = \mathrm{Im}\,\Pi$, since $\Pi(0)$ is real.

Again, one must stress that the peculiar relation (48.24) is not accidental. It is called the **dispersion relation** (please do not confuse it with the same term that is used e.g. in optics for a relation of the wavelength to frequency) and it is a special consequence of Cauchy's theorem in complex analysis, which in the considered case relies on specific analytic properties of the function $\overline{\Pi}(q^2)$ (or simply $\Pi(q^2)$) extended to the complex plane. Unfortunately, within this introductory text there is not enough space for discussing such a problem in more detail; so, we adopt a pragmatic approach by using the dispersion relation (48.24) when it is helpful and refer the interested reader e.g. to the book [7]. In any case, let us add a technical remark. We know that $\mathrm{Im}\,\overline{\Pi}(q^2) = \mathrm{Im}\,\Pi(q^2)$; in fact, the relation (48.16) clearly indicates *a priori* that the imaginary part of $\Pi(q^2)$ has nothing to do with UV divergences. If one takes for granted the form of dispersion relation, one might try to write down such a representation for $\Pi(q^2)$ itself, namely

$$\Pi(q^2) \overset{?}{=} \frac{1}{\pi} \int_{4m^2}^{\infty} \frac{\mathrm{Im}\,\Pi(t)}{t - q^2} dt \tag{48.25}$$

289

(for the moment, we suppress the symbol $i0$). However, $\operatorname{Im}\Pi(t)$ is asymptotically (i.e. for $t \to \infty$) constant (see (48.14)), so the integral in (48.25) diverges logarithmically. This is an alternative manifestation of the UV divergence of the fermion loop (UV divergence "in disguise"). One subtraction (which does not change the analytic properties of $\Pi(q^2)$) is needed to remove the divergence; in particular, subtracting a divergent constant

$$\Pi(0) = \frac{1}{\pi} \int_{4m^2}^{\infty} \frac{\operatorname{Im}\Pi(t)}{t}\, dt\,, \qquad (48.26)$$

one gets

$$\overline{\Pi}(q^2) = \Pi(q^2) - \Pi(0) = \frac{1}{\pi} \int_{4m^2}^{\infty} \left( \frac{\operatorname{Im}\Pi(t)}{t - q^2} - \frac{\operatorname{Im}\Pi(t)}{t} \right) dt$$

$$= \frac{1}{\pi} q^2 \int_{4m^2}^{\infty} \frac{\operatorname{Im}\Pi(t)}{t(t - q^2)}\, dt\,, \qquad (48.27)$$

and the relation (48.24) is thereby recovered. Thus, the convergent integral representation (48.24) or (48.27) is rightly called "dispersion relation with one subtraction". The technique of dispersion relations is quite useful in some practical calculations, as we will see in what follows.

In the rest of this chapter we are going to discuss a simple example of one-loop QED radiative correction, namely the correction to the Coulomb potential due to the vacuum polarization. It is one of the earliest and best known results in QED; the original papers concerning this were published in mid 1930s by Edwin A. Uehling [49] and independently by Robert Serber [48] (of course, without Feynman diagrams that were born much later). So, let us consider the scattering of a charged spin-$\frac{1}{2}$ fermion (electron) in an external electrostatic field described by the potential $V(\vec{x})$; it means that we employ QED incorporating both quantized and classical electromagnetic fields. In the lowest (first) order, such a scattering process is described by the familiar tree diagram shown in Fig. 48.2. Note that $q = p' - p = (0, \vec{q})$, since the energy is

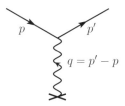

**Fig. 48.2:** Tree-level graph for the electron scattering in an external electromagnetic field.

conserved, in contrast to the momentum. The corresponding matrix element is

$$\mathcal{M}^{(1)} = e\,\bar{u}(p')\gamma_0 u(p) U(\vec{q})\,, \qquad (48.28)$$

where $U(\vec{q})$ is the Fourier transform of the potential, i.e.

$$U(\vec{q}) = \int d^3x\, e^{-i\vec{q}\cdot\vec{x}} V(\vec{x})\,. \qquad (48.29)$$

The contribution of the vacuum polarization is given by the 3rd order diagram shown in Fig. 48.3, where the fermion bubble is now represented by the renormalized quantity $\left(-i\overline{\Pi}_{\mu\nu}(q)\right)$

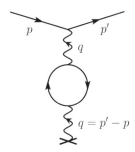

Vacuum polarization contribution to electron scattering in an external field.

involving the subtracted form factor $\overline{\Pi}(q^2)$. Using the standard Feynman rules and taking into account the transversality of $\overline{\Pi}_{\mu\nu}(q)$, as well as the notorious identity $\bar{u}(p')\slashed{q}u(p) = 0$, one gets for the corresponding matrix element

$$\mathcal{M}^{(3)} = e\,\bar{u}(p')\gamma_0 u(p) U(\vec{q}) \left(-\overline{\Pi}(q^2)\right). \tag{48.30}$$

Thus,

$$\mathcal{M}^{(1)} + \mathcal{M}^{(3)} = e\,\bar{u}(p')\gamma_0 u(p) U(\vec{q}) \left(1 - \overline{\Pi}(q^2)\right), \tag{48.31}$$

where, of course, $q^2 = -|\vec{q}|^2$, since $q = (0, \vec{q})$. Eq. (48.31) means that the sum of diagrams in Figs. 48.2 and 48.3 leads to an "effective potential" whose Fourier transform is

$$U_{\text{eff}}(\vec{q}) = U(\vec{q}) \left(1 - \overline{\Pi}(-|\vec{q}|^2)\right). \tag{48.32}$$

The Coulomb potential is, in our system of units,

$$V(\vec{x}) = \frac{1}{4\pi} \frac{e}{r}, \quad r = |\vec{x}|, \tag{48.33}$$

so that its Fourier transform $U(\vec{q})$ becomes

$$U(\vec{q}) = \frac{e}{|\vec{q}|^2}. \tag{48.34}$$

For $\overline{\Pi}(-|\vec{q}|^2)$, the dispersion relation (48.23) now comes in handy. It reads

$$\overline{\Pi}(-|\vec{q}|^2) = -\frac{\alpha}{3\pi}|\vec{q}|^2 \int_{4m^2}^{\infty} \frac{dt}{t} \sqrt{1 - \frac{4m^2}{t}} \left(1 + \frac{2m^2}{t}\right) \frac{1}{t + |\vec{q}|^2}, \tag{48.35}$$

and from (48.32) along with (48.34) we thus have

$$U_{\text{eff}}(\vec{q}) = \frac{e}{|\vec{q}|^2}\left[1 + \frac{\alpha}{3\pi}|\vec{q}|^2 \int_{4m^2}^{\infty} \frac{dt}{t} \sqrt{1 - \frac{4m^2}{t}} \left(1 + \frac{2m^2}{t}\right) \frac{1}{t + |\vec{q}|^2}\right]. \tag{48.36}$$

Now, our task is to carry out the inverse Fourier transformation, i.e. evaluate $V_{\text{eff}}(\vec{x})$ according to

$$V_{\text{eff}}(\vec{x}) = \int \frac{d^3q}{(2\pi)^3} U_{\text{eff}}(\vec{q})\, e^{i\vec{q}\cdot\vec{x}}. \tag{48.37}$$

For this purpose, one needs the formula

$$\int \frac{d^3q}{(2\pi)^3} e^{i\vec{q}\cdot\vec{x}} \frac{1}{t + |\vec{q}|^2} = \frac{1}{4\pi} \frac{e^{-r\sqrt{t}}}{r} \tag{48.38}$$

(its verification is left as a homework for any diligent reader). Thus, one gets

$$V_{\text{eff}}(r) = \frac{1}{4\pi} \frac{e}{r} \left( 1 + \frac{2\alpha}{3\pi} \int_1^\infty du\, e^{-2mru} \left( 1 + \frac{1}{2u^2} \right) \frac{\sqrt{u^2 - 1}}{u^2} \right), \tag{48.39}$$

where we have introduced, instead of $t$, a dimensionless integration variable $u$ such that $t = 4m^2u^2$.

So, the result of our calculation can be summarized in an elegant form by writing it as

$$V_{\text{eff}}(r) = \frac{e}{4\pi r} Q(r), \tag{48.40}$$

where $Q(r)$ is an "effective charge" depending on the distance,

$$Q(r) = 1 + \frac{2\alpha}{3\pi} \int_1^\infty du\, e^{-2mru} \left( 1 + \frac{1}{2u^2} \right) \frac{\sqrt{u^2 - 1}}{u^2}. \tag{48.41}$$

It is most instructive to find out what is the limiting behaviour of $Q(r)$ for small and large $r$. It turns out that for leading terms one gets

$$Q(r) = 1 + \frac{\alpha}{3\pi} \left( \ln \frac{1}{(mr)^2} + \dots \right), \quad mr \ll 1 \tag{48.42}$$

and

$$Q(r) = 1 + \frac{\alpha}{4\sqrt{\pi}(mr)^{3/2}} e^{-2mr} + \dots, \quad mr \gg 1. \tag{48.43}$$

Thus, it is seen that at large distances, one has the Coulomb potential with the usual charge $e$, while at small distances the effective charge grows logarithmically and goes to infinity for $r \to 0$. A derivation of the results (48.42) and (48.43) is left as a challenge for calculus aficionados.

Intuitively, such a picture now justifies the term "vacuum polarization" that we have been using from Chapter 38 on. The idea is that the QED vacuum behaves, in a sense, like a polarizable medium that contributes to "screening" of the charge at larger distances by means of a "cloud of virtual electron–positron pairs" appearing in the loops. The relation between the values of $Q(r)$ at large and small distances is thus analogous to the relation between renormalized and bare electromagnetic coupling: recall that according to (46.37) one has $e^2 = (1 - \Pi(0))e_0^2$, with $\Pi(0) > 0$. Thus, $e^2 < e_0^2$, i.e. the vacuum polarization $\Pi(0)$ causes the **screening** of the bare coupling constant in QED.

# Chapter 49

# Calculable quantities:
# UV finite without counterterms

We already know that in QED there are one-loop Feynman diagrams that are UV finite, and thus do not generate renormalization counterterms. They correspond precisely to situations, in which the would-be counterterms with dimension greater than four could destroy the perturbative renormalizability of the theory in higher orders. In this context, a particularly interesting case is the fermion box, discussed in detail in Chapter 43. Such a diagram is the only contribution to an intriguing process, namely the photon–photon elastic scattering (or, as it is often called, the **light-by-light scattering**). Such a process certainly does not exist at the classical level, i.e. it is a purely quantum effect. A detailed calculation of the contribution of the box diagram shown in Fig. 43.7 is unfortunately very tedious; explicit results are available e.g. in the book [7]. Nevertheless, one can get at least a reasonable estimate of the cross section of the photon–photon scattering in the low-energy limit, e.g. for visible light. The technique appropriate for such a purpose is based on the idea of an **effective Lagrangian**; so, the discussion of the scattering process in question provides us with an opportunity to illustrate the power and modest beauty of this rather general method.

The basic idea is in fact quite simple. If one considers photon energies ($E_\gamma$) much smaller than the rest mass ($m$) of the virtual charged particle (the electron, for definiteness) inside the loop, one can assume that the contribution of the closed-loop diagram is approximately equal to the contribution of a 1st order (tree-level) graph corresponding to an effective Lagrangian for direct interaction of four photons. Now, before proceeding further, a remark is in order here. An assumption of the above-mentioned kind should sound familiar to anybody acquainted with fundamentals of weak interaction theory. There, the original Fermi-type model of a direct four-fermion interaction is a low-energy effective theory corresponding to the deeper underlying theory (the standard electroweak model) involving intermediate vector bosons (IVB) as the "force carriers". The "matching condition" for those two theories to be equivalent in the low-energy limit is $G_F \propto g^2/m_W^2$, where $G_F$ is the Fermi coupling constant, $g$ is a dimensionless coupling for the IVB interactions and $m_W$ is the IVB mass.

Returning to the problem of the photon–photon scattering in QED, our assumption corresponds, schematically, to the approximate equality depicted in Fig. 49.1 for $E_\gamma \ll m$; the main problem is how to estimate the interaction strength $G$ corresponding to the heavy dot in the tree diagram appearing there.

To begin with, one must take into account that an effective four-photon Lagrangian should be gauge invariant (since the underlying QED Lagrangian is so), i.e. it must be made of a product

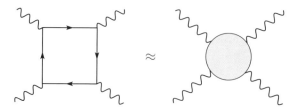

**Fig. 49.1:** Schematic depiction of the origin of an effective Lagrangian for the light-by-light scattering in QED.

of four electromagnetic field tensors $F_{\mu\nu}$; appropriate forms would be e.g.

$$F_{\alpha\beta}F^{\beta\mu}F_{\mu\nu}F^{\nu\alpha} , \tag{49.1}$$

or

$$(F_{\mu\nu}F^{\mu\nu})^2 . \tag{49.2}$$

We will discuss the relevant $FFFF$ structures later on; for now, it is sufficient to notice that the mass dimension of a term like this is eight. This in turn means that the effective coupling $G$ has the dimension minus four (since the full Lagrangian has the dimension four). Obviously, the coupling $G$ must be made of the electron mass (in general, the mass of the charged particle in the loop) and, because of the equality depicted in Fig. 49.1, it must be proportional to $e^4$ (i.e. $\alpha^2$). Thus, the matching condition reads

$$G = \text{const.} \frac{\alpha^2}{m^4} \tag{49.3}$$

(notice that the last relation is analogous to the above-mentioned matching condition for the Fermi constant). Now, estimating the scattering cross section is an easy task. The contribution of the tree-level effective diagram in Fig. 49.1 is proportional to $G$, so that the cross section is proportional to $G^2$. The dimension of cross section is (length)$^2$, i.e. (mass)$^{-2}$, and apart from $G^2$, it is proportional to an appropriate power of photon energy. So, it is clear that the cross section we are trying to estimate behaves like

$$\sigma(E_\gamma)\big|_{E_\gamma \ll m} = \text{const.} \frac{\alpha^4}{m^8} E_\gamma^6 . \tag{49.4}$$

This is the key result: in the low-energy limit, the light-by-light scattering cross section is proportional to $E_\gamma^6$! Note that our derivation has been rather simple, two basic ingredients being **gauge invariance** and **dimensional arguments**, but the result (49.4) is in fact quite strong. Concerning the numbers, for visible light, $E_\gamma \doteq 2 - 3\,\text{eV}$; thus, taking into account that the electron mass is $m \doteq 0.5\,\text{MeV}$, $\alpha = 1/137$ and invoking the conversion constant $\hbar c = 197\,\text{MeV fm}$ with $1\,\text{fm} = 10^{-13}\,\text{cm}$, one has $1\,\text{MeV}^{-2} \doteq 4 \times 10^{-22}\,\text{cm}^2$. Then one gets

$$\sigma(\gamma\gamma \to \gamma\gamma)\big|_{\substack{\text{visible} \\ \text{light}}} \simeq 10^{-67}\,\text{cm}^2 . \tag{49.5}$$

So, for the visible light the cross section is terribly small and thus it is not surprising that it has not been observed yet.

Let us now review some more detailed results available in the literature. The effective low-energy four-photon Lagrangian resulting from the calculation of the box diagram in Fig. 49.1

can be written as a linear combination of the expressions (49.1) and (49.2). In fact, it is more common to use another basis, consisting of $(F^2)^2 = (F_{\mu\nu}F^{\mu\nu})^2$ and $(F \cdot \widetilde{F})^2 = (F_{\mu\nu}\widetilde{F}^{\mu\nu})^2$, where $\widetilde{F}_{\mu\nu}$ is the dual of $F_{\mu\nu}$, i.e.

$$\widetilde{F}_{\mu\nu} = \frac{1}{2}\,\epsilon_{\mu\nu\rho\sigma}F^{\rho\sigma}\,. \tag{49.6}$$

With some effort, utilizing the properties of the Levi-Civita symbol $\epsilon_{\mu\nu\rho\sigma}$, one finds out that

$$\left(F \cdot \widetilde{F}\right)^2 = -2\left(F^2\right)^2 + 4\left\langle FFFF\right\rangle\,, \tag{49.7}$$

where $\langle FFFF \rangle$ is a shorthand notation for the expression (49.1). A detailed evaluation of the box diagram (including, of course, all permutations of external photon lines) leads to the result

$$\mathscr{L}_{\text{eff}} = \frac{\alpha^2}{90m^4}\left[\left(F^2\right)^2 + \frac{7}{4}\left(F \cdot \widetilde{F}\right)^2\right]\,. \tag{49.8}$$

This can be recast in a form that is most usual in current literature, namely

$$\mathscr{L}_{\text{eff}} = \frac{2\alpha^2}{45m^4}\left[\left(\vec{E}^2 - \vec{B}^2\right)^2 + 7\left(\vec{E} \cdot \vec{B}\right)^2\right]\,, \tag{49.9}$$

where $\vec{E}$ and $\vec{B}$ denote the strength of the electric and magnetic field, respectively. To arrive at (49.9) from (49.8), one employs the identities

$$\vec{E}^2 - \vec{B}^2 = -\frac{1}{2}\,F_{\mu\nu}F^{\mu\nu}\,, \qquad \vec{E} \cdot \vec{B} = \frac{1}{4}\,F_{\mu\nu}\widetilde{F}^{\mu\nu}\,. \tag{49.10}$$

The result (49.9) is called the **Euler–Heisenberg Lagrangian**, since, historically, it was derived first in mid 1930s by Werner Heisenberg and his doctoral student Hans Euler (needless to say, before the advent of Feynman diagrams). Since then, it was recovered by other authors using different techniques, including, of course, the direct evaluation of the box diagram. Nice reviews of this time-honoured subject can be found in ref. [50]. A lot of useful information is contained also in the book [15] (including some remarkable details of Euler's biography).

It is perhaps worth mentioning that an effective Lagrangian of the Euler–Heisenberg type can also be evaluated for other QED models. In particular, for scalar QED the contributing one-loop diagrams are shown in Chapter 47 (see Fig. 47.8); the sum of their contributions leads to the effective low-energy Lagrangian of the form

$$\mathscr{L}_{\text{eff}}^{(\text{scalar})} = \frac{\alpha^2}{1440\,m_s^4}\left[7\left(F^2\right)^2 + \left(F \cdot \widetilde{F}\right)^2\right]\,, \tag{49.11}$$

where $m_S$ is the mass of the charged scalar in the relevant closed loops. By the way, this result has been recovered in the paper [51]. In this paper, one can also find the result for QED of massive charged spin-1 particles (i.e. electromagnetic interactions of $W^\pm$ bosons within SM). The calculation is indeed a formidable task and the result is

$$\mathscr{L}_{\text{eff}}^{(\text{vector})} = \frac{\alpha^2}{160\,m_W^4}\left[29\left(F^2\right)^2 + 27\left(F \cdot \widetilde{F}\right)^2\right] \tag{49.12}$$

(note that in such a case the relevant QED one-loop diagrams are topologically the same as in scalar QED, with the obvious replacement of propagators and the rule for vertices).

Let us return to the "textbook case" of spinor QED. When one employs the effective Euler–Heisenberg Lagrangian (49.9), the evaluation of the low-energy cross section for photon–photon scattering is straightforward, though somewhat tedious. For unpolarized photons one gets the result (see e.g. [6])

$$\sigma(\gamma\gamma \to \gamma\gamma)\big|_{E_\gamma \ll m} = \frac{1}{2\pi}\frac{7784}{89100}\frac{\alpha^4}{m^8}E_\gamma^6 . \qquad (49.13)$$

Note that the numerical factor in (49.13) turns out to be roughly 0.01, so one may say that our estimate based on the simple form (49.4) is not so bad.

If one goes beyond the low-energy region, the cross section $\sigma(\gamma\gamma \to \gamma\gamma)$ depends quite strongly on the photon energy. The analytic expression is, of course, rather complicated (see e.g. the book [7]), but the plot of the energy dependence of $\sigma$ is quite instructive and the reader is encouraged to look it up in Chapter 12 of [7] (Fig. 24 therein). It is remarkable that the cross section in question, which is so tiny for the visible light, becomes quite sizable (at least by the particle physics standards) in the region of photon energies of the order of 1 MeV. The maximum value of $\sigma$ is about $10^{-30}$ cm$^2$ and corresponds to the photon collision energy in the c.m. system close to the threshold for production of an electron–positron pair (i.e. for the inelastic process $\gamma\gamma \to e^- e^+$). At this threshold, the plot of $\sigma(E_\gamma)$ has a cusp singularity corresponding to the discontinuous derivative of $\sigma$. For $E_\gamma \to \infty$ the cross section decreases rapidly to zero.

Finally, let us remark that a measurement of such a cross section is an extremely challenging task, for obvious reasons. Nevertheless, quite recently, the collaboration ATLAS at LHC (CERN) announced an observation of events of the desired type, in the so-called ultraperipheral collisions of heavy ions; in fact, what is observed is a collision of "quasi-real" virtual photons constituting the electromagnetic fields of colliding lead ions (see [52]).

The matrix element for the photon–photon scattering is, as we have emphasized, UV finite in QED and no particular counterterm is needed. Generally, such a quantity is called "calculable".[29] There is another prominent example of a quantity of such a type, namely the magnetic moment of the electron (or muon, if you want). The lowest-order prediction for this follows already from the Dirac equation considered as an equation of relativistic quantum mechanics (see Chapter 2). Within QED, there is a calculable radiative correction to the lowest-order result. The successful calculation of such an $\mathcal{O}(\alpha)$ correction to the electron magnetic moment had become a breakthrough argument in favour of QED at the end of 1940s and quantum field theory then entered its modern era, in a sense its "golden age".[30] The evaluation of the so-called **Schwinger correction** mentioned above is the subject of the next chapter.

---

[29]In Czech: "spočitatelná, resp. "vyčíslitelná" veličina.

[30]It is fair to note that the term "golden age" should be taken with a grain of salt. In particular, while there had been many precision calculations within renormalized QED, late 1950s and 1960s brought a sort of crisis of the QFT status, which was mostly due to unsatisfactory description of strong and weak interactions by means of the available QFT models. Thus, many theorists in the particle physics community rejected at that time QFT as a fundamental concept and looked for radically different alternatives. Nevertheless, the ultimate change of the paradigm in favour of QFT came in the early 1970s with the advent of the non-Abelian gauge theory of weak and strong interactions, i.e. the highly successful scheme now commonly called the standard model of particle physics. Concerning this, the reader may find relevant information in several books cited here, see e.g. [14, 22, 27] and [31].

# Chapter 50

# Schwinger correction

As we have recalled at the end of the preceding chapter, one of the highlights of the original Dirac theory is a prediction of the value of the spin magnetic moment of the electron; we have discussed various aspects of this achievement is some detail in chapters 2 and 13. In particular, in Chapter 13 we have stressed that the original result

$$\mu_e = \mu_B = \frac{e}{2m} \tag{50.1}$$

(where $\mu_B$ is the common symbol for the Bohr magneton) is in fact a "conditional" prediction, as it depends on an additional assumption concerning the interaction of the Dirac particle with an external electromagnetic field: for deriving the result (50.1), it is assumed that the interaction is "minimal", i.e. it is defined by means of the familiar replacement $\partial_\mu \to \partial_\mu - ieA_\mu$ in the equation for the free particle. Apparently, such a recipe is appropriate for a pointlike particle, like the electron or muon (but, as we know, it would be insufficient for a phenomenological description of the proton or neutron).

In any case, the theoretical derivation of the result (50.1) was a stunning success in its time, since it is indeed very close to the experimental value of $\mu_e$. However, some high-precision measurements carried out in 1940s (see Chapter 13) have shown that there is in fact a small deviation from the Dirac prediction, with the relative magnitude of about 0.001. So, it has become an obvious challenge for theorists to explain such a tiny discrepancy within the framework of QED, which at that time was developing rapidly. The first successful calculation of the QED correction to the Dirac prediction is due to J. Schwinger, who had done it in 1948 without using Feynman diagrams (he never used them). In what follows, we are going to derive the famous Schwinger's result with the help of standard techniques of Feynman diagrams.

First of all, we have to find out how to extract the desired magnetic moment from the matrix element for the electron scattering in an external field. QED Feynman diagrams of the first and third order that might be of interest are shown in Fig. 50.1 (though we have already discussed the diagrams (a) and (b) in Chapter 26; here we have summarized them all for reader's convenience).

Before proceeding to the diagram calculation, let us outline our basic strategy. We will consider the electron scattering in a static, spatially homogeneous magnetic field, i.e. use the four-potential of the form

$$A_\mu(x) = \left(0, A_j(\vec{x})\right), \tag{50.2}$$

with the vector $A_j(\vec{x})$ chosen so that the field strength

$$B_j = \epsilon_{jkl}\partial_k A_l(\vec{x}) \tag{50.3}$$

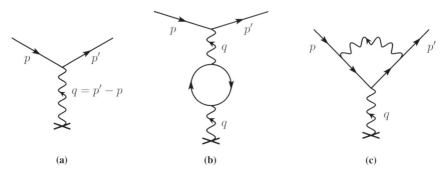

**Fig. 50.1:** (a) 1st order diagram for electron scattering in an external electromagnetic field; (b) and (c) are 3rd order diagrams involving both external and quantized field.

is a constant. Since the spin magnetic moment is a static characteristic of a particle, we will examine the relevant matrix element for non-relativistic, quasi-static electrons; then, one may envisage obtaining a form involving the scalar product of the $\vec{B}$ with the vector of the would-be magnetic moment, proportional to Pauli spin matrices, The corresponding coefficient of proportionality is then identified as the value of the electron magnetic moment in question.

So, now we are ready to evaluate the relevant diagrams in Fig. 50.1, keeping in mind the hint sketched above. For the contribution of the diagram (a) one has

$$\mathcal{M}_a = e\,\bar{u}(p')\gamma^j u(p)\widetilde{A}_j(\vec{q})\,, \tag{50.4}$$

where $\widetilde{A}_j(\vec{q})$ is the Fourier transform of the vector potential in (50.2) (note that in what follows, tilde always denotes the Fourier transform). For our purpose it will be useful to employ the Gordon decomposition (see (C.28) in Appendix C)

$$\bar{u}(p')\gamma^j u(p) = \frac{1}{2m}\bar{u}(p')\big[(p+p')^j + i\sigma^{jk}q_k\big]u(p)\,, \tag{50.5}$$

separating the "convective" and "spin" parts of the Dirac current. Let us recall that $\sigma_{jk} = \frac{i}{2}[\gamma_j,\gamma_k]$, and this is simply related to the spin matrices $\Sigma$ through the identity (see (C.36))

$$\sigma_{jk} = \epsilon_{jkl}\Sigma_l\,. \tag{50.6}$$

Needless to say, in the standard representation of $\gamma$-matrices one has

$$\vec{\Sigma} = \begin{pmatrix} \vec{\sigma} & 0 \\ 0 & \vec{\sigma} \end{pmatrix}\,, \tag{50.7}$$

where $\vec{\sigma}$ are the Pauli matrices.

Let us now work out the part of the matrix element (50.4) involving the spin matrices $\sigma^{jk}$. We have

$$\widetilde{A}_j(\vec{q}) = \int d^3x\, e^{-i\vec{q}\cdot\vec{x}} A_j(\vec{x})\,,$$

and thus

$$q_k\widetilde{A}_j(\vec{q}) = i\int d^3x\, e^{-i\vec{q}\cdot\vec{x}}\partial_k A_j(\vec{x})\,. \tag{50.8}$$

Then, using the identities (50.5) and (50.6) (note also that $\sigma^{jk} = \sigma_{jk}$), the expression (50.4) is recast as

$$\mathcal{M}_a = \frac{e}{2m}\bar{u}(p')\big[(p+p')^j + i\epsilon_{jkl}\Sigma_l q_k\big]u(p)\widetilde{A}_j(\vec{q})\,. \tag{50.9}$$

298

Furthermore, from (50.8), along with (50.3), one gets

$$i\widetilde{B}_l(\vec{q}) = \epsilon_{lkj} q_k \widetilde{A}_j(\vec{q}) . \tag{50.10}$$

Thus, the matrix element (50.9) becomes finally

$$\mathcal{M}_a = \frac{e}{2m} \bar{u}(p') u(p)(p+p')^j \widetilde{A}_j(\vec{q})$$
$$+ \frac{e}{2m} \bar{u}(p')\left(\vec{\Sigma} \cdot \vec{\widetilde{B}}\right) u(p) \tag{50.11}$$

(when passing from (50.9), (50.10) to (50.11) please don't forget that $\epsilon_{jkl} = -\epsilon_{lkj}$).

For a constant $\vec{B}$, its Fourier transform is proportional to $\delta^{(3)}(\vec{q})$ (thus, in such a scattering process both energy and momentum are conserved, as it must be for any static homogeneous external field); having it in mind, in what follows we will write simply $\vec{\Sigma} \cdot \vec{B}$ in the second term in the expression (50.11). Now, in the spirit of the strategy outlined above, we restrict ourselves to quasi-static electrons. Appealing again to reader's previous training in diracology, let us recall that in such a case, $u(p)$ (as well as $u(p')$) is reduced to

$$u(p) = \begin{pmatrix} w \\ 0 \end{pmatrix}, \tag{50.12}$$

i.e. only the upper two components survive, and the two-component column $w$ embodies the two possible electron spin states. Note that we have in mind the standard representation of $\gamma$-matrices; so, we are going to utilize the form (50.7) for $\vec{\Sigma}$. Moreover, since

$$\gamma_0 = \begin{pmatrix} \mathbb{1} & 0 \\ 0 & -\mathbb{1} \end{pmatrix}$$

in the standard representation, the matrix products in (50.11) obviously become

$$\bar{u}(p') u(p) = w_f^\dagger w_i ,$$
$$\bar{u}(p') \vec{\Sigma} \cdot \vec{B} u(p) = w_f^\dagger \vec{\sigma} \cdot \vec{B} w_i , \tag{50.13}$$

where $i$ and $f$ are usual labels for the initial and final state, respectively.

In fact, experimental measurements of the magnetic moment are based on detecting spin-flip transitions from the initial to final state. This means that the first term in the matrix element (50.11), being proportional to $w_f^\dagger w_i$, is irrelevant for our purpose; we are left with the non-trivial spin contribution

$$\mathcal{M}_a = \frac{e}{2m} w_f^\dagger \vec{\sigma} \cdot \vec{B} w_i , \tag{50.14}$$

and there one can see the value of the electron magnetic moment corresponding precisely to the original Dirac prediction (50.1). This is just the anticipated result.

Note that, in musical terms, the tempo we have chosen for the preceding exposition was "andante", or rather "largo", but it has not been a loss of time or energy, because in most of the available textbooks the details of the relevant calculation are usually skipped; so, our lengthy discussion has been presented here for the reader's convenience.

Now, we would like to calculate the $\mathcal{O}(\alpha)$ radiative correction to the lowest-order value (50.1) that might originate in one-loop diagrams in Figs. 50.1(b),(c). To this end, we will need a formula describing the general form of the matrix element in question. It turns out that the matrix element for the electron scattering in an arbitrary external field $A_\mu(x)$ can be written, in general, as follows:

$$\mathcal{M} = e\,\bar{u}(p')\left[F_1(q^2)\gamma^\mu + \frac{i}{2m}\sigma^{\mu\nu}q_\nu F_2(q^2)\right]u(p)\widetilde{A}_\mu(q)\,, \tag{50.15}$$

where $F_1$ and $F_2$ are two independent form factors. Note that the form (50.15) does not depend on the perturbation expansion. Obviously, the tree-level matrix element (50.4) has such a form, with $F_1(q^2) = 1$ and $F_2(q^2) = 0$. If we stick to the on-shell renormalization scheme, the form factor $F_1(q^2)$ is normalized by the condition $F_1(0) = 1$; this corresponds to the subtraction in the vertex function $\Gamma_\mu(p', p)$ at $q = 0$ (see (47.10)). The derivation of the formula (50.15) can be found in the Appendix F. Here, let us emphasize its important consequences for our calculation of the desired radiative correction.

We already know that for obtaining the value of the magnetic moment, the kinematical configuration with four-momentum transfer $q = (0, \vec{q})$, $\vec{q} = 0$ is considered. Now, the contribution of the diagram (b) in Fig. 50.1 is just the tree-level matrix element multiplied by $\overline{\Pi}(q^2) = \Pi(q^2) - \Pi(0)$ (cf. (48.30)), so it vanishes for $q = 0$; thus, the vacuum polarization graph (b) does not contribute to the radiative correction we are looking for. As for the diagram (c), there is a subtraction at $q = 0$ for the term involving $\gamma^\mu$, so the contribution to the form factor $F_1$ from this diagram vanishes for $q = 0$. It means that the form factor $F_1(q^2)$, normalized by $F_1(0) = 1$, gives just the tree-level value (50.1). On the other hand, the term in (50.15) involving the form factor $F_2$ can contribute even for $q = 0$, since, as we know from the previous discussion of the tree diagram (a), the factor $q_\nu$ is absorbed in the definition of the magnetic field strength. In fact, it should be clear now that for working out the contribution of the $F_2$ term in (50.15) one can essentially repeat the steps that followed the relation (50.5); the only extra ingredient is the additional factor $F_2(0)$ in the contribution to the magnetic moment. Thus, we end up with the general formula

$$\mu_e = \frac{e}{2m}\left[1 + F_2(0)\right]. \tag{50.16}$$

To summarize our preceding discussion: First, the diagram (b) does not contribute at all. Second, the contribution of the diagram (c) involving $\gamma_\mu$ is irrelevant, as it vanishes for $q = 0$ (due to the on-shell subtraction). Finally, when computing the vertex function $\Gamma_\mu(p', p)$, the term proportional to $\sigma^{\mu\nu}q_\nu$ is UV convergent (let us recall that the only UV divergence in $\Gamma_\mu(p', p)$ is proportional to $\gamma_\mu$) and thus **the form factor $F_2(q^2)$ is calculable.**[31] The only problem that remains to be clarified is how to get all relevant terms involving the structure $\sigma^{\mu\nu}q_\nu$ in a practical evaluation of the diagram (c). The answer is quite easy. Because of Lorentz covariance, an explicit calculation of the vertex function $\Gamma_\mu(p', p)$ brings naturally terms proportional to $\gamma_\mu$, $p_\mu$ and $p'_\mu$. While the $\gamma_\mu$ terms can be thrown away, those involving $p_\mu$ and $p'_\mu$ are expected to appear eventually in the combination $p_\mu + p'_\mu$, and this in turn can be converted to $\sigma_{\mu\nu}q^\nu$ by means of the Gordon identity. Thus, the strategy of the evaluation of the relevant part of the vertex diagram (c) is clearly set, and we may now go *in medias res* to start the calculation, which is quite tedious, but its result is nice and truly rewarding.

The contribution of the "interior" of the diagram (c) can be written as (cf. also (41.19))

$$i\Gamma_\mu(p', p) = e^3 \int \frac{\mathrm{d}^4 l}{(2\pi)^4}\, \frac{\gamma_\alpha(\slashed{l} + \slashed{p}' + m)\gamma_\mu(\slashed{l} + \slashed{p} + m)\gamma^\alpha}{[(l + p')^2 - m^2][(l + p)^2 - m^2](l^2 - \lambda^2)}\,. \tag{50.17}$$

Note that we do not invoke here any UV regularization, since we need only the above-mentioned finite part of the integral (50.17). The first few steps of the evaluation of the expression (50.17)

---

[31]This, of course, is gratifying, since a counterterm for the contribution proportional to $\sigma^{\mu\nu}q_\mu$ would have the structure $\bar{\psi}\sigma_{\mu\nu}\psi F^{\mu\nu}$, and such an UV divergent term would spoil the renormalizability (cf. Chapter 44).

are routine: introducing Feynman parametrization, shifting properly the loop momentum $l$ and using the symmetric integration, one gets

$$i\Gamma_\mu(p',p)\Big|_{\text{UVfin.}} = 2e^3 \int\limits_0^1 dx \int\limits_0^{1-x} dy \int \frac{d^4l}{(2\pi)^4} \frac{N_\mu}{(l^2 - C)^3} , \tag{50.18}$$

where the numerator $N_\mu$ reads

$$N_\mu = \gamma_\alpha(-x\slashed{p} + (1-y)\slashed{p}' + m)\gamma_\mu((1-x)\slashed{p} - y\slashed{p}' + m)\gamma^\alpha , \tag{50.19}$$

and

$$C = (x+y)^2 m^2 + (1-x-y)\lambda^2 - xy\,q^2 . \tag{50.20}$$

Note that in arriving at the formula (50.20) we have set $p^2 = m^2$, $p'^2 = m^2$. (Needless to say, the reader is encouraged to recover the above expressions independently.) Now, we in fact need the relevant expression for $\Gamma_\mu(p',p)$ sandwiched between $\bar{u}(p')$ and $u(p)$, as this enters the matrix element (50.15). To work it out, we utilize the algebra of $\gamma$-matrices, in particular the "chain identities" (see Appendix C, formulae (C.24)) for $n = 4$, and also Dirac equations $(\slashed{p} - m)u(p) = 0$ and $\bar{u}(p')(\slashed{p}' - m) = 0$. At the same time, we discard systematically the terms involving just $\gamma_\mu$ alone, as these are irrelevant. The calculation is quite lengthy and boring; an intermediate result is

$$\bar{u}(p')N_\mu u(p) = \bar{u}(p')u(p)\Big\{ [\,4x(1-x) - 4(1-x)(1-y) + 4(1-2x)\,]\,m\,p_\mu$$
$$+ [\,4y(1-y) - 4(1-x)(1-y) + 4(1-2y)\,]\,m\,p'_\mu\Big\} + \ldots \tag{50.21}$$

where the ellipsis means the irrelevant terms.

As a next step, we may employ our master formula for the loop-momentum integration (see (39.8)). In our case, this gives

$$\int \frac{d^4l}{(2\pi)^4} \frac{1}{(l^2 - C)^3} = -\frac{i}{32\pi^2}\frac{1}{C} . \tag{50.22}$$

From (50.20) it is obvious that $C$ is symmetric under the interchange of $x$ and $y$.

Now, one may notice that the expression (50.21) has a remarkable symmetry: the coefficients at $p_\mu$ and $p'_\mu$ can be made equal upon the interchange $x \leftrightarrow y$ in one of them. To see that such an interchange is legitimate, one should realize that the integration over the Feynman parameters has the following simple property:

$$\int\limits_0^1 dx \int\limits_0^{1-x} dy\, f(x,y) = \int\limits_0^1 dx \int\limits_0^{1-x} dy\, f(y,x) \tag{50.23}$$

(the proof is easy). Taking all this into account, it is clear that under the integral over $x$ and $y$ one may identify the coefficients at $p_\mu$ and $p'_\mu$ and the expression (50.21) thus becomes, effectively,

$$\bar{u}(p')N_\mu u(p) \quad\longrightarrow\quad 4m(p_\mu + p'_\mu)[x(1-x) + 1 - 2x - (1-x)(1-y)]\bar{u}(p')u(p)$$
$$= 4m(p_\mu + p'_\mu)(y - xy - x^2)\bar{u}(p')u(p) , \tag{50.24}$$

where we have already omitted the irrelevant terms. This is the anticipated result: $p_\mu$ and $p'_\mu$ enter in the combination $p_\mu + p'_\mu$, which can be immediately turned into $\sigma_{\mu\nu}q^\nu$ (with the help of

301

Gordon identity) and thus fits in the general form (50.15). Explicitly, the Gordon identity tells us that

$$(p_\mu + p'_\mu)\bar{u}(p')u(p) = 2m\,\bar{u}(p')\gamma_\mu u(p) - i\,\bar{u}(p')\sigma_{\mu\nu}q^\nu u(p)\,. \tag{50.25}$$

One thus has, preserving only the relevant term,

$$\bar{u}(p')\Gamma_\mu(p',p)u(p) = e^3\frac{i}{4\pi^2}\,m\,\bar{u}(p')\sigma_{\mu\nu}q^\nu u(p)\int_0^1 dx\int_0^{1-x} dy\,\frac{y - xy - x^2}{(x+y)^2 m^2 - xy\,q^2}\,. \tag{50.26}$$

Note that to obtain the last result, we have used Eqs. (50.18), (50.20), (50.22), (50.24) and (50.25); in the denominator of the integrand we have also discarded the IR regulator $\lambda$, as the integral in question has no IR divergence. The identification of the form factor $F_2(q^2)$ is now easy: the contribution of the diagram (c) to the scattering matrix element in question is

$$\mathcal{M}_c = \bar{u}(p')\Gamma_\mu(p',p)u(p)\widetilde{A}^\mu(q)\,,$$

and comparing the structure of the expression (50.26) with the general form (50.15), one gets

$$F_2(q^2) = \frac{e^2}{2\pi^2}\,m^2\int_0^1 dx\int_0^{1-x} dy\,\frac{y - xy - x^2}{(x+y)^2 m^2 - xy\,q^2}\,. \tag{50.27}$$

By means of appropriate substitutions, the last expression can be reduced to a simple one-dimensional integral

$$F_2(q^2) = \frac{\alpha}{2\pi}\int_0^1 du\,\frac{1}{1 - \dfrac{q^2}{m^2}\,u(1-u)}\,, \tag{50.28}$$

where we have also used $e^2 = 4\pi\alpha$. (The passage from (50.27) to (50.28) is left to enthusiasts as a nice exercise in computing integrals.) Thus, the correction to the electron magnetic moment is, according to the key formula (50.16),

$$F_2(0) = \frac{\alpha}{2\pi}\,. \tag{50.29}$$

This is the celebrated **Schwinger correction**, published first in 1948. Obviously, J. Schwinger was proud of this achievement and so it is also placed on his tombstone (see Fig. 50.2). Let us add that this computational success (that matched the experimental results) was certainly a part of the reason why he received the Nobel Prize in 1965 together with R. Feynman and S. Tomonaga (the precise citation of the Nobel committee was "for their fundamental work in quantum electrodynamics, with deep-ploughing consequences for the physics of elementary particles").

So, the QED prediction for the electron spin magnetic moment is, more generally,

$$\mu_e = \mu_B\left(1 + \frac{\alpha}{2\pi} + \ldots\right), \tag{50.30}$$

where the ellipsis stands for contributions of the order $\mathcal{O}(\alpha^2)$ and higher. Of course, the evaluation of those higher-order contributions is rather laborious indeed; nevertheless, as of today, theoretical calculations have been done up to the order $(\alpha/\pi)^4$ (both numerical and analytic) and, recently, also the correction of the order $(\alpha/\pi)^5$ has been obtained (numerically).

Such a theoretical precision matches the accuracy of the best current experiments and the electron magnetic moment is in fact a quantity most precisely ever measured. For an illustration, let us mention explicitly the order $\mathcal{O}(\alpha^2)$. The corresponding result had been obtained in 1957 independently by Charles Sommerfield and André Petermann. Using their result, one may write, approximately,

$$\mu_e = \mu_B\left[1 + \frac{\alpha}{2\pi} - 0.328\left(\frac{\alpha}{\pi}\right)^2 + \mathcal{O}(\alpha^3)\right],\tag{50.31}$$

where the coefficient at $(\alpha/\pi)^2$ arises as

$$\frac{197}{144} + \frac{\pi^2}{12} - \frac{\pi^2}{2}\ln 2 + \frac{3}{4}\zeta(3) \doteq -0.328,\tag{50.32}$$

with $\zeta(3) = \sum_{n=1}^{\infty} 1/n^3$ being the relevant value of the Riemann zeta-function ($\zeta(3) \doteq 1.2$ is also known as the Apéry's constant). The structure of the expression (50.32) indicates that obtaining it must be a tough job.

## Epilogue

Coming back to the Schwinger correction (50.29) that we have been able to recover explicitly, after some effort that is not overly time-consuming, we can see that numerically, $\alpha/2\pi \doteq 0.001$ if we set $\alpha = 1/137$. One may now recall the introductory Chapter 2, where we have reproduced the original Dirac's prediction $\mu_e = \mu_B$. Thus, one might say, with some self-irony, that after going through the remaining 48 chapters (which correspond roughly to six months of a lecture course within the academic year), our knowledge of relativistic quantum theory has improved by about one per mille. Well, of course, I am kidding. In fact, we have seen that one has to master a lot of material, in order to be able to compute a tiny QED radiative correction like (50.29). One may conclude that QED is a great theory indeed, which fully deserved the Nobel Prize in its time. Now it is a subsection of the more comprehensive standard model of elementary particle physics.

# Appendix A

# Basic properties
# of Lorentz transformations

The Lorentz transformation of spacetime coordinates is written as

$$x'^{\mu} = \Lambda^{\mu}{}_{\nu} x^{\nu} \tag{A.1}$$

or

$$x'_{\mu} = \Lambda_{\mu}{}^{\nu} x_{\nu} , \tag{A.2}$$

where, in accordance with the rules for raising and lowering the indices, one has

$$\Lambda_{\mu}{}^{\nu} = g_{\mu\rho} \Lambda^{\rho}{}_{\sigma} g^{\sigma\nu} . \tag{A.3}$$

When the coordinates $x^{\mu}$ are ordered in a column, the relation (A.1) is recast in the matrix form simply as $x' = \Lambda x$ (with $\mu$ and $\nu$ being the row and column index, respectively); for convenience, the matrix defined by the elements appearing in (A.2) may be then denoted as $\overline{\Lambda}$. The formula (A.3) is thus tantamount to the matrix identity

$$\overline{\Lambda} = g \cdot \Lambda \cdot g , \tag{A.4}$$

where $g$ stands for the $4 \times 4$ matrix representing the metric of the Minkowski space (which is the same for the components $g_{\mu\nu}$ and $g^{\mu\nu}$); recall that our convention is $g = \text{diag}(+1, -1, -1, -1)$.

The invariance of the spacetime interval under Lorentz transformations means that $x'^{\mu} x'_{\mu} = x^{\rho} x_{\rho}$. Thus one gets

$$\Lambda^{\mu}{}_{\rho} \Lambda_{\mu}{}^{\sigma} x^{\rho} x_{\sigma} = \delta^{\sigma}_{\rho} x^{\rho} x_{\sigma} , \tag{A.5}$$

and this implies the pseudoorthogonality relation

$$\Lambda^{\mu}{}_{\rho} \Lambda_{\mu}{}^{\sigma} = \delta^{\sigma}_{\rho} . \tag{A.6}$$

In the matrix form it reads

$$\Lambda^{\mathsf{T}} \cdot \overline{\Lambda} = \mathbb{1} , \tag{A.7}$$

where $\Lambda^{\mathsf{T}}$ is the transpose of $\Lambda$. Since we are working with finite dimensional matrices, the relation (A.7) implies also

$$\overline{\Lambda} \cdot \Lambda^{\mathsf{T}} = \mathbb{1} , \tag{A.8}$$

which, in compliance with the above definitions, means

$$\Lambda_{\nu}{}^{\rho} \Lambda^{\mu}{}_{\rho} = \delta^{\mu}_{\nu} . \tag{A.9}$$

305

Now, taking into account that $\delta_\nu^\mu = g_\nu^\mu$ and raising the index $\nu$, from (A.9) one gets

$$\Lambda^\mu{}_\rho \Lambda^{\nu\rho} = g^{\mu\nu} ,$$

and this can be rewritten as

$$g^{\mu\nu} = \Lambda^\mu{}_\rho \Lambda^\nu{}_\sigma g^{\rho\sigma} . \tag{A.10}$$

However, (A.10) is just the transformation law for a 2nd order tensor. Thus, the elementary considerations presented above lead to the conclusion that the metric components $g^{\mu\nu}$ in fact constitute a 2nd order tensor under Lorentz transformations (this justifies the standard term metric tensor). Obviously, it is a rather exceptional tensor, whose salient feature is that its components are the same in all Lorentz reference frames. In other words, $g^{\mu\nu}$ is a purely numerical Lorentz tensor (recall that the Kronecker delta $\delta_{jk}$ has an analogous property with respect to three-dimensional rotations).

The above identities (A.4), (A.7) have a simple consequence that is worth mentioning here. Employing them, one has

$$\Lambda^{\mathsf{T}} \cdot g \cdot \Lambda \cdot g = \mathbb{1} . \tag{A.11}$$

Taking the determinant of both sides of the identity (A.11), having in mind that $\det g = -1$ and $\det \Lambda^{\mathsf{T}} = \det \Lambda$, one gets

$$(\det \Lambda)^2 = 1 . \tag{A.12}$$

Thus one arrives at the well-known fact that for any Lorentz transformation one has

$$\det \Lambda = \pm 1 . \tag{A.13}$$

At this point, it is useful to recall the standard terminology concerning the whole set (in fact group) of Lorentz transformations. Among the matrices $\Lambda$ satisfying the pseudoorthogonality relations (A.7), (A.8), there are such that preserve the direction of time, as well as those that reverse it. Transformations of the first type are called **orthochronous**. Concerning the second type, the basic example is provided by the pure time reversal denoted conventionally as $T$, for which $\Lambda = \Lambda_T = \mathrm{diag}(-1, 1, 1, 1, )$. According to the sign of $\det \Lambda$ shown in (A.13), one distinguishes between **proper** ($\det \Lambda = +1$) and **improper** ($\det \Lambda = -1$) Lorentz transformations. Under such as criterion, the time reversal matrix $\Lambda_T$ certainly represents an improper transformation. Another obvious example is the **spatial inversion** (**space reflection**, or the **parity** transformation $P$), for which $\Lambda = \Lambda_P = \mathrm{diag}(1, -1, -1, -1)$. One the other hand, the continuous Lorentz transformations, such as "boosts" and spatial rotations, are proper (more about these shortly).

The sign ambiguity embodied in (A.13) enables one to distinguish between tensors and pseudotensors. Let us first recall some familiar examples. A quantity is a true scalar if it is invariant under both proper Lorentz transformations and parity, while a pseudotensor changes its sign under parity. Concerning vectors and pseudovectors, both behave in the same way under proper transformations (according to the paradigm (A.1)); with respect to parity, a true vector $V^\mu$ transforms as $V^\mu \rightarrow (V^0, -\vec{V})$, while a pseudovector (axial vector) $A^\mu$ transforms as $A^\mu \rightarrow (-A^0, \vec{A})$. To summarize and generalize the above examples, for a pseudotensor, an additional factor of $\det \Lambda$ should be involved in its transformation law, apart from the standard chain of matrix elements of $\Lambda$.

There is an important example of a pseudotensor that is just a collection of numbers, i.e. its components are the same in all Lorentz reference frames: it is the familiar **Levi-Civita symbol** $\varepsilon_{\mu\nu\rho\sigma}$, known also as the **permutation symbol**. Just to be sure, let us recall that $\varepsilon_{\mu\nu\rho\sigma}$ is $\pm 1$ according to the signature of the particular permutation $\mu\nu\rho\sigma$ with respect to the basic set 0123 (our convention is such that $\varepsilon_{0123} = +1$). So, let us see that $\varepsilon_{\mu\nu\rho\sigma}$ is indeed a Lorentz pseudotensor. To this end, one may start with an appropriate representation of the determinant of $\Lambda$, namely

$$\det \Lambda = \Lambda^0{}_\mu \Lambda^1{}_\nu \Lambda^2{}_\rho \Lambda^3{}_\sigma (-\varepsilon^{\mu\nu\rho\sigma}) \tag{A.14}$$

(note that the minus sign in (A.14) is due to our convention, according to which $\varepsilon^{0123} = -1$). Now, since a transposition of the rows in a determinant changes its sign, one may also write

$$\Lambda^\alpha{}_\mu \Lambda^\beta{}_\nu \Lambda^\gamma{}_\rho \Lambda^\delta{}_\sigma (-\varepsilon^{\mu\nu\rho\sigma}) = \det \Lambda \cdot (-\varepsilon^{\alpha\beta\gamma\delta}) . \tag{A.15}$$

Thus we get

$$\varepsilon^{\alpha\beta\gamma\delta} = \det \Lambda \, \Lambda^\alpha{}_\mu \Lambda^\beta{}_\nu \Lambda^\gamma{}_\rho \Lambda^\delta{}_\sigma \varepsilon^{\mu\nu\rho\sigma} , \tag{A.16}$$

where we have also utilized the simple fact that $(\det \Lambda)^{-1} = \det \Lambda$, due to (A.13). The pseudotensor property of the four-dimensional Levi-Civita symbol is thereby proven. It means that one has at disposal two purely numerical Lorentz tensors, namely $g_{\mu\nu}$ and $\varepsilon_{\mu\nu\rho\sigma}$, which thus naturally appear in many formulae for various quantities of tensor character encountered in field theory models.

Let us now proceed to the formal description of the continuous (proper) Lorentz transformations. The corresponding matrices $\Lambda$ form a six-parameter Lie group; it means that a particular $\Lambda$ may be written in the exponential form

$$\Lambda = \exp\left(-\frac{i}{2}\omega^{\alpha\beta} I_{\alpha\beta}\right) , \tag{A.17}$$

where $\omega^{\alpha\beta} = -\omega^{\beta\alpha}$ stand for the relevant parameters and $I_{\alpha\beta} = -I_{\beta\alpha}$ are $4 \times 4$ matrix generators (which form a basis of the corresponding Lie algebra). Of course, the summation in the exponent runs over $\alpha, \beta = 0, 1, 2, 3$. An attentive reader should notice that the antisymmetry property of $\omega^{\alpha\beta}$ guarantees that there are just six independent values of these parameters. As usual, one may now consider an infinitesimal form of (A.17), which reads

$$\Lambda = \mathbb{1} - \frac{i}{2}\Delta\omega^{\alpha\beta} I_{\alpha\beta} , \tag{A.18}$$

where we have distinguished the infinitesimal parameters $\Delta\omega^{\alpha\beta}$ from the original finite ones. In fact, one would like to represent such an infinitesimal $\Lambda$ in a natural way as a small deviation from unit matrix; for the matrix elements this would mean

$$\Lambda^\mu{}_\nu = \delta^\mu{}_\nu + \Delta\omega^\mu{}_\nu . \tag{A.19}$$

It is not difficult to show that one can arrive at (A.19) if the generators in (A.18) are chosen as

$$\left(I_{\alpha\beta}\right)^\mu{}_\nu = i\left(g^\mu{}_\alpha g_{\beta\nu} - g^\mu{}_\beta g_{\alpha\nu}\right) \tag{A.20}$$

(let us stress that one should not confuse the role of the Greek indices $(\alpha, \beta)$ and $(\mu, \nu)$; $\alpha, \beta$ are labelling the six generators, while $\mu, \nu$ mark the elements of a given matrix $I_{\alpha\beta}$). Indeed, employing (A.20) in (A.18), one gets

$$\begin{aligned}
\Lambda^\mu{}_\nu &= \left(\mathbb{1} - \frac{i}{2}\Delta\omega^{\alpha\beta} I_{\alpha\beta}\right)^\mu{}_\nu \\
&= \delta^\mu{}_\nu - \frac{i}{2}\Delta\omega^{\alpha\beta} i\left(g^\mu{}_\alpha g_{\beta\nu} - g^\mu{}_\beta g_{\alpha\nu}\right) \\
&= \delta^\mu{}_\nu + \frac{1}{2}\Delta\omega^{\mu\beta} g_{\beta\nu} - \frac{1}{2}\Delta\omega^{\alpha\mu} g_{\alpha\nu} \\
&= \delta^\mu{}_\nu + \frac{1}{2}\Delta\omega^\mu{}_\nu + \frac{1}{2}\Delta\omega^{\mu\alpha} g_{\alpha\nu} \\
&= \delta^\mu{}_\nu + \Delta\omega^\mu{}_\nu ,
\end{aligned} \tag{A.21}$$

and that's it.

With the explicit form of the generators at hand, one may verify that the following commutation relation is valid:

$$[I_{\mu\nu}, I_{\rho\sigma}] = i \left( g_{\mu\sigma} I_{\nu\rho} + g_{\nu\rho} I_{\mu\sigma} - g_{\mu\rho} I_{\nu\sigma} - g_{\nu\sigma} I_{\mu\rho} \right) \tag{A.22}$$

(the corresponding calculation is left to a diligent reader as an instructive exercise). Further, it is useful to express the generators $I_{\alpha\beta}$ in an explicit matrix form. Utilizing the formula (A.20) one gets

$$I_{01} = -i \begin{pmatrix} 0 & 1 & 0 & 0 \\ 1 & 0 & 0 & 0 \\ 0 & 0 & 0 & 0 \\ 0 & 0 & 0 & 0 \end{pmatrix}, \quad I_{02} = -i \begin{pmatrix} 0 & 0 & 1 & 0 \\ 0 & 0 & 0 & 0 \\ 1 & 0 & 0 & 0 \\ 0 & 0 & 0 & 0 \end{pmatrix}, \quad I_{03} = -i \begin{pmatrix} 0 & 0 & 0 & 1 \\ 0 & 0 & 0 & 0 \\ 0 & 0 & 0 & 0 \\ 1 & 0 & 0 & 0 \end{pmatrix},$$

$$I_{23} = -i \begin{pmatrix} 0 & 0 & 0 & 0 \\ 0 & 0 & 0 & 0 \\ 0 & 0 & 0 & 1 \\ 0 & 0 & -1 & 0 \end{pmatrix}, \quad I_{31} = -i \begin{pmatrix} 0 & 0 & 0 & 0 \\ 0 & 0 & 0 & -1 \\ 0 & 0 & 0 & 0 \\ 0 & 1 & 0 & 0 \end{pmatrix}, \quad I_{12} = -i \begin{pmatrix} 0 & 0 & 0 & 0 \\ 0 & 0 & 1 & 0 \\ 0 & -1 & 0 & 0 \\ 0 & 0 & 0 & 0 \end{pmatrix}. \tag{A.23}$$

It is worth noting that the matrices (A.23) are traceless; this in turn means that for $\Lambda$ given by (A.17) one has $\det \Lambda = 1$, due to the known identity

$$\mathrm{Tr} \ln M = \ln \det M \tag{A.24}$$

valid for any regular matrix $M$. In this way it is confirmed that a matrix $\Lambda$ in (A.17) represents a proper Lorentz transformation.

When one examines the transformations defined by (A.17) in detail, it is not difficult to arrive at the conclusion that $I_{0j}$, $j = 1, 2, 3$, generate Lorentz boosts (i.e. correspond to relative uniform motions of two reference frames), while $I_{jk}$, $j, k = 1, 2, 3$, generate spatial rotations. The point is that due to some obvious properties of the matrices (A.23) with respect to multiplication, the exponential (A.17) can be evaluated explicitly for the individual one-parameter subgroups of transformations generated by $I_{01}$, $I_{02}$, etc. and obtain e.g. the formulae shown in Chapter 4 (cf. (4.25), (4.34)); any hard-working reader is encouraged to perform such an exercise and recover the relevant hyperbolic or trigonometric functions from the Taylor expansion of the exponential (A.17) in question.

Next, let us introduce a convenient notation

$$I_{01} = K_1, \qquad I_{02} = K_2, \qquad I_{03} = K_3,$$
$$I_{23} = J_1, \qquad I_{31} = J_2, \qquad I_{12} = J_3. \tag{A.25}$$

Then it is straightforward to show that the commutation relations (A.22) are tantamount to

$$[J_j, J_k] = i\varepsilon_{jkl} J_l,$$
$$[J_j, K_k] = i\varepsilon_{jkl} K_l, \tag{A.26}$$
$$[K_j, K_k] = -i\varepsilon_{jkl} J_l.$$

Notice that the rotation generators $J_j$, $j = 1, 2, 3$, emerge indeed with the expected commutators (corresponding to an angular momentum). The formulae (A.26) are instrumental in the discussion of representations of the Lorentz algebra (see Appendix B).

308

# Appendix B

# Representations of Lorentz group

We provide here an elementary treatment of finite-dimensional representations of the proper Lorentz group; in fact, relying on the basic relation (A.17), we examine the representations of the **Lorentz algebra**, i.e. we are aiming at a systematic description of all possible matrix generators satisfying the commutation relations (A.26) (or, equivalently, (A.22)).

There is a simple trick that enables one to solve such a problem. Instead of $\vec{J}$ and $\vec{K}$ appearing in (A.25), (A.26), let us introduce their linear combinations

$$\vec{M} = \frac{1}{2}(\vec{J} + i\vec{K}), \qquad \vec{N} = \frac{1}{2}(\vec{J} - i\vec{K}) \tag{B.1}$$

(note that we have used here a shorthand vector notation, so that $\vec{J}$ stands for $J_j$, $j = 1, 2, 3$, etc.). It is straightforward to show that the commutation relations (A.26) imply

$$\begin{aligned} [M_j, M_k] &= i\varepsilon_{jkl}M_l\,, \\ [N_j, N_k] &= i\varepsilon_{jkl}N_l\,, \\ [M_j, N_k] &= 0\,. \end{aligned} \tag{B.2}$$

Thus, $\vec{M}$ and $\vec{N}$ are triplets of matrices representing two independent angular momenta; this also means that the Lorentz algebra is equivalent to that of the product group $SU(2) \times SU(2)$ (in a better mathematical notation it can be viewed as $su(2) \oplus su(2)$). However, the construction of the finite-dimensional representations of the $SU(2)$ algebra is known very well from the quantum mechanical theory of angular momentum (spin). Using a generic notation $\vec{J}$ for the spin components, the corresponding matrices can be labelled by a number $j$ that is a non-negative integer or half-integer and they are of the order $2j + 1$. As the knowledgeable reader may recall, the explicit form of such spin matrices can be obtained by means of the technique of the raising and lowering ("ladder") operators $J_\pm = J_1 \pm iJ_2$. The label $j$ corresponds to the eigenvalue $j(j + 1)$ of the quadratic Casimir operator $\vec{J}^2 = J_1^2 + J_2^2 + J_3^2$, which commutes with any component $J_j$, $j = 1, 2, 3$ (and in a given representation it is a multiple of the unit matrix).

Thus, one may classify the representations of the matrices $\vec{M}$, $\vec{N}$ by means of pairs $(j_1, j_2)$ with $j_1$, $j_2$ obeying the above familiar rules. Consequently, such a labelling may be employed for the representations of the generators $\vec{J}$, $\vec{K}$, which are expressed in terms of $\vec{M}$, $\vec{N}$ as

$$\vec{J} = \vec{M} + \vec{N}\,, \qquad \vec{K} = \frac{1}{i}(\vec{M} - \vec{N})\,. \tag{B.3}$$

Since $\vec{M}$, $\vec{N}$ are independent (commuting) angular momenta, the Lorentz group representation corresponding to a given pair $\vec{J}$, $\vec{K}$ (i.e. $\vec{M}$, $\vec{N}$) is described by the direct product of the matrices corresponding to $\vec{M}$ and $\vec{N}$, and thus the dimension of the representation in question is $(2j_1 + 1)(2j_2 + 1)$.

The above described construction may be elucidated by means of some examples of the lowest-dimensional representations. The simplest case is the trivial (scalar) representation $(0, 0)$, in which the matrices $\vec{M}$, $\vec{N}$ are zero (and, of course, the same is then true for $\vec{J}$ and $\vec{K}$). In the context of relativistic quantum mechanics or field theory this corresponds to the Klein–Gordon equation that describes spinless particles. Next, one may consider the representations $(0, \frac{1}{2})$ and $(\frac{1}{2}, 0)$. It is quite easy to realize what is their matrix content. Indeed, spin $\frac{1}{2}$ is described by the familiar Pauli matrices (more precisely, by $\frac{1}{2}\vec{\sigma}$), so that one has

$$(0, \tfrac{1}{2}) : \qquad \vec{M} = 0, \qquad \vec{N} = \frac{1}{2}\vec{\sigma}, \qquad\qquad (B.4)$$

$$(\tfrac{1}{2}, 0) : \qquad \vec{M} = \frac{1}{2}\vec{\sigma}, \qquad \vec{N} = 0. \qquad\qquad (B.5)$$

Thus, according to (B.3), one gets

$$(0, \tfrac{1}{2}) : \qquad \vec{J} = \frac{1}{2}\vec{\sigma}, \qquad \vec{K} = \frac{i}{2}\vec{\sigma}, \qquad\qquad (B.6)$$

$$(\tfrac{1}{2}, 0) : \qquad \vec{J} = \frac{1}{2}\vec{\sigma}, \qquad \vec{K} = -\frac{i}{2}\vec{\sigma}. \qquad\qquad (B.7)$$

Both these representations are two-dimensional, and one may thus expect that they are implemented in the description of the Weyl equation(s) (cf. Chapter 9). To check it, let us look at the equation (9.8), i.e. $\sigma^\mu \partial_\mu \psi = 0$, which is the same as

$$i\frac{\partial \psi}{\partial t} = -i(\vec{\sigma} \cdot \vec{\nabla})\psi \qquad\qquad (B.8)$$

(cf. (9.6)). In our straightforward analysis of its relativistic covariance we have found out that the relevant transformation of the wave function is

$$S = \exp\left(-\frac{i}{4}\omega^{\alpha\beta}W_{\alpha\beta}\right), \qquad\qquad (B.9)$$

with (see (9.23))

$$
\begin{array}{lll}
W_{12} = \sigma_3, & W_{31} = \sigma_2, & W_{23} = \sigma_1, \\
W_{01} = i\sigma_1, & W_{02} = i\sigma_2, & W_{03} = i\sigma_3.
\end{array}
\qquad (B.10)
$$

Now, comparing (B.9) with the generic form (A.17) of a representation of the Lorentz group, and taking into account the correspondence (A.25), one would like to match the generators (B.10) with $\vec{J}$ and $\vec{K}$ in such a way that

$$
\begin{array}{lll}
\dfrac{1}{2}W_{12} = J_3, & \dfrac{1}{2}W_{31} = J_2, & \dfrac{1}{2}W_{23} = J_1, \\[2mm]
\dfrac{1}{2}W_{01} = K_1, & \dfrac{1}{2}W_{02} = K_2, & \dfrac{1}{2}W_{03} = K_3.
\end{array}
\qquad (B.11)
$$

From (B.6) it is seen that such a coincidence occurs indeed for the representation $(0, \frac{1}{2})$. In a similar way, one would be able to show that the relevant transformation law for the Weyl equation of the second type, i.e. (9.7), corresponds to the representation $(\frac{1}{2}, 0)$ specified in

(B.7). The representations $(0, \frac{1}{2})$ and $(\frac{1}{2}, 0)$ are irreducible in the usual sense and, despite having the same dimension, they are inequivalent; this is due to the simple fact that there is no similarity transformation between $\vec{\sigma}$ and $-\vec{\sigma}$. Concerning the common terminology, these two representations correspond to 2-component **relativistic Weyl spinors**.

As we know, Dirac equation involving a non-zero mass term is written for a four-component field or wave function, so the question is what is the corresponding representation of the Lorentz group, from the point of view of the classification described above. The answer is that it is the direct sum $(\frac{1}{2}, 0) \oplus (0, \frac{1}{2})$ (note such a representation is reducible). Let us convince ourselves that it is indeed so. Using (B.6) and (B.7), the Lorentz generators corresponding to $(\frac{1}{2}, 0) \oplus (0, \frac{1}{2})$ may be written as

$$\vec{J} = \begin{pmatrix} \frac{1}{2}\vec{\sigma} & 0 \\ 0 & \frac{1}{2}\vec{\sigma} \end{pmatrix}, \qquad \vec{K} = \begin{pmatrix} -\frac{i}{2}\vec{\sigma} & 0 \\ 0 & \frac{i}{2}\vec{\sigma} \end{pmatrix}. \tag{B.12}$$

On the other hand, looking back at the results of Chapter 4, the transformation matrix for the solution of Dirac equation is given by

$$S = \exp\left(-\frac{i}{4}\omega^{\alpha\beta}\sigma_{\alpha\beta}\right), \qquad \sigma_{\alpha\beta} = \frac{i}{2}[\gamma_\alpha, \gamma_\beta]. \tag{B.13}$$

Thus, similarly as in (B.11), one would like to arrive at the identification

$$\frac{1}{2}\sigma_{23} = J_1, \qquad \frac{1}{2}\sigma_{31} = J_2, \qquad \frac{1}{2}\sigma_{12} = J_3,$$

$$\frac{1}{2}\sigma_{01} = K_1, \qquad \frac{1}{2}\sigma_{02} = K_2, \qquad \frac{1}{2}\sigma_{03} = K_3 \tag{B.14}$$

for a particular realization of the $\gamma$-matrices appearing in (B.13). It turns out that this is indeed possible; a subtle point is that the relevant set of $\gamma$-matrices is not the most common standard representation (3.37), but the choice (3.38) that is called **spinor** (or **chiral**) representation. Indeed, using (3.38) one gets

$$\sigma_{01} = i\gamma_0\gamma_1 = \begin{pmatrix} -i\sigma_1 & 0 \\ 0 & i\sigma_1 \end{pmatrix},$$

$$\sigma_{02} = i\gamma_0\gamma_2 = \begin{pmatrix} -i\sigma_2 & 0 \\ 0 & i\sigma_2 \end{pmatrix},$$

$$\sigma_{03} = i\gamma_0\gamma_3 = \begin{pmatrix} -i\sigma_3 & 0 \\ 0 & i\sigma_3 \end{pmatrix},$$

$$\sigma_{12} = i\gamma_1\gamma_2 = \begin{pmatrix} \sigma_3 & 0 \\ 0 & \sigma_3 \end{pmatrix}, \tag{B.15}$$

$$\sigma_{31} = i\gamma_3\gamma_1 = \begin{pmatrix} \sigma_2 & 0 \\ 0 & \sigma_2 \end{pmatrix},$$

$$\sigma_{23} = i\gamma_2\gamma_3 = \begin{pmatrix} \sigma_1 & 0 \\ 0 & \sigma_1 \end{pmatrix},$$

where we have utilized some well-known properties of Pauli matrices, in particular $\sigma_1\sigma_2 = i\sigma_3$, $\sigma_2\sigma_3 = i\sigma_1$, $\sigma_3\sigma_1 = i\sigma_2$. Thus, the pattern of matching (B.14) is verified. Obviously, such a coincidence also justifies the usual term "spinor representation" for (3.38) and the Dirac field or a wave function transforming according to (B.13) is rightly called **bispinor** (or **Dirac spinor**).

The alternative label "chiral" for (3.38) is related to the fact that the corresponding matrix $\gamma_5$ has then the block diagonal form

$$\gamma_5 = \begin{pmatrix} \mathbb{1} & 0 \\ 0 & -\mathbb{1} \end{pmatrix}.$$

As a last example of a low-lying representation of Lorentz algebra we will consider the case $(\frac{1}{2}, \frac{1}{2})$. The dimension of such a representation is four, so one may guess that it could be equivalent to the Lorentz transformations of a four-vector (e.g. spacetime coordinates, in particular), with the relevant generators given by the "canonical" matrices shown in (A.23), (A.25). It is indeed so, but some commentary is in order here. For $j_1 = j_2 = \frac{1}{2}$, the matrix triplets $\vec{M}$ and $\vec{N}$ can be expressed directly in terms of Pauli matrices, but to construct the desired $(\frac{1}{2}, \frac{1}{2})$ representation of $\vec{J}$ and $\vec{K}$ one should find some appropriate four-dimensional equivalents for $\vec{M}$ and $\vec{N}$. It is not immediately obvious how to guess it out of hand, but one may proceed the other way round; substituting (A.23) into (B.1) one gets

$$M_1 = \frac{1}{2}\begin{pmatrix} 0 & 1 & 0 & 0 \\ 1 & 0 & 0 & 0 \\ 0 & 0 & 0 & -i \\ 0 & 0 & i & 0 \end{pmatrix}, \quad M_2 = \frac{1}{2}\begin{pmatrix} 0 & 0 & 1 & 0 \\ 0 & 0 & 0 & i \\ 1 & 0 & 0 & 0 \\ 0 & -i & 0 & 0 \end{pmatrix}, \quad M_3 = \frac{1}{2}\begin{pmatrix} 0 & 0 & 0 & 1 \\ 0 & 0 & -i & 0 \\ 0 & i & 0 & 0 \\ 1 & 0 & 0 & 0 \end{pmatrix} \quad \text{(B.16)}$$

and

$$N_1 = \frac{1}{2}\begin{pmatrix} 0 & -1 & 0 & 0 \\ -1 & 0 & 0 & 0 \\ 0 & 0 & 0 & -i \\ 0 & 0 & i & 0 \end{pmatrix}, \quad N_2 = \frac{1}{2}\begin{pmatrix} 0 & 0 & -1 & 0 \\ 0 & 0 & 0 & i \\ -1 & 0 & 0 & 0 \\ 0 & -i & 0 & 0 \end{pmatrix}, \quad N_3 = \frac{1}{2}\begin{pmatrix} 0 & 0 & 0 & -1 \\ 0 & 0 & -i & 0 \\ 0 & i & 0 & 0 \\ -1 & 0 & 0 & 0 \end{pmatrix}. \quad \text{(B.17)}$$

One may check readily that the squares of the matrices (B.16), (B.17) are proportional to the $4 \times 4$ unit matrix and the Casimir operators $\vec{M}^2$ and $\vec{N}^2$ become

$$\vec{M}^2 = \frac{3}{4} \cdot \mathbb{1}, \qquad \vec{N}^2 = \frac{3}{4} \cdot \mathbb{1}. \quad \text{(B.18)}$$

It is reassuring, due to the familiar arithmetic identity $\frac{3}{4} = \frac{1}{2}(\frac{1}{2} + 1)$; the representation contents $(\frac{1}{2}, \frac{1}{2})$ of the standard four-vector transformations generated by (A.23) is thereby confirmed. Note also that unlike the four-dimensional bispinors discussed previously, the representation $(\frac{1}{2}, \frac{1}{2})$ is irreducible.

We would like to conclude this appendix with a challenge for a truly diligent and determined reader. Obviously, there must be also 3-dimensional representations $(1, 0)$ and $(0, 1)$ and the question is, where one may encounter them. Well, quite correctly one should stick to the rule formulated in the well-known play by J. Cimrman "The soothsayer" (in Czech: "Vizionář"), namely "We may not even indicate...". Nevertheless, so as to make the life of a potential explorer easier, let us point towards the electromagnetic tensor $F_{\mu\nu}$, more precisely to the field strengths $\vec{E}$ and $\vec{B}$. So, good luck!

# Appendix C

# Review of "diracology"

The purpose of this appendix is to collect in one place some important identities for the Dirac gamma matrices and solutions of the Dirac equation, which are employed frequently in the main text; thus we have taken the liberty to use the rather informal expression "diracology" in the title.

To begin with, let us recall the basic anticommutation relation

$$\{\gamma_\mu, \gamma_\nu\} = 2g_{\mu\nu} \tag{C.1}$$

and the definition of the fully anticommuting matrix $\gamma_5$, which in our convention reads

$$\gamma_5 = i\gamma^0\gamma^1\gamma^2\gamma^3 . \tag{C.2}$$

So, $\gamma_5$ satisfies

$$\{\gamma_\mu, \gamma_5\} = 0 , \qquad \mu = 0, 1, 2, 3 , \qquad \gamma_5^2 = 1 . \tag{C.3}$$

Note that the right-hand side of (C.1) in fact means $2g_{\mu\nu} \cdot 1$ with $1$ being the $4 \times 4$ unit matrix, but we are going to use the shorthand notation like (C.1) whenever it does not lead to confusion. Taking into account that $g_\mu^\mu = 4$, from (C.1) one gets immediately

$$\gamma_\mu\gamma^\mu = 4 , \tag{C.4}$$

and a sequence of "chain identities" then follows:

$$\gamma_\alpha\gamma_\mu\gamma^\alpha = -2\gamma_\mu ,$$
$$\gamma_\alpha\gamma_\mu\gamma_\nu\gamma^\alpha = 4g_{\mu\nu} , \tag{C.5}$$
$$\gamma_\alpha\gamma_\lambda\gamma_\mu\gamma_\nu\gamma^\alpha = -2\gamma_\nu\gamma_\mu\gamma_\lambda .$$

The derivation of these formulae is straightforward: one starts with the basic anticommutation relation (C.1) and then (C.3) may be utilized; the procedure is recursive. Finding the next term in the sequence (C.5) is left to the reader as an instructive exercise. Note that another pertinent label for the identities of the type (C.5) would be "sandwich relations".

Next, there are highly useful identities for traces of products of $\gamma$-matrices. First of all, utilizing the properties (C.3) of the matrix $\gamma_5$, it is easy to show that the trace of a product of an odd number of $\gamma$-matrices is zero; in a neat shorthand notation we will write

$$\mathrm{Tr}(\text{odd} \#) = 0 . \tag{C.6}$$

As for the products involving an even number of $\gamma$-matrices one has, e.g.

$$\mathrm{Tr}(\gamma_\mu\gamma_\nu) = 4g_{\mu\nu} ,$$
$$\mathrm{Tr}(\gamma_\mu\gamma_\nu\gamma_\rho\gamma_\sigma) = 4(g_{\mu\nu}g_{\rho\sigma} - g_{\mu\rho}g_{\nu\sigma} + g_{\mu\sigma}g_{\nu\rho}) . \tag{C.7}$$

The proof of such formulae is straightforward and relies just on employing (C.1) in an appropriate way; again, the procedure is recursive, as indicated in Chapter 3. The overall factor 4 comes from Tr $\mathbb{1}$; this is immediately obvious from the first identity (C.7). Another familiar form of the relations (C.7) reads

$$
\begin{aligned}
\mathrm{Tr}(\slashed{a}\slashed{b}) &= 4a \cdot b \,, \\
\mathrm{Tr}(\slashed{a}\slashed{b}\slashed{c}\slashed{d}) &= 4\left[(a \cdot b)(c \cdot d) - (a \cdot c)(b \cdot d) + (a \cdot d)(b \cdot c)\right],
\end{aligned}
\tag{C.8}
$$

with $\slashed{a} = a^\mu \gamma_\mu$, etc. Of course, this is obtained readily from (C.7) and the definition of the scalar products, $a \cdot b = g_{\mu\nu} a^\mu b^\nu$, etc.

The formulae (C.7) have a clear tensor character; it is not surprising, since the products of the metric tensor components are the only algebraic expressions that may descend from anticommutators of $\gamma$-matrices. In fact, there is another argument in favour of such a tensor structure. To see this, one may recall the relation (4.9) involving the bispinor transformation matrix $S$. Then one has

$$
\begin{aligned}
\mathrm{Tr}(\gamma^\mu \gamma^\nu \cdots \gamma^\tau \gamma^\omega) &= \mathrm{Tr}(S S^{-1} \gamma^\mu \gamma^\nu \cdots \gamma^\tau \gamma^\omega) \\
&= \mathrm{Tr}(S^{-1} \gamma^\mu \gamma^\nu \cdots \gamma^\tau \gamma^\omega S) \\
&= \mathrm{Tr}(S^{-1} \gamma^\mu S S^{-1} \gamma^\nu S \cdots S^{-1} \gamma^\tau S S^{-1} \gamma^\omega S) \\
&= \Lambda^\mu{}_\alpha \Lambda^\nu{}_\beta \cdots \Lambda^\tau{}_\gamma \Lambda^\omega{}_\delta \, \mathrm{Tr}(\gamma^\alpha \gamma^\beta \cdots \gamma^\tau \gamma^\omega) \,.
\end{aligned}
\tag{C.9}
$$

Thus it is clear that any such trace is a (purely numerical) tensor under Lorentz transformations; obviously, such a quantity can be constructed just from components of the metric tensor. By the way, the relation (C.9) also provides an alternative proof of the fact that $\mathrm{Tr}(\mathrm{odd}\ \#) = 0$, since one obviously cannot construct a tensor with an odd number of indices by using only components of the metric tensor carrying two indices. It is worth noting that the structure of the formulae for the traces of the type $\mathrm{Tr}(\mathrm{even}\ \#)$ is quite uniform (one just has to keep in mind that the number of relevant terms grows rapidly with increasing number of $\gamma$-matrices inside the trace, cf. (3.23)). In addition to the formulae shown above, one may also add the "palindromic" (or "backwards") relation

$$
\mathrm{Tr}(\gamma_\alpha \gamma_\beta \cdots \gamma_\tau \gamma_\omega) = \mathrm{Tr}(\gamma_\omega \gamma_\tau \cdots \gamma_\beta \gamma_\alpha)
\tag{C.10}
$$

that can be proved by employing the properties of the matrix of charge conjugation (cf. Chapter 5).

Let us now consider the trace identities involving $\gamma_5$. Apart from the situations that may be reduced to $\mathrm{Tr}(\mathrm{odd}\ \#) = 0$, the most frequently used identities are as follows:

$$
\begin{aligned}
\mathrm{Tr}\,\gamma_5 &= 0 \,, \\
\mathrm{Tr}(\gamma_\mu \gamma_\nu \gamma_5) &= 0 \,, \\
\mathrm{Tr}(\gamma_\mu \gamma_\nu \gamma_\rho \gamma_\sigma \gamma_5) &= 4i\, \varepsilon_{\mu\nu\rho\sigma} \,.
\end{aligned}
\tag{C.11}
$$

A proof of the first two relations is quite straightforward. As for the last formula in (C.11), it is not difficult to find out that the result for such a trace is totally antisymmetric in the indices $\mu$, $\nu$, $\rho$, $\sigma$; thus, it must be proportional to the Levi-Civita symbol $\varepsilon_{\mu\nu\rho\sigma}$. The overall factor $4i$ is then fixed easily by an explicit calculation for $(\mu\nu\rho\sigma) = (0123)$; one gets, using the definition (C.2),

$$
\mathrm{Tr}(\gamma_0 \gamma_1 \gamma_2 \gamma_3 \gamma_5) = \mathrm{Tr}(i\gamma_5 \cdot i\gamma_5) = 4i
\tag{C.12}
$$

(let us stress that our convention in (C.11) is $\varepsilon_{0123} = +1$).

Now, obtaining a continuation of the series (C.11) is not so straightforward as in the case of $\gamma$-matrix products without $\gamma_5$. For instance, if one wants to derive a formula for

$\mathrm{Tr}(\gamma_\lambda\gamma_\mu\gamma_\nu\gamma_\rho\gamma_\sigma\gamma_\tau\gamma_5)$, the procedure relying on sequential anticommutations of $\gamma_\lambda$ toward the end of the whole chain yields, because of the presence of the extra $\gamma_5$, just the so-called **Schouten identity**

$$g_{\mu\lambda}\varepsilon_{\nu\rho\sigma\tau} - g_{\nu\lambda}\varepsilon_{\mu\rho\sigma\tau} + g_{\rho\lambda}\varepsilon_{\mu\nu\sigma\tau} - g_{\sigma\lambda}\varepsilon_{\mu\nu\rho\tau} + g_{\tau\lambda}\varepsilon_{\mu\nu\rho\sigma} = 0 \tag{C.13}$$

instead of the desired trace formula. However, one may proceed differently. As we know from Chapter 3, sixteen $4 \times 4$ matrices $\Gamma_A$ defined by means of products of $\gamma$-matrices (see (3.24) through (3.28)) form a basis in the space of $4 \times 4$ matrices. So, one may expand e.g. the product $\gamma_\lambda\gamma_\mu\gamma_\nu$ in such a basis and utilize the known properties of traces of $\gamma$-matrix products. In this way it becomes clear that there are only two types of non-trivial contributions to such an expansion; explicitly, one has

$$\gamma_\lambda\gamma_\mu\gamma_\nu = (g_{\lambda\mu}g_{\nu\omega} - g_{\lambda\nu}g_{\mu\omega} + g_{\lambda\omega}g_{\mu\nu})\gamma^\omega - i\varepsilon_{\lambda\mu\nu\omega}\gamma^\omega\gamma_5 . \tag{C.14}$$

Then, using (C.14), one gets

$$\mathrm{Tr}(\gamma_\lambda\gamma_\mu\gamma_\nu\gamma_\rho\gamma_\sigma\gamma_\tau\gamma_5) = 4i(g_{\lambda\mu}\varepsilon_{\nu\rho\sigma\tau} - g_{\lambda\nu}\varepsilon_{\mu\rho\sigma\tau} + g_{\mu\nu}\varepsilon_{\lambda\rho\sigma\tau} + g_{\sigma\tau}\varepsilon_{\lambda\mu\nu\rho} - g_{\rho\tau}\varepsilon_{\lambda\mu\nu\sigma} + g_{\rho\sigma}\varepsilon_{\lambda\mu\nu\tau}) . \tag{C.15}$$

The expressions appearing in (C.11) and (C.15) are pseudotensors due to the presence of the Levi-Civita symbol (cf. Appendix A). It is not difficult to realize that such an algebraic structure is in fact a simple consequence of the properties of the $\gamma_5$ matrix. To see this, one may utilize the same procedure that has led to (C.9). For a proper Lorentz transformation, the matrix $S$ in (4.9) is generated by $\sigma_{\mu\nu} = \frac{i}{2}[\gamma_\mu, \gamma_\nu]$, which commutes with $\gamma_5$, and thus we get

$$\mathrm{Tr}(\gamma^\mu\gamma^\nu \cdots \gamma^\tau\gamma^\omega\gamma_5) = \Lambda^\mu{}_\alpha\Lambda^\nu{}_\beta \cdots \Lambda^\tau{}_\gamma\Lambda^\omega{}_\delta \, \mathrm{Tr}(\gamma^\alpha\gamma^\beta \cdots \gamma^\gamma\gamma^\delta\gamma_5) . \tag{C.16}$$

However, for the parity transformation one has $S = S_P = \gamma_0$ (see Chapter 5), so that

$$S_P^{-1}\gamma_5 S_P = -\gamma_5 . \tag{C.17}$$

Thus, a general result may be written as

$$\mathrm{Tr}(\gamma^\mu\gamma^\nu \cdots \gamma^\tau\gamma^\omega\gamma_5) = \det\Lambda \, \Lambda^\mu{}_\alpha\Lambda^\nu{}_\beta \cdots \Lambda^\tau{}_\gamma\Lambda^\omega{}_\delta \, \mathrm{Tr}(\gamma^\alpha\gamma^\beta \cdots \gamma^\gamma\gamma^\delta\gamma_5) , \tag{C.18}$$

which is just the envisaged pseudotensor transformation law (cf. Appendix A). On the basis of the above symmetry argument it is obvious that any trace of the considered type must be expressed by means of products of a certain number of components of the metric tensor and a single Levi-Civita symbol. Now one also arrives at an alternative elegant proof of the identity $\mathrm{Tr}(\gamma_\mu\gamma_\nu\gamma_5) = 0$: such an expression would be a Lorentz pseudotensor with two indices, but, obviously, there is no such a thing (to get it from $\varepsilon_{\mu\nu\rho\sigma}$, one would have to contract two indices with the help of the metric tensor, but this gives zero due to the antisymmetry of the Levi-Civita symbol).

Some of the identities shown above can be generalized in a straightforward way to an $n$-dimensional spacetime (such a generalization may be useful, among other things, within the method of dimensional regularization for closed-loop diagrams). To this end, let us reconsider the construction of the matrix basis described in Chapter 3, which relies on forming all possible independent products of $\gamma$-matrices. In an $n$-dimensional spacetime one has $n$ $\gamma$-matrices satisfying the relation (C.1). Thus, they possess basically the same properties as in the case $n = 4$, and the total number of their linearly independent products, starting with $\mathbb{1}$ and ending up with $\gamma_0\gamma_1 \cdots \gamma_{n-1}$, is equal to

$$\binom{n}{0} + \binom{n}{1} + \ldots + \binom{n}{n-1} + \binom{n}{n} = 2^n . \tag{C.19}$$

According to the statements formulated in Chapter 3 (which hold for a general $n$ as well), these $2^n$ matrices constitute a basis in the matrix space in question, and this in turn means that they can be represented in the form $2^{n/2} \times 2^{n/2}$ (for the purpose of our discussion we may restrict ourselves to even-dimensional spacetimes, $n = 2k$, $k = 1, 2, \ldots$). For $n = 4$ we recover our good old $4 \times 4$ matrices; for $n = 2$ one should use two $\gamma$-matrices $2 \times 2$, for $n = 6$ one has six basic $\gamma$-matrices $8 \times 8$, etc. Note that we may introduce immediately also a fully anticommuting extra matrix $\gamma_{n+1}$ (as an analogue of $\gamma_5$ in four dimensions), defined conventionally as e.g.

$$\gamma_{n+1} = i\gamma^0\gamma^1 \cdots \gamma^{n-1} . \tag{C.20}$$

Now, how about a generalization of the identities shown above? Concerning the traces, one gets readily $\mathrm{Tr}(\text{odd}\,\#) = 0$ and for $\mathrm{Tr}(\text{even}\,\#)$ we obtain, in analogy with (C.7),

$$\mathrm{Tr}(\gamma_\mu\gamma_\nu) = 2^{n/2}g_{\mu\nu} ,$$
$$\mathrm{Tr}(\gamma_\mu\gamma_\nu\gamma_\rho\gamma_\sigma) = 2^{n/2}(g_{\mu\nu}g_{\rho\sigma} - g_{\mu\rho}g_{\nu\sigma} + g_{\mu\sigma}g_{\nu\rho}) , \tag{C.21}$$

etc. The point is that the derivation of (C.21) proceeds in much the same way as in the case $n = 4$, except that here $\mathrm{Tr}\,\mathbb{1} = 2^{n/2}$.

As for the "chain" or "sandwich" identities generalizing (C.3), (C.5), one has to use $g^\mu_\mu = n$, so that the anticommutation relation (C.1) then yields

$$\gamma_\mu\gamma^\mu = n ,$$
$$\gamma_\alpha\gamma_\mu\gamma^\alpha = (2 - n)\gamma_\mu ,$$
$$\gamma_\alpha\gamma_\mu\gamma_\nu\gamma^\alpha = 4g_{\mu\nu} + (n - 4)\gamma_\mu\gamma_\nu , \tag{C.22}$$
$$\gamma_\alpha\gamma_\lambda\gamma_\mu\gamma_\nu\gamma^\alpha = -2\gamma_\nu\gamma_\mu\gamma_\lambda - (n - 4)\gamma_\lambda\gamma_\mu\gamma_\nu .$$

As we have already noted before, formulae for traces involving $\gamma_{n+1}$ do not have such a regular structure as (C.21), because of the specific properties of the Levi-Civita pseudotensor. For an illustration, let us display at least one particular example (in which, however, some salient features of the problem are manifested). So, for $n = 6$ one gets

$$\mathrm{Tr}\,\gamma_7 = 0 ,$$
$$\mathrm{Tr}(\gamma_\mu\gamma_\nu\gamma_7) = 0 ,$$
$$\mathrm{Tr}(\gamma_\mu\gamma_\nu\gamma_\rho\gamma_\sigma\gamma_7) = 0 , \tag{C.23}$$
$$\mathrm{Tr}(\gamma_\mu\gamma_\nu\gamma_\rho\gamma_\sigma\gamma_\tau\gamma_\omega\gamma_7) = 8i\,\varepsilon_{\mu\nu\rho\sigma\tau\omega} ,$$

if $\gamma_7$ is defined according to (C.19) and, in compliance with our earlier convention, we set $\varepsilon_{012345} = +1$. An attentive reader may notice that a generic feature of the formulae like (C.23) is the increasing number of vanishing traces of the type $\mathrm{Tr}(\text{even}\,\#\,\gamma_{n+1})$ due to the total antisymmetry of the Levi-Civita symbol.

Let us now return to four dimensions. Apart from the identities shown previously, there is a triplet of elegant formulae that are highly useful in calculations of decay rates and scattering cross sections in models involving Dirac fields. These read

$$\mathrm{Tr}(\slashed{a}\gamma_\mu\slashed{b}\gamma_\nu) \cdot \mathrm{Tr}(\slashed{c}\gamma^\mu\slashed{d}\gamma^\nu) = 32\,[(a \cdot c)(b \cdot d) + (a \cdot d)(b \cdot c)] ,$$
$$\mathrm{Tr}(\slashed{a}\gamma_\mu\slashed{b}\gamma_\nu\gamma_5) \cdot \mathrm{Tr}(\slashed{c}\gamma^\mu\slashed{d}\gamma^\nu\gamma_5) = 32\,[(a \cdot c)(b \cdot d) - (a \cdot d)(b \cdot c)] , \tag{C.24}$$
$$\mathrm{Tr}(\slashed{a}\gamma_\mu\slashed{b}\gamma_\nu) \cdot \mathrm{Tr}(\slashed{c}\gamma^\mu\slashed{d}\gamma^\nu\gamma_5) = 0 .$$

The proof of the first relation is based on a straightforward application of (C.7) and the third identity is obtained easily from (C.7) and (C.11). To prove the second formula, one employs (C.11) and an identity for the Levi-Civita tensor, namely

$$\varepsilon^{\mu\nu\alpha\beta}\varepsilon_{\rho\sigma\alpha\beta} = -2(\delta^\mu_\rho\delta^\nu_\sigma - \delta^\mu_\sigma\delta^\nu_\rho) . \tag{C.25}$$

Note that in the present text we occasionally use the nickname "formulae 32" for the triplet (C.24).

Next, we are going to discuss the **Gordon identity** (often called also **Gordon decomposition**). Its basic ingredients are bispinor amplitudes of Dirac plane waves $u(p)$, $v(p)$ satisfying the familiar equations

$$(\not p - m)u(p) = 0, \qquad (\not p + m)v(p) = 0, \tag{C.26}$$

and, equivalently,

$$\bar u(p)(\not p - m) = 0, \qquad \bar v(p)(\not p + m) = 0. \tag{C.27}$$

(we suppress here the spin labels). A basic Gordon identity reads

$$\bar u(p)\gamma_\mu u(p') = \frac{1}{2m}\bar u(p)\left[(p + p')_\mu + i\sigma_{\mu\nu}(p - p')^\nu\right]u(p'), \tag{C.28}$$

with $\sigma_{\mu\nu} = \frac{i}{2}[\gamma_\mu, \gamma_\nu]$. The derivation of this formula is quite easy if one hits on a clever trick that consists in representing zero in a rather sophisticated form, namely

$$0 = \bar u(p)\left[(\not p - m)\gamma_\mu + \gamma_\mu(\not p' - m)\right]u(p'). \tag{C.29}$$

From (C.29) one gets readily

$$\bar u(p)\gamma_\mu u(p') = \frac{1}{2m}\bar u(p)\left(\not p\gamma_\mu + \gamma_\mu\not p'\right)u(p'), \tag{C.30}$$

and the matrix products in parentheses are then conveniently rewritten in terms of anticommutators and commutators (recall that in general one has $AB = \frac{1}{2}\{A, B\} + \frac{1}{2}[A, B]$). Thus we have

$$\not p\gamma_\mu = \frac{1}{2}\{\not p, \gamma_\mu\} + \frac{1}{2}[\not p, \gamma_\mu] = p_\mu + \frac{1}{2}p^\nu[\gamma_\nu, \gamma_\mu] = p_\mu + i\sigma_{\mu\nu}p^\nu, \tag{C.31}$$

and similarly

$$\gamma_\mu\not p' = \frac{1}{2}\{\gamma_\mu, \not p'\} + \frac{1}{2}[\gamma_\mu, \not p'] = p'_\mu - i\sigma_{\mu\nu}p'^\nu. \tag{C.32}$$

So, substituting (C.31) and (C.32) into (C.30), one recovers (C.29). Now it is also clear that there are three additional identities of the type (C.29) involving $v$ and/or $\bar v$ along with (or instead of) $u$ and $\bar u$. To derive these, one may start with the trick (C.29), where an expression $\not p - m$ or $\not p' - m$ is replaced by $(-\not p - m)$ or $(-\not p' - m)$ if $\bar v$ and/or $v$ enters the game. Thus it is clear that the "master identity" (C.28) generates three extra variants simply by changing appropriately the sign of $p$ and/or $p'$. In such a way, one arrives at

$$\bar u(p)\gamma_\mu v(p') = \frac{1}{2m}\bar u(p)\left[(p - p')_\mu + i\sigma_{\mu\nu}(p + p')^\nu\right]v(p'),$$
$$\bar v(p)\gamma_\mu u(p') = \frac{1}{2m}\bar v(p)\left[(-p + p')_\mu + i\sigma_{\mu\nu}(-p - p')^\nu\right]u(p'), \tag{C.33}$$
$$\bar v(p)\gamma_\mu v(p') = \frac{1}{2m}\bar v(p)\left[(-p - p')_\mu + i\sigma_{\mu\nu}(-p + p')^\nu\right]v(p').$$

Note that the forms (C.28), (C.33) may be viewed as a decomposition of the "Dirac current" into its "convective" and "spin" part; the common terminology reflects just this observation.

Finally, let us comment briefly on a general definition of the spin matrices and on the intriguing relation (7.3) involving $\gamma_5$. In Chapter 2 we have introduced the triplet $\Sigma_j$, $j = 1, 2, 3$, defined directly in terms of Pauli matrices (see (2.10)). Such a straightforward definition corresponds to the standard representation of $\gamma$-matrices. In fact, it is not difficult to figure out a natural definition of $\vec{\Sigma}$, which is independent of a specific representation of $\gamma$-matrices. It reads

$$\Sigma_j = \frac{1}{2}\varepsilon_{jkl}\sigma_{kl}, \qquad \text{i.e.} \quad \sigma_{jk} = \varepsilon_{jkl}\Sigma_l \tag{C.34}$$

(let us recall that the position of the indices is irrelevant — up or down, it does not matter). From (C.34) one gets readily

$$\Sigma_1 = \sigma_{23}, \qquad \Sigma_2 = \sigma_{31}, \qquad \Sigma_3 = \sigma_{12}. \tag{C.35}$$

As we know from Chapter 4, the matrices $\frac{1}{2}\sigma_{\alpha\beta}$ satisfy commutation relations of the Lorentz algebra (see (4.22) and (4.24)). Further, the fundamental Lorentz generators $I_{kl}$ labelled appropriately as

$$J_1 = I_{23}, \qquad J_2 = I_{31}, \qquad J_3 = I_{12} \tag{C.36}$$

correspond to components of an angular momentum, in view of the relations (A.26) in Appendix A. Thus it is clear that the $4 \times 4$ matrices $\vec{S} = \frac{1}{2}\vec{\Sigma}$ defined through Eq. (C.34) satisfy the commutation relations of angular momentum (spin).

Now it should also be obvious that the above-mentioned relation (7.3) is representation independent. Indeed, suing the general definition (C.34) and the fundamental theorem on $\gamma$-matrices, one can see immediately that $\vec{\Sigma}$, $\vec{\alpha}$ and $\gamma_5$ are changed via a similarity transformation when passing from one representation to another. Therefore, the validity of Eq. (7.3) in the standard representation guarantees that it holds in any other one as well.

# Appendix D

# More about spin states of Dirac field

The purpose of this appendix is to show that the one-particle states of the quantized Dirac field, $b^\dagger(p, s)|0\rangle$ or $d^\dagger(p, s)|0\rangle$ are indeed eigenstates of the helicity. Our treatment is adapted, with minor modifications, from the book [18].

To begin with, one has to establish some identities that are instrumental in achieving the desired goal. First, employing the basic anticommutation relations (19.15), one obtains

$$\{\psi(x), \psi(y)\} = 0 \,,$$
$$\{\psi(\vec{x}, t), \psi^\dagger(\vec{y}, t)\} = \delta^{(3)}(\vec{x} - \vec{y}) \,. \tag{D.1}$$

Note that the first identity (D.1) is an obvious consequence of (19.15). A proof of the second relation requires some calculation, which is not difficult and hopefully can be left to the diligent reader as an instructive exercise.

Next, utilizing the results of Chapter 15, the operators of the momentum and angular momentum of the quantized Dirac field may be written as (cf. (15.52) through (15.54))

$$\vec{P} = \int d^3x \, \psi^\dagger(x) \left(-i\vec{\nabla}\right)\psi(x) \,,$$
$$\vec{M} = \int d^3x \, \psi^\dagger(x) \left(\vec{L} + \frac{1}{2}\vec{\Sigma}\right)\psi(x) \,, \tag{D.2}$$

where

$$\vec{L} = -i(\vec{x} \times \vec{\nabla}) \tag{D.3}$$

and $\frac{1}{2}\vec{\Sigma}$ are the familiar spin matrices. Let us also recall that in our notation, $\nabla^j$, $j = 1, 2, 3$, means $\partial/\partial x^j$. A crucial consequence of the above results is represented by the following commutators

$$[\vec{P}, \psi(x)] = i\vec{\nabla}\psi(x) \,,$$
$$[\vec{M}, \psi(x)] = \left(i(\vec{x} \times \vec{\nabla}) - \frac{1}{2}\vec{\Sigma}\right)\psi(x) \,. \tag{D.4}$$

The above commutation relations are proved easily. Indeed, one may utilize the elementary identity

$$[AB, C] = A\{B, C\} - \{A, C\}B \,, \tag{D.5}$$

319

and this yields, e.g. for the first commutator in (D.4),

$$[\vec{P}, \psi(x)] = \left[ \int d^3y\, \psi^\dagger(\vec{y}, t)(-i\vec{\nabla})\psi(\vec{y}, t), \psi(\vec{x}, t) \right]$$

$$= \int d^3y\, \left( \psi^\dagger(\vec{y}, t)(-i\vec{\nabla}_y)\{\psi(\vec{y}, t), \psi(\vec{x}, t)\} - \{\psi^\dagger(\vec{y}, t), \psi(\vec{x}, t)\}\left(-i\vec{\nabla}\psi(\vec{y}, t)\right) \right) \tag{D.6}$$

(note that we have been allowed to use the same value of $t$ in all field operators appearing in (D.6), since $\vec{P}$ is time independent). From (D.6) and (D.1) one then gets readily the first formula (D.4). The second relation (D.4) is obtained in the same way.

Now, we may take Hermitian conjugation of (D.4) and let the result act on the vacuum $|0\rangle$, taking into account that

$$\vec{P}|0\rangle = 0\,, \qquad \vec{M}|0\rangle = 0 \tag{D.7}$$

(because of normal ordering). One thus gets, after a simple manipulation,

$$\vec{P}\psi^\dagger(x)|0\rangle = i\vec{\nabla}\psi^\dagger(x)|0\rangle\,,$$
$$\vec{M}\psi^\dagger(x)|0\rangle = \left( i(\vec{x}\times\vec{\nabla})\psi^\dagger + \psi^\dagger\frac{1}{2}\vec{\Sigma} \right)|0\rangle\,. \tag{D.8}$$

As a last item (last but not least), we will need a formula expressing the creation operator $b^\dagger(p, s)$ (or $d^\dagger(p, s)$) in terms of $\psi(x)$. To this end, let us recall the expansion of $\psi(x)$ in the annihilation and creation operators (see (19.6)); we reproduce it here for reader's convenience:

$$\psi(x) = \int d^3k\, N_k \sum_r \left[ b(k, r)u(k, r)\, e^{-ikx} + d^\dagger(k, r)v(k, r)\, e^{ikx} \right]. \tag{D.9}$$

Multiplying (D.9) by $e^{ipx}$ and integrating over $d^3x$, one gets the notorious delta functions that enable one to replace the variable $\vec{k}$ with $\vec{p}$ or $-\vec{p}$, respectively. Next, one may multiply the resulting expression by $u^\dagger(p, s)$ from the left and employ then the relations (19.9), (19.10) (these follow from Gordon identities discussed in Appendix C, see (C.28), (C.33)). In this way, the term in (D.9) that involves $d^\dagger$ along with $v$ drops out and we are left just with the annihilation operator $b(p, s)$. One thus has

$$b(p, s) = N_p \int d^3x\, e^{ipx} u^\dagger(p, s)\psi(x)\,. \tag{D.10}$$

Taking Hermitian conjugate of (D.10) we obtain finally

$$b^\dagger(p, s) = N_p \int d^3x\, e^{-ipx}\psi^\dagger(x)u(p, s)\,. \tag{D.11}$$

Now we are in a position to take up the problem of helicity. We would like to examine the action of the scalar product $\vec{P}\cdot\vec{M}$ on the one-particle state $b^\dagger(p, s)|0\rangle$, i.e.

$$|a\rangle = \vec{P}\cdot\vec{M}\, b^\dagger(p, s)|0\rangle\,. \tag{D.12}$$

In this context, it is useful to realize that $\vec{P}\cdot\vec{M} = \vec{M}\cdot\vec{P}$, since

$$[M^j, P^j] = 0 \tag{D.13}$$

(such a vanishing commutator has an obvious analogue in ordinary quantum mechanics, where one has, in general, $[M^j, P^k] = i\varepsilon^{jkl} P^l$). So, we are going to evaluate

$$|a\rangle = \vec{M} \cdot \vec{P} b^\dagger(p, s)|0\rangle = \vec{M} \cdot \vec{p} \, b^\dagger(p, s)|0\rangle . \tag{D.14}$$

Employing the representation (D.11), one has

$$|a\rangle = \vec{p} \cdot \vec{M} N_p \int d^3x \, e^{-ipx} \psi^\dagger(x)|0\rangle u(p, s) . \tag{D.15}$$

Since $\vec{M}$ is independent of spacetime coordinates, (D.15) can be recast as

$$|a\rangle = \vec{p} N_p \int d^3x \, e^{-ipx} \vec{M} \psi^\dagger(x)|0\rangle u(p, s)$$

$$= \vec{p} N_p \int d^3x \, e^{-ipx} \left( i(\vec{x} \times \vec{\nabla}) \psi^\dagger + \psi^\dagger \frac{1}{2} \vec{\Sigma} \right) |0\rangle u(p, s) , \tag{D.16}$$

where we have utilized the second identity in (D.8). Let us first consider the "orbital part" of (D.16), which involves the operator

$$\mathcal{O} = p^j \int d^3x \, e^{-ipx} (\vec{x} \times \vec{\nabla})^j \psi^\dagger(x)$$

$$= p^j \varepsilon^{jrs} \int d^3x \, e^{-ipx} x^r \nabla^s \psi^\dagger(x) . \tag{D.17}$$

Performing the integration by parts (and discarding the surface term), (D.17) is recast as

$$\mathcal{O} = -p^j \varepsilon^{jrs} \int d^3x \, \nabla^s \left( e^{-ipx} x^r \right) \psi^\dagger(x)$$

$$= -p^j \varepsilon^{jrs} \int d^3x \left( i p^s \, e^{-ipx} x^r + e^{-ipx} \delta^{rs} \right) \psi^\dagger(x) . \tag{D.18}$$

Now it is clear that $\mathcal{O}$ vanishes, because of the total antisymmetry of the Levi-Civita symbol $\varepsilon^{jrs}$. Thus, the contribution of the orbital part of the scalar product $\vec{M} \cdot \vec{P}$ in (D.14) is seen to be identically zero, in accordance with an intuitive expectation (note that classically, orbital angular momentum is orthogonal to the linear momentum, $\vec{L} \cdot \vec{P} = 0$).

All this means that the expression (D.16) is reduced to

$$|a\rangle = N_p \int d^3x \, e^{-ipx} \psi^\dagger(x)|0\rangle \frac{1}{2} (\vec{p} \cdot \vec{\Sigma}) u(p, s) . \tag{D.19}$$

However, according to the definition of $u(p, s)$ as the amplitude of a plane wave carrying a definite helicity one has

$$\vec{n} \cdot \vec{S} \, u(p, s) = h \, u(p, s) , \tag{D.20}$$

where $\vec{n} = \vec{p}/|\vec{p}|$, $\vec{S} = \frac{1}{2} \vec{\Sigma}$ and $h$ is the corresponding helicity ($h = \pm \frac{1}{2}$). Thus, using (D.20) and (D.11) we can see that

$$|a\rangle = h \, |\vec{p}| b^\dagger(p, s)|0\rangle ,$$

i.e.

$$\frac{\vec{p}}{|\vec{p}|} \cdot \vec{M} b^\dagger(p, s)|0\rangle = h \, b^\dagger(p, s)|0\rangle , \tag{D.21}$$

and this is what we wanted to show. The analysis of the state $d^\dagger(p, s)|0\rangle$ may proceed in much the same way.

# Appendix E

# Photon propagator
# in a general covariant gauge

---

The "Fermi trick" (31.26) may be generalized by means of a simple modification of the gauge-fixing term, using

$$\mathscr{L} = -\frac{1}{4}F_{\mu\nu}F^{\mu\nu} - \frac{1}{2\alpha}(\partial \cdot A)^2 \,, \tag{E.1}$$

with $\alpha$ being an arbitrary real parameter. In such a case one has

$$\frac{\partial \mathscr{L}}{\partial(\partial_\rho A_\sigma)} = -F^{\rho\sigma} - \frac{1}{\alpha}(\partial \cdot A)g^{\rho\sigma} \,, \tag{E.2}$$

and the equation of motion then becomes

$$\Box A^\sigma + \left(\frac{1}{\alpha} - 1\right)\partial_\rho\partial^\sigma A^\rho = 0 \,,$$

or

$$\left[g_\rho^\sigma \Box + \left(\frac{1}{\alpha} - 1\right)\partial^\sigma\partial_\rho\right]A^\rho = 0 \,. \tag{E.3}$$

Thus, Eq. (E.3) for $\alpha \neq 1$ differs from d'Alembert equation and the procedure of canonical operator quantization is therefore considerably more complicated than in the case $\alpha = 1$. However, since we are primarily interested in the corresponding propagator, one may try a heuristic method relying on a general observation that the propagator of a quantized field is a particular Green's function of the equation of motion in question (cf. our previous findings concerning scalar field). Now, for $\alpha = 1$ one obviously gets

$$\Box \mathscr{D}_\nu^\mu = g_\nu^\mu \delta^{(4)}(x) \,, \tag{E.4}$$

simply because

$$[A_\mu(x), \dot{A}_\nu(y)]_{\text{E.T.}} = -ig_{\mu\nu}\delta^{(4)}(x - y)$$

(see (31.30)). Thus, for Eq. (E.3) one may try

$$\left[g_\lambda^\mu \Box + \left(\frac{1}{\alpha} - 1\right)\partial^\mu\partial_\lambda\right]\mathscr{D}_\nu^\lambda(x) = g_\nu^\mu \delta^{(4)}(x) \,. \tag{E.5}$$

Differential equation (E.5) can be solved in usual way by means of Fourier transformation. Writing

$$\mathscr{D}_{\mu\nu}(x) = \int \frac{\mathrm{d}^4 q}{(2\pi)^4}D_{\mu\nu}(q)\,e^{iqx} \,, \tag{E.6}$$

322

from (E.5) one then gets an algebraic equation for $D_{\mu\nu}(q)$, namely

$$\left[ -g^\mu_\lambda q^2 - \left( \frac{1}{\alpha} - 1 \right) q^\mu q_\lambda \right] D^\lambda_\nu(q) = g^\mu_\nu \,. \tag{E.7}$$

This can be written in matrix form as

$$L \cdot D = \mathbb{1} \,, \tag{E.8}$$

where the matrix elements of $L$ are just

$$L^\mu_\lambda = -g^\mu_\lambda q^2 - \left( \frac{1}{\alpha} - 1 \right) q^\mu q_\lambda \tag{E.9}$$

and $\mathbb{1}$ is the $4 \times 4$ unit matrix. Finding $D$ thus amounts to inverting $L$, i.e. $D = L^{-1}$. The most efficient way of solving (E.8) consists in decomposing the matrices $L$ and $D$ (as well as $\mathbb{1}$) in terms of two independent projection operators. Since the matrices in question are also Lorentz tensors, one may formally raise and lower the matrix indices whenever it may be convenient. The projection operators we have in mind are

$$\begin{aligned} (P_\mathrm{T})_{\mu\nu} &= g_{\mu\nu} - \frac{q_\mu q_\nu}{q^2} \,, \\ (P_\mathrm{L})_{\mu\nu} &= \frac{q_\mu q_\nu}{q^2} \,, \end{aligned} \tag{E.10}$$

where the labels T and L stand for transverse and longitudinal, respectively. It is easy to see that $P_\mathrm{T}$ and $P_\mathrm{L}$ are (mutually orthogonal) projection operators: it holds

$$(P_\mathrm{T})^2 = P_\mathrm{T} \,, \qquad (P_\mathrm{L})^2 = P_\mathrm{L} \,, \qquad P_\mathrm{T} \cdot P_\mathrm{L} = 0 \,, \qquad P_\mathrm{T} + P_\mathrm{L} = \mathbb{1} \,. \tag{E.11}$$

Now, writing $L$ and $D$ as

$$\begin{aligned} L &= A P_\mathrm{T} + B P_\mathrm{L} \,, \\ D &= X P_\mathrm{T} + Y P_\mathrm{L} \,, \end{aligned} \tag{E.12}$$

where $A, B, X, Y$ are some numerical coefficients, and using (E.11), Eq. (E.8) is recast as

$$A X \, P_\mathrm{T} + B Y \, P_\mathrm{L} = P_\mathrm{T} + P_\mathrm{L} \,, \tag{E.13}$$

and thus

$$X = \frac{1}{A} \,, \qquad Y = \frac{1}{B} \,. \tag{E.14}$$

In this way, the problem of inverting the $4 \times 4$ matrix $L$ is reduced to inverting numbers $A$ and $B$, a trivial operation. From (E.9) it is seen that

$$L = -q^2 P_\mathrm{T} - \frac{1}{\alpha} q^2 P_\mathrm{L} \,, \tag{E.15}$$

so

$$D = \frac{1}{-q^2} P_\mathrm{T} + \frac{1}{-\frac{1}{\alpha} q^2} P_\mathrm{L} \,,$$

i.e., using (E.10), one gets finally

$$D_{\mu\nu}(q) = \frac{1}{q^2}\left[-g_{\mu\nu} + (1-\alpha)\frac{q_\mu q_\nu}{q^2}\right]. \tag{E.16}$$

The expression (E.16) represents a class of propagators of the electromagnetic field, corresponding to **covariant gauges** (often called "$\alpha$-gauges"). Admittedly, it is somewhat vague statement at the present heuristic level, but this subtle issue goes beyond the scope of our elementary approach. Of course, one thus obtains just the relevant functional dependence of $D_{\mu\nu}(q)$, while the replacement $1/q^2 \to 1/(q^2 + i\varepsilon)$ for the true Feynman propagator is done by hand (well, but we know that at least for $\alpha = 1$ we are able to do it correctly).

There are some prominent values of $\alpha$, related to famous names:

- $\alpha = 1$: **Feynman gauge**,
- $\alpha = 0$: **Landau gauge**.

Note that $\alpha = 0$ does not make sense in the Lagrangian (E.1), but for the propagator it's OK. Finally, notice that for $\alpha \to \infty$ one recovers the original gauge invariant Maxwell Lagrangian, but in this case the matrix $L$ is singular (it is reduced to $P_\mathrm{T}$) and has no inverse. This is precisely the point of **gauge fixing** — for the gauge invariant Lagrangian itself the problem of finding the propagator is ill-defined.

# Appendix F

# Electromagnetic form factors of electron

In this appendix, we discuss briefly the general formula (50.15). As we have already mentioned before, at the tree level it is valid automatically, with $F_1(q^2) = 1$, $F_2(q^2) = 0$. Now the question is what one gets in higher orders for the matrix product

$$\mathcal{M}_\mu = \bar{u}(p')\Gamma_\mu(p',p)u(p)\,, \tag{F.1}$$

where $\Gamma_\mu(p',p)$ denotes the vertex function represented by Feynman diagrams with two external electron lines and one external photon line (pictorially, see Fig. F.1). Needless to say, $p$ and $p'$

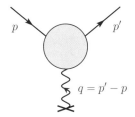

**Fig. F.1:** Schematic depiction of the QED vertex function that embodies the electron magnetic moment.

are taken to be on the mass shell, $p^2 = p'^2 = m^2$, while the external photon line is in general off-shell. Within the covariant perturbation expansion we are using, the $\mathcal{M}_\mu$ should be a Lorentz four-vector, and it is a function of two independent four-momenta $p$ and $p'$.

Thus, one may guess immediately that the most general form of the $\mathcal{M}_\mu$ could be

$$\mathcal{M}_\mu = \bar{u}(p')\left[A_1(q^2)\gamma_\mu + A_2(q^2)p_\mu + A_3(q^2)p'_\mu\right]u(p)\,. \tag{F.2}$$

As regards the invariant amplitudes (form factors) $A_j(q^2)$, $j = 1, 2, 3$, these might depend on $p^2$, $p'^2$ and the scalar product $p \cdot p'$. But $p^2 = p'^2 = m^2$, so there is just one independent kinematical invariant, e.g. $q^2 = (p'-p)^2 = 2m^2 - 2p \cdot p'$. A remark is perhaps in order here. When the loop integrations inside the blob in Fig. F.1 are carried out, one gets some Lorentz invariant form factors and the resulting $\Gamma_\mu(p',p)$ incorporates diverse products of $\gamma$-matrices, which enter the game either as $\gamma_\mu$, or slashed combinations $\not{p}$ and $\not{p}'$. To simplify the products of $\gamma$-matrices, one can employ their basic anticommutation relations, which lead e.g. to $\not{p}\gamma_\mu = 2p_\mu - \gamma_\mu\not{p}$, $\not{p}\not{p}' = 2p \cdot p' - \not{p}'\not{p}$, etc. In this way, one may eventually encounter just a finite number of $\gamma$-matrix products, such as

$$\not{p}\gamma_\mu,\ p_\mu\not{p}',\ \not{p}\gamma_\mu\not{p}',\ \ldots \tag{F.3}$$

The readers are encouraged to activate their imagination and try to find all possible relevant $\gamma$-matrix products involved here. By the way, a detailed explicit evaluation of $\Gamma_\mu(p', p)$ at one-loop level, sketched in Chapter 50, is quite instructive in this context. Higher-order diagrams can of course produce long chains of $\gamma$-matrices, but one can always employ the anticommutation relations, and move the matrices inside the chain in such a way that one eventually gets $(\not p)^2 = p^2 = m^2$ and similarly for $(\not p')^2$. Moreover, one should put factors $\not p$ and $\not p'$ in the right order, such that $\not p'$ stands on the left and $\not p$ on the right; one may then utilize Dirac equations $\bar u(p')\not p' = m\,\bar u(p')$ and $\not p u(p) = m\,u(p)$. A typical example of the above-mentioned manipulations is as follows: one may get easily, upon appropriate anticommutations of $\gamma$-matrices,

$$\not p \gamma_\mu \not p' = 2p_\mu \not p' + 2p'_\mu \not p - 2p \cdot p' \gamma_\mu - \not p' \gamma_\mu \not p .$$

In such a way, one can justify the simple structure (F.2).

Well, after such a long explanatory comment, let us take the form (F.2) for granted and proceed further. For our purpose, it is more convenient to recast the expression in terms of the combinations $p'_\mu + p_\mu$ and $p'_\mu - p_\mu$, so that we write Eq. (F.2) in an equivalent form

$$\mathcal{M}_\mu = \bar u(p')\big[A(q^2)\gamma_\mu + B(q^2)(p'_\mu + p_\mu) + C(q^2)(p'_\mu - p_\mu)\big]u(p). \tag{F.4}$$

Now we may use the Ward–Takahashi (WT) identity (see Chapter 42, the formula (42.20)), which in our present notation reads simply

$$q^\mu \mathcal{M}_\mu = 0 . \tag{F.5}$$

Using the decomposition (F.4) and the identities $\bar u(p')\not q u(p) = 0$, $(p'_\mu + p_\mu)(p'^\mu - p^\mu) = 0$, the WT relation (F.5) is reduced to $q^2 C(q^2) = 0$, or,

$$C(q^2) = 0 . \tag{F.6}$$

Thus, the form (F.4) becomes

$$\mathcal{M}_\mu = \bar u(p')\big[A(q^2)\gamma_\mu + B(q^2)(p'_\mu + p_\mu)\big]u(p), \tag{F.7}$$

and with the help of the Gordon identity (50.25) this can be immediately rewritten as a combination of terms involving $\gamma_\mu$ and $\sigma_{\mu\nu}q^\nu$. The formula (50.15) is thereby proven.

Finally, let us remark that we have demonstrated the validity of the WT identity at the one-loop level, but in fact it is quite general, as a consequence of the gauge invariance of QED. A detailed discussion of this topic can be found e.g. in the book [6], Chapter 8, section 8.4.1. Thus, we may conclude that the form (50.15) is valid to any order of perturbation expansion.

# Bibliography

Below I have listed some books that I was using occasionally when preparing my lectures during the past years. Thus, the reader may find there more details concerning the topics discussed in the present text. One item of the list is rather special, so it is worth mentioning explicitly; I have in mind [23] and the reason is as follows. In my experience, the mathematically minded students sometimes complain that the standard QFT methods are somewhat sloppy (though efficient) and strive for due rigorousness. The remarkable book [23], written by a professional mathematician (with deep respect for physics) responds, at least partly, to such needs. So, I have included it in the list, for the reader's convenience. Needless to say, apart from the limited set of textbooks and monographs displayed here, there are many other good books covering the huge and fascinating area of the quantum field theory. Further, the list of relevant literature continues with references to original and review papers dealing with some particular themes treated in these lecture notes. The selection of cited papers has been rather minimalistic; some of them reflect the history of QFT and particle physics, while the other ones might (along with the comprehensive books) arouse the reader's interest and open them new horizons in the rich QFT landscape. In particular, the papers cited in ref. [56] contain a remarkable defence of the "conventional QFT" (i.e. the approach pursued also in the present lecture notes) in contrast to a rigorous axiomatic theory, which might be a "holy grail" for a mathematically minded reader. One more remark is perhaps in order here. It is clear that the present text is oriented primarily to possible applications in particle physics. In fact, the scope of quantum field theory is much broader. In particular, QFT methods are highly efficient also in the condensed matter physics or nuclear physics, i.e. in the theory of many-body systems in general. The reader interested in these aspects of quantum field theory may find some relevant information e.g. in the book [10].

[1] J. Bjorken, S. Drell: *Relativistic quantum mechanics*, McGraw Hill Book Company, New York 1964.
[2] J. Bjorken, S. Drell: *Relativistic quantum fields*, Mc Graw Hill Book Company, New York 1965.
[3] A. Messiah: *Quantum mechanics*, Dover Publications Inc., Mineola, New York 1999.
[4] P. Strange: *Relativistic quantum mechanics*, Cambridge Univ. Press, Cambridge 1998.
[5] W. Greiner: *Relativistic quantum mechanics*, Springer-Verlag, Berlin 1994.
[6] C. Itzykson, J. B. Zuber: *Quantum field theory*, Mc Graw Hill Book Company, New York 1980.
[7] V. B. Berestetskii, E. M. Lifshitz and L. P. Pitaevskii: *Quantum electrodynamics (Landau & Lifshitz course of theoretical physics, Vol. 4)*, Butterworth–Heinemann, Oxford 1999.
[8] S. Schweber: *An introduction to relativistic quantum field theory*, Dover Publications, Mineola, New York 1989.
[9] S. Schweber: *QED and the men who made it*, Princeton University Press, Princeton 1994.
[10] S. J. Chang: *Introduction to quantum field theory*, World Scientific, Singapore 1990.
[11] A. Das: *Lectures on quantum field theory*, World Scientific, Singapore 2008.
[12] G. Sterman: *An introduction to quantum field theory*, Cambridge University Press, Cambridge 1993.

[13]  S. Weinberg: *The quantum theory of fields I*, Cambridge University Press 1995.

[14]  M. Peskin, D. Schroeder: *Introduction to quantum field theory*, Westview Press 1995.

[15]  W. Greiner, J. Reinhardt: *Quantum electrodynamics*, Springer-Verlag, Berlin 1994.

[16]  N. N. Bogoliubov, D. V. Shirkov: *Quantum fields*, Addison-Wesley, Boston 1982.

[17]  N. Nakanishi, I. Ojima: *Covariant operator formalism of gauge theories and quantum gravity*, World Scientific, Singapore 1990.

[18]  T. D. Lee: *Particle physics and introduction to field theory*, Harwood Academic Publishers, London 1988.

[19]  L. Ryder: *Quantum field theory*, Cambridge University Press 1996.

[20]  I. Duck, E. C. G. Sudarshan: *Pauli and the spin-statistics theorem*, World Scientific, Singapore 1997.

[21]  R. Streater, A. Wightman: *PCT, spin and statistics, and all that*, Princeton University Press, Princeton 1980.

[22]  J. Hořejší: *Fundamentals of electroweak theory*, Karolinum Press, Prague 2003;
P. Langacker: *The standard model and beyond*, 2nd edition, CRC Press, Boca Raton 2017.

[23]  G. Folland: *Quantum field theory: A tourist guide for mathematicians*, American Mathematical Society, Providence, Rhode Island 2008.

[24]  B. Thaller: *The Dirac equation*, Springer-Verlag, Berlin 1992.

[25]  J. Collins: *Renormalization*, Cambridge University Press, Cambridge 1984.

[26]  J. Formánek: *Úvod do relativistické kvantové mechaniky a kvantové teorie pole I* (in Czech), Karolinum, Praha 2004.

[27]  M. Schwartz: *Quantum field theory and the standard model*, Cambridge University Press, Cambridge 2014.

[28]  L. M. Brown (editor): *Renormalization: From Lorentz to Landau (and beyond)*, Springer-Verlag, Berlin 1993.

[29]  L. Álvarez-Gaumé, M. Vázquez-Mozo: *An invitation to quantum field theory*, Springer-Verlag, Berlin 2012.

[30]  E. Manoukian: *Renormalization*, Academic Press, New York 1983.

[31]  K. Huang: *Quarks, leptons & gauge fields*, 2nd edition, World Scientific, Singapore 1992.

[32]  P. A. M. Dirac: *The quantum theory of the electron*, Proc. Roy. Soc. London **A 117** (1928) 610.

[33]  P. A. M. Dirac: *A theory of electrons and protons*, Proc. Roy. Soc. London **A 126** (1930) 360.

[34]  S. Watanabe: *Chirality of K particle*, Phys. Rev. **106** (1957) 1306.

[35]  R. P. Feynman, M. Gell-Mann: *Theory of Fermi interaction*, Phys. Rev. **109** (1958) 193.

[36]  N.Dombey, A. Calogeracos: *Seventy years of the Klein paradox*, Phys. Reports **315** (1999) 41.

[37]  P. Krekora et al.: *Klein paradox in spatial and temporal resolution*,
Phys. Rev. Lett. **92** (2004) 040406.

[38]  L. Lamata et al.: *Dirac equation and quantum relativistic effects in a single trapped ion*, Phys. Rev. Lett. **98** (2007) 253005;
R. Gerritsma et al.: *Quantum simulation of the Dirac equation*, Nature **463** (2010) 68;
Zhi-Yong Wang, Cai-Dong Xiong: *Zitterbewegung by quantum field-theory considerations*, Phys. Rev. **A 77** (2008) 045402;
P. Krekora et al.: *Relativistic electron localization and the lack of Zitterbewegung*,
Phys. Rev. Lett. **93** (2004) 043004.

[39]  W. Dittrich: *On the Pauli–Weisskopf anti-Dirac paper*, Eur. Phys. J. **H 40** (2015) 261.

[40]  F. J. Dyson: *The radiation theories of Tomonaga, Schwinger, and Feynman*,
Phys. Rev. **75** (1949) 486.

[41]  G. 't Hooft, M. J. G. Veltman: *Regularization and renormalization of gauge fields*,
Nucl. Phys. **B 44** (1972) 189.

[42]  C. G. Bollini, J. J. Giambiagi: *Dimensional renormalization: The number of dimensions as a regularizing parameter*, Nuovo Cim. **B 12** (1972) 20.

[43] W. Pauli, F. Villars: *On the invariant regularization in relativistic quantum theory*, Rev. Mod. Phys. **21** (1949) 434.

[44] J. Hořejší, J. Novotný, O. I. Zavialov: *Dimensional regularization in four dimensions*, Czech. J. Phys. **B 39** (1989); *Dimensional regularization of the VVA triangle graph as a continuous superposition of of Pauli–Villars regularizations*, Phys. Lett. **B 213** (1988) 173.

[45] J. C. Ward: *An identity in quantum electrodynamics*, Phys. Rev. **78** (1950) 182.

[46] Y. Takahashi: *On the generalized Ward identity*, Nuovo Cimento **6** (1957) 371.

[47] F. J. Dyson: *The S-matrix in quantum electrodynamics*, Phys. Rev. **75** (1949) 1736.

[48] R. Serber: *A note on positron theory and proper energies*, Phys. Rev. **49** (1936) 545.

[49] E. A. Uehling: *Polarization effects in the positron theory*, Phys. Rev. **48** (1935) 55.

[50] G. Dunne: *Heisenberg-Euler effective Lagrangians: basics and extensions*, arXiv:hep-th/0406216; *The Heisenberg–Euler effective action: 75 years on*, Int. J. Mod. Phys. **A 27** (2012) 1260004.

[51] F. Přeučil, J. Hořejší: *Effective Euler–Heisenberg Lagrangians in models of QED*, J. Phys. **G 45** (2018) 085005.

[52] M. Aaboud et al. (ATLAS Collaboration): *Evidence for light-by-light scattering in heavy ion collisions with the ATLAS detector at the LHC*, Nature Phys. **13** (2017) 852.

[53] J. Schwinger: *On quantum electrodynamics and the magnetic moment of the electron*, Phys. Rev. **73** (1948) 416.

[54] A. S. Blum: *The state is not abolished, it withers away: How quantum field theory became a theory of scattering*, Stud. Hist. Phil. Sci. **B 60** (2017) 46.

[55] B. Delamotte: *A hint of renormalization*, Am. J. Phys. **72** (2004) 170.

[56] D. Wallace: *In defence of naiveté: The conceptual status of Lagrangian quantum field theory*, Synthese **151** (2006) 33; *Taking particle physics seriously: A critique of the algebraic approach to quantum field theory*, Stud. Hist. Phil. Sci. **B 42** (2011) 116.

# Index

331

light-by-light scattering, 293
local U(1) symmetry, 132
longitudinal polarization, 81
loop diagrams, 217, 221
loop momentum, 223, 249, 263, 301
Lorentz algebra, 100, 309
Lorentz transformations, 28, 305
Lorentz-invariant phase space, 145
Lorenz condition, 3, 190, 195
Lorenz-like constraint, 126

Majorana representation of Dirac matrices, 22, 34
Mandelstam invariants, 148, 156, 182
mass dimension, 264, 277, 294
mass renormalization, 270, 273
mass shell, 115, 325
matching condition, 293
Maxwell field, 98, 104, 186
metric tensor, 4, 20, 306
minimal electromagnetic interaction, 15
minimal subtraction, 283
Minkowski space, 18, 82, 188, 305
momentum cut-off, 229, 240
Mott formula, 159, 161
Møller scattering, 214

negative energy solutions, 14, 37, 68, 86–87
negative norm squared, 193
neutrino, 62, 140, 156
Noether's identity, 97
Noether's theorem, 95
non-Abelian gauge theory, 296
non-covariant photon propagator, 190
non-renormalizable theory, 277
normal ordering, 113
normal product, 207

occupation numbers, 114
on-shell scheme, 283
one-particle-irreducible Feynman diagram, 261
optical theorem, 288
orbital angular momentum, 13
orthochronous Lorentz transformations, 306
oscillator decomposition, 115

palindromic relation, 36, 314
parity, 30, 60

parity violation, 31, 62
particle interpretation, 111, 122–124, 127, 189
Pauli equation, 16
Pauli exclusion principle, 86, 125
Pauli matrices, 11, 13, 16, 57, 60
Pauli term, 85, 265, 277
Pauli–Jordan function, 196
Pauli–Villars regularization, 237
photon propagator, 190, 199, 322
photon–photon scattering, 256, 261, 281, 296
Planck constant, 4
plane waves, 6, 38
Poincaré invariance, 104
polarization sum, 82, 175, 185, 188
polarization vector, 79, 187
positron, 86, 139, 158, 181, 184
power counting, 264
probability current, 8, 12, 28
probability density, 6, 9, 12, 63
Proca equation, 79, 91, 126
Proca field, 90, 126, 154, 174
propagator of Dirac field, 170, 172, 223
propagator of massive vector field, 174, 177
propagator of scalar field, 163
proper Lorentz transformations, 306
pseudoscalar, 31
pseudovector, 32, 306

QED with massive photon, 181, 265
quantized field, 107, 110, 111, 115, 163
quantum chromodynamics, 281
quantum electrodynamics, 1, 16, 174, 181, 211, 302
quasi-real virtual photons, 296

radiation gauge, 187, 197
radiative corrections, 278, 284
relativistic covariance, 9, 23, 58, 78
renormalization, 225, 267, 271–277
renormalization constant, 275
renormalization counterterms, 272–277
representations of Lorentz group, 25, 309
right-handed particle, 52
rotations, 28, 81, 99

$S$-matrix, 1, 136, 139, 143, 163, 212, 216
$S$-matrix unitarity, 288

Notes: